Movement-Aware Applications for Sustainable Mobility: Technologies and Approaches

Monica Wachowicz
Wageningen University and Research Centre, The Netherlands

Information Science REFERENCE
INFORMATION SCIENCE REFERENCE
Hershey · New York

Director of Editorial Content:	Kristin Klinger
Director of Book Publications:	Julia Mosemann
Acquisitions Editor:	Lindsay Johnston
Development Editor:	Joel Gamon
Publishing Assistant:	Sean Woznicki
Typesetter:	Myla Harty
Quality control:	Jamie Snavely
Cover Design:	Lisa Tosheff
Printed at:	Yurchak Printing Inc.

Published in the United States of America by
Information Science Reference (an imprint of IGI Global)
701 E. Chocolate Avenue
Hershey PA 17033
Tel: 717-533-8845
Fax: 717-533-8661
E-mail: cust@igi-global.com
Web site: http://www.igi-global.com/reference

Library of Congress Cataloging-in-Publication Data

Movement-aware applications for sustainability mobility : technologies and approaches / Monica Wachowicz, editor.
p. cm.

Includes bibliographical references and index.
Summary: "This book will provide insight on the issues and repercussions of collecting and analysing the movement of people using techniques such as privacy preserving data mining, ontologies, space-time modeling and visualization"--Provided by publisher.

ISBN 978-1-61520-769-5 (hbk.) -- ISBN 978-1-61520-770-1 (ebook) 1. Location-based services. 2. Geographic information systems. 3. Global Positioning System. 4. Geospatial data--Social aspects. 5. Transportation--Social aspects. I. Wachowicz, Monica.
TK5105.65.M68 2010
388.01'1--dc22

2009036053

British Cataloguing in Publication Data
A Cataloguing in Publication record for this book is available from the British Library.

Editorial Advisory Board

Table of Contents

Section 2
Exploring New Methods on the Analysis of Movement Behaviour

Section 3
Privacy Issues and New Perspectives on Movement-Aware Services

Detailed Table of Contents

The impact of urban development and climate change has created the impetus to monitor changes in the environment, particularly, the behaviour, habitat and movement of fauna species. The aim of this chapter is to present the design and development of a sensor network based on Smartphones to automatically collect and analyse acoustic and visual data for environmental monitoring purposes. Due to the communication and sophisticated programming facilities offered by Smartphones, software tools can be developed to allow data to be collected, partially processed and sent to a remote server over the network for storage and further processing. This sensor network which employs a client-server architecture has been deployed in three applications: monitoring a rare bird species near Brisbane Airport, study of koalas behaviour at St Bees Island, and detection of fruit flies. The users of this system include scientists (e.g. ecologists, ornithologists, computer scientists) and community groups participating in data collection or reporting on the environment (e.g. students, bird watchers). The chapter focuses on the following aspects of our research: issues involved in using Smartphones as sensors; the overall framework for data acquisition, data quality control, data management and analysis; current and future applications of the Smartphone-based sensor network, and our future research directions.

In this chapter the authors analyze the concept and definitions of anonymity in the modern connected world. In particular, they explore if modern technology renders complete anonymity impossible and if a new definition of anonymity needs to be adopted. They examine examples of anonymous use of technology that illustrate the complexity of the concept of anonymity and demonstrate that access to anonymity is not uniform for data owners with regard to multiple data controllers and audiences in complex systems and processes. They evaluate legal definitions of "anonymity" and "anonymous data" as well as the right to anonymity provided in the European directives and by some European statutes, observing that anonymity cannot be absolute, that only "relative" anonymity is realistic in the present technological environment, and that different degrees of anonymity exist. The authors address the issue of measuring these degrees or levels of anonymity in complex systems, in order to provide a new foundation for a nuanced and comprehensive understanding of anonymity. They conclude that the concept of relative anonymity can become the basis for a new and more effective approach to personal data protection.

The growing amount of ubiquitously available location based information offers great chances for the development of mobile pedestrian navigation services. Yet, providing spot-on real-time adequate information remains challenging, as relevant information needs to be filtered and group-related behaviour patterns and preferences are still insufficiently known. In this chapter, the authors focus on two aspects. Firstly, they consider the issue of how to comprehensively investigate human spatio-temporal behaviour patterns in order to enable the provision of customised navigation information, and secondly, they discuss potential impacts of ubiquitous wayfinding systems on human perception of space and possible resulting navigation behaviour modifications.

The relatively new ability to rapidly transfer digital information to people as they move through cities opens up exciting possibilities for services that alleviate some of the issues that dense urban centers face. This chapter examines the potential for one such service – named FriendFreight – to reduce some of the negative effects of goods transportation in cities. FriendFreight operates through exploiting the real-time location information of people and goods and the ability for members of a community to deliver items for others while moving through the city themselves. It aims to lower the number of 'unnecessary trips' that people make to obtain some small goods - groceries, books, documents and dry-cleaning - and reduce what the authors define as *travel demand* in the city. However, the success of such a service relies not only on accessing real time location information but also on an understanding of how and why people might deliver goods for each other. Thus, this chapter explores two things: how incentives, trust and reciprocity can be built in services that harness digital information; and how the feasibility of a service like FriendFreight can be established given a particular real world context - Copenhagen and bicycles. Through this, the authors show that access to real-time location and movement information can open up innovative ways of tackling problems in cities from the 'bottom-up' but that essential to this is the nurturing of trust between users of the service. They also demonstrate that it is possible to achieve a significant reduction in travel demand through using FriendFreight for certain types of goods in the context of Copenhagen.

This chapter presents a model that addresses some shortcomings associated with limitations in the way that electronic identities are used in an application environment. This model, Snowflake, allows the exchange of validation data in order to establish trust when identity attributes are used in applications. Snowflake leverages upon groups of validation authorities as electronic intermediaries to facilitate application interoperation. The application area of choice is the electronic reporting of fish catches by a vessel's master in line with the European Union legal framework. Grouping together the services of multiple validation authorities maximises the utilisation of resources available especially in environments that do not shy out in sharing resources and data such and the electronic fish logbook provides a good platform in this respect.

Section 4
Selected Readings

Through the advancement of electronics technologies, human motion analysis applications span many domains. Existing commercially available magnetic, mechanical and optical systems for motion capture and analyses are far from being able to operate in natural scenarios and environments. The current shortcoming of requiring the subject to wear sensors and markers on the body has prompted development directed towards a marker-less setup using computer vision approaches. However, there are still many challenges and problems in computer vision methods such as inconsistency of illumination, occlusion and lack of understanding and representation of its operating scenario. The authors present a video-based marker-less motion capture method that has the potential to operate in natural scenarios such as occlusive and cluttered scenes. In specific applications in sports biomechanics and education, which are stimulated by the usage of interactive digital media and augmented reality, accurate and reliable capture of human motion are essential.

The Speckled Computing project is a large multisite research project based in Scotland, UK. The aim of the project is to investigate, prototype, and produce tiny (1mm3) computational devices, called Specks, that can be configured into wireless sensor networks, called SpeckNets. Our particular interest is in how people might interact in such environments, what interaction tools they require, and what characteristics are required to be provided by the operating system of the Specks. Interaction in these environments places the human physically inside an information space. At one time, the human may be interacting with one Speck, at another with a hundred, and at another with several thousand. Moreover, the Specks themselves have no input method, apart from their sensors, and no output display. We explore these issues through taking some theories of distributed information spaces, some design principles from information visualization, and report on some empirical studies of prototypes and simulations that have been developed.

Chapter 15
Modelling Pedestrian Movement to Measure On-Street Crime Risk 243
Spencer Chainey, UCL Jill Dando Institute of Crime Science, UK
Jake Desyllas, Intelligent Space Partnership, UK

This chapter presents results for the first large-scale analysis of street crime rates that utilizes accurate on-street pedestrian population estimates. Pedestrian counts were generated at the street segment level for an area in central London (UK) using a modeling process that utilized key indicators of pedestrian movement and sample observations. Geocoded street crime positioned on street segments then allowed for street crime rates to be calculated for the entire central London study area's street network. These street crime rate measures were then compared against street crime volume patterns (e.g., hotspot maps of street crime density) and street crime rate statistics and maps that were generated from using the residential population as the denominator. The research demonstrates the utility of pedestrian modeling for generating better and more realistic measures for street crime rates, suggesting that if the residential population is used as a denominator for local level street crime analysis it may only misinform and mislead the interpretation and understanding of on- to pedestrians. The research also highlights the importance of crime rate analysis for understanding and explaining crime patterns, and suggests that with accurate analysis of crime rates, policing, and crime prevention initiatives can be improved.

Chapter 16
A Framework for Data Warehousing and Mining in Sensor Stream Application Domains 264
Nan Jiang, Cedarville University, USA

The recent advances in sensor technologies have made these small, tiny devices much cheaper and convenient to use in many different applications, for example, the weather and environmental monitoring applications, the hospital and factory operation sites, sensor devices on the traffic road and moving vehicles and so on. The data collected from sensors forms a sensor stream and is transferred to the server to perform data warehousing and mining tasks for the end user to perform data analysis. Several data preprocessing steps are necessary to enrich the data with domain information for the data warehousing and mining tasks in the sensor stream applications. This chapter presents a general framework for domain-driven mining of sensor stream applications. The proposed framework is able to enrich sensor

streams with additional domain information that meets the application requirements. Experimental studies of the proposed framework are performed on real data for two applications: a traffic management and an environmental monitoring site.

Foreword

Occasionally you find yourself reading a sentence that you realise would not have made any sense only a year or two ago. So it is with the title of this excellent book: after all, what are 'movement-aware applications' and what is 'sustainable mobility'? It is only the precipitous speed of technological development that means such concepts are needed to help us make sense of developments taking place in this field. Location technologies are driving changes so far-reaching that they will eventually lead to both profound social change and the restructuring of our transport systems and city fabric. As with the wristwatch and the commoditisation of time 19th century, so it will be with location on your mobile device in the 21st century. Greater knowledge of places and distance will begin to drive decision-making about lifestyles, including travel choices, shopping options and leisure pursuits. A dramatic reduction in the time we spend being lost or misjudging trip times will give us back time for the things that matter most to us.

We can only speculate on what these changes will mean for mobility in the medium to long term, but some implications stand out. For example, real time transport information will likely lead to the redistribution of congestion e.g. onto roads parallel to limited access highways. Knowledge of traffic speeds by transport mode in cities might be used to re-price public transport by time-taken rather than by distance. Mobile social networks might be used to support demand-responsive transport services. Augmented reality pedestrian navigation has the potential to become a major driver of shopping behaviour. There are scientific applications too such as animal tracking and health monitoring. Perhaps in the long term we will see personal carbon budgets administered through movement-aware applications.

I'm delighted to write this foreword as this book does an important job in gathering together research on these topics into one place. The initiative shown by Monica Wachowicz in bringing this group of authors together is both insightful and timely, as it gives interested parties a synoptic view of the research in this field. It is also important as an alternative vision of mobility to that developed by intelligent transport systems (ITS) sector, which has tended to see mobility innovation as supply-driven. This book offers alternative thinking on demand-driven mobility initiatives and is a must read for all those interested in the future of mobility and transportation.

Jonathan Raper
November 2009

Jonathan Raper *joined the Department in 1999 as Professor of Geographic Information Science and was Head of Department from 2003-2006. He founded the geographic information centre ("giCentre at City") in 1999, which is now one of the UK's premier centres of expertise in geographic information systems, geovisualisation and location based services (LBS) research.*

Jonathan in known internationally for his work on GIS and LBS, having published around 150 research articles and seven books in these fields, and supervised a dozen completed Ph.Ds. He was awarded the Royal Geographical Society Gill Memorial Award for research in GIS in 1995 and the Kitchin Award by the Saddlers Company for significant contributions to City University in 2008. Jonathan's main research work in the last decade has been the design, development and implications of location-based services. He also conducts research in geographic information policy, especially locational privacy, the philosophy of spatial and temporal representation and the handling and analysis of environmental information. He is a Fellow of the Royal Geographical Society/Institute of British Geographers and a member of Association of British Geographers and the Chartered Institute of Library and Information Professionals. He is a member of the Steering Committee of the Location and Timing Knowledge Transfer Network and is the Editor-in-Chief of the Journal of Location Based Services. Jonathan is now a part-time Professor at City and has founded a University spinoff company called Placr, which develops mobile geospatial solutions.

Preface

Sustainability is just one word and yet there exists over 300 definitions. The term was originally used in 1987 by the World Commission on Environment and Development, which coined what has become the most often-quoted definition of sustainable development as "forms of progress that meet the needs of the present without compromising the ability of future generations to meet their needs." In particular, sustainable mobility refers to the needs of society to move freely, gain access, communicate, trade and establish relationships without sacrificing other essential human or ecological requirements today or in the future (Mobility Project 2030, World Business Council for Sustainable Development). This can only be achieved by establishing a set of principles that provide a framework for policy goals that will change over time, in response to the priorities in the economy (e.g. access to jobs and economic resiliency), social/equity (e.g. mobility choices, health societies, and community legacy) and environment (e.g. climate change, pollution, energy use, landscape, and resource efficiency). The strategies derived from these principles would not only be about the modes people are using, nor only about transportation. They will evolve from the knowledge about the collective movement patterns that are evidence of the human behaviour at different scales.

The proliferation of mobile technologies for "everywhere, anytime" services and applications is already helping in the fulfilment of some of these strategies in travel behaviour, nature preservation, and health monitoring. Recently, Gartner Inc. has identified eight mobile technologies that will evolve significantly through 2010, and will have an impact on short-term sustainable mobility strategies and policies. They are Bluetooth 3.0, Mobile User Interfaces (UIs), Location Sensing, 802.11n, Display Technologies, Mobile Web and Widgets, Cellular Broadband, and Near Field Communication. This book focuses on the use of location sensing technology and its wider applicability in supporting sustainable mobility. Each chapter of the book presents current research on developing innovative approaches for gathering, representing, storing and analysing movement data sets being generated from location sensing technology today.

This book also introduces the latest findings on movement analysis that will open up tremendous opportunities for new movement-aware applications of public utility and large societal and economic impact. Examples of these applications are described in this book and they are strongly related to sustainable mobility, such as health, transportation and spatial planning. They also show the important global trends in location sensing technology, as well as the fostering of cross-disciplinary research on Geographic Information Science. Therefore, we have put together an interdisciplinary and cross-border book with highly qualified authors in Geography, Geomatics, Spatial Planning, Law, Transport Management Mathematics, and Computational Sciences. Specifically, this book offers a unique opportunity for learning about how different techniques are being used to develop new movement-aware applications based on different perspectives from data providers, scientists, application developers, and users.

The inspiration for this book came from the new insights that have emerged from the synergy of the outcomes of two workshops that I have organised within a two-year period. The first workshop was on the Knowledge Discovery for Sustainable Mobility: Challenges and repercussions of discovering behavioural patterns from people in motion, hosted by the Waag Society in Amsterdam, The Netherlands, on September 11th and 12th, 2006. This workshop aimed at getting a community input and feedback on the issues and repercussions of discovering behavioural patterns from people in motion using privacy preserving data mining, spatio-temporal ontologies, and visualisation techniques that could enable innovative movement-aware applications. Specifically, the workshop provided a forum for discussion based on different perspectives from data providers, scientists, application developers, and users. The second workshop was on Mobility Issues in Spatial Data Infrastructures, organised in conjunction with the GEOIDE Annual Conference on June 9th 2007, Halifax, Canada. The aim of this workshop was to address the methodological and technological issues on the gathering and analysis of movement data sets, and its involvement into infrastructures. It provided a basis on which to compare their applicability in a specific context and case studies of how location information is being used in spatial data infrastructure settings.

This book was written to provide a single reference for movement dynamics. The possible range of topics to cover is nearly limitless, therefore we have tried to cover the most important ones, moving from mobile technologies, innovative approaches, new applications and privacy issues. This will be an opportunity for students, researchers and practitioners to acquire new concepts in the production, awareness and sharing (use) of sensors, data, and information products for supporting sustainable mobility.

ORGANISATION OF THIS BOOK

This book is organised into four parts, each devoted to a particular aspect of movement-aware applications for sustainable mobility. Section 1 covers the state-of-art in location sensing technology for tracking and sensing the movement of people with interesting examples on transportation management and health monitoring. Section 2 explores innovative methods developed so far for analysing the human and animal movements. Section 3 is devoted to the implementation of movement-aware services, and the related issues of measuring different levels of anonymity. The main focus is on addressing both technical and legal implications of location sensing technology. Section 4 contains some addtional selected readings. Each section of the book consists of four chapters, which one covering relevant background information and some advanced topics, giving full proofs, methodological procedures, and application results in order to help the reader appreciate the utility of the work published in this book. Authors of the chapters have tried as much as possible to introduce the technical background necessary for readers with no expertise in location sensing technology and movement-aware applications.

Monica Wachowicz
Wageningen University and Research Centre, The Netherlands

Acknowledgment

This book would not have been possible without the contribution of many people. First, I would like to thank the authors and the reviewers for working very hard in improving their chapters during the peer-review process. Their efforts have definitely contributed to the overall quality of this book. Second, I would like to thank the Waag Society and the GEOIDE Network, for hosting the workshops.

Finally, my additional thanks to the funding agencies that made this work possible by supporting the People in Motion Project (Decision Making Subsidies for Investing in Knowledge Infrastructure Framework, Space for Geo-Information Programme), the Geographic Privacy-aware Knowledge Discovery and Delivery Project (Sixth EU Framework Programme, Future and Emerging Technologies) and the Mobility, Data Mining, and Privacy Project (Seventh EU Framework Programme, Future and Emerging Technologies).

Monica Wachowicz
Wageningen University and Research Centre, The Netherlands

Section 1
State-of-the-Art in Tracking and Sensing

Chapter 1
Data Collection on Personal Movement Using Mobile ICTs:
Old Wine in New Bottles?

Martin Lee-Gosselin
Université Laval, Canada

Sean T. Doherty
Wilfrid Laurier University, Canada

Amer Shalaby
University of Toronto, Canada

ABSTRACT

This chapter examines the current status of data collection methods employing location/time-aware devices to observe evolving patterns of spatio-temporal behaviour, including patterns that are affected by ICTs. Drawing mostly on transport research, it is suggested that two streams of development have emerged: a "passive" stream that maximises the automatic interpretation of positioning data, and an "active" stream, that is using increasingly sophisticated mobile computing devices and/or the Internet to engage respondents in the validation, interpretation and enhancement of their own data. Recent and future developments are described that promise to go beyond simply using technologies to carry out conventional travel surveys: rather, some new classes of data may be obtained, notably because of common or overlapping interests with other fields, such as public health research. There are, however, some ethical and public acceptability constraints that must be respected.

INTRODUCTION

Amidst growing evidence that Information and Communication Technologies (ICTs) may be having profound effects on the organisation of human activities in space and time (Dijst, 2004), survey researchers have focussed on the challenges of collecting data that is sufficiently detailed, and with sufficiently long observation periods, for the study of such effects. Conventional methods, notably time-use and activity-travel diaries, can provide valuable micro-behavioural data, especially if a trained interviewer questions a respondent in-depth around a "skeleton" of self-observations, such as a simplified travel log. But this is rarely done for

DOI: 10.4018/978-1-61520-769-5.ch001

more than one or two days, and is considered too burdensome for periods of more than seven days, even though diaries have occasionally been used with success to collect minimal spatio-temporal data over multi-week periods, such as six weeks. Of course, increasing respondent burden aggravates many standard problems of surveys, especially recruitment and selection bias, item non-response, and invalid response. It is therefore no surprise that survey methodologists are increasingly exploring the use of ICTs as part of their *toolkit* to reduce respondent burden while collecting person-based data on spatio-temporal behaviour. The length of observation and the level of detail sought is compatible with a holistic, multi-scaled, and interpersonal view of the spatial extent of human activities, and with detecting phenomena such as the temporal and spatial fragmentation of activities, coordination with others and (ideally) multitasking.

In this paper, we examine two major streams of ICT-enabled activity and travel data collection methods that have emerged. The first is essentially *passive* and envisages the largely automatic interpretation of spatio-temporal monitoring data. The second *active* stream is highly interactive. It uses portable and web-based computing technologies to allow subjects to validate automatic interpretations, typically every 1-2 days; it also allows for more immediate interaction with subjects, such as live queries directly following a passively tracked specific event (e.g. a fall by a pedestrian). We give an overview of the status of these two streams, and then look forward. We draw primarily on the experience of a multi-university team in Canada that, since 1995, has experimented with mostly off-the-shelf ICTs to devise new ways of measuring personal activity and travel patterns in urban areas. We start with a brief summary of the earliest methods. This is followed by a discussion of an overall architecture for ICT-enabled activity and travel data collection methods, and then a section each is devoted to the "passive" and "active" streams. A sixth section speculates about a number of new measurement concepts that seem ripe for development. In a final section, we ask ourselves how far we have come in using some ICTs to help study phenomena that are influenced by all ICTs, discuss the probable limits of the usefulness of these methods, and examine some of the risks that may be entailed for society and for researchers.

THE INITIATION OF ICT-AIDED DATA COLLECTION METHODS

While the use of ICTs to understand vehicular traffic is a relatively mature art, the same cannot be said of related methods to understand human activity in time and space. For more than 20 years, increasingly miniaturised electronic sensor and computing technologies have made it possible to collect timed microbehavioural data on human subjects and to record the data in portable loggers. For example, in the public health and sports medicine literatures, one finds citations of research into free activity using portable accelerometers going back to 1984 or earlier (Matthew, 2005), and as we discuss below, the measurement of physical activity is a rapidly expanding field of research that is taking advantage of the potential of multiple sensors.

In the travel behaviour field, since the mid 1990s, attention has inevitably focussed on the potential of location-aware systems such as Global Positioning Systems (GPS), triangulation on FM radio stations, or the automatic reporting of the location of mobile telephones in the "cells" employed by GSM (Global System for Mobile communications). We have seen a diversity of efforts to take advantage of such systems to collect temporal and spatial data on human mobility with unprecedented levels of detail and accuracy and, moreover, to collect data in an automatic or semi-automatic manner over much longer periods of observation than are covered by the vast majority of travel and activity diary methods.

Initially, these were mostly limited to travel in private motor vehicles (for example, Murakami, Wagner, & Neumeister, 1999; Ueno et al, 1999), because the power requirements of equipment in continuous use could easily be met with a connection to vehicle electrics, the problems of reception were minimised, and the linking of movement to ground features was simplified by staying on road networks. Nevertheless, even in the 1990s, some experiments took place to use GPS to survey personal mobility in all modes of transport and off road networks (such as Kalfs & Saris, 1997). A review of the emergence of location/time-aware devices in activity and travel survey methods will be found in Lee-Gosselin and Harvey (2006).

Substantial progress has been made, but arguably the potential for such data collection has, until now, been under-realised for two reasons. First, the hard and soft technologies of geolocation, data storage, data transfer and power management have only very recently reached the stage that robust, pocketable devices have been feasible and affordable. Secondly, it has been a case of "old wine in new bottles": the focus has often been on doing better the types of data measurement that were well-established before relevant ICTs were available, or on validating traditional measures. While there is nothing wrong in this, radically new types of measurement can also be imagined.

A DESIGN FRAMEWORK FOR ICT-AIDED PERSONAL ACTIVITY AND TRAVEL SURVEYS

It is quite normal at the early stages of adoption of a new practice, and especially a practice employing a new technology, that one or two applications dominate the image of that practice. As experience is gained, applications can be differentiated and technical requirements are understood with more nuance. Such was the case of ICT-aided activity and travel surveys which, in the early stages, were overwhelmingly focused on two opportunities to improve on conventional survey methods: (a) overcoming partially missing data, notably by detecting short and some non-routine trips that are often missed by diaries; and (b) recording route choice by private vehicles together with the time and distance travelled on different classes of road, something that is infeasible with non-ICT methods. Both of these attracted much interest because of their potential to enhance the art of travel demand forecasting, which was (and to a large extent still is) invested above all in the understanding and management of vehicle flows on urban road networks. However, a full appreciation of the potential of ICT-aided surveys for both urban and long-distance travel requires a framework for the multiple components of a survey design.

Figure 1 shows one such framework: it is a proposed architecture for activity and travel surveys that use ICT technologies to provide tracking of the timing and location of individual people or vehicles using location/time-aware devices (LADs). It is structured to show the flow of stages in the design and implementation of a survey, although for current purposes, we will assume that the design of sampling and recruitment, which must take into account the additional issues that are associated with asking respondents to use tracking technologies, has been decided.

At the left, there are three stages that are common to all LAD-aided surveys. Respondent training (1) can involve quite detailed tasks for the correct deployment of equipment and sometimes for the transfer of data and/or its periodic validation and enhancement. However, technological developments are tending to lower the level of effort that these tasks require. Fully informed consent is also critically important for this class of surveys, including instruction on how to suspend tracking should the respondent wish to do so.

The choice of data collection technology (2) depends, in part, on the type of mobility involved: for example, FM radio station triangulation or GSM telephone cell identification may be insuf-

Figure 1. Stages of location/time aware device aided surveys. Adapted, updated and extended from Lee-Gosselin and Harvey (2006).

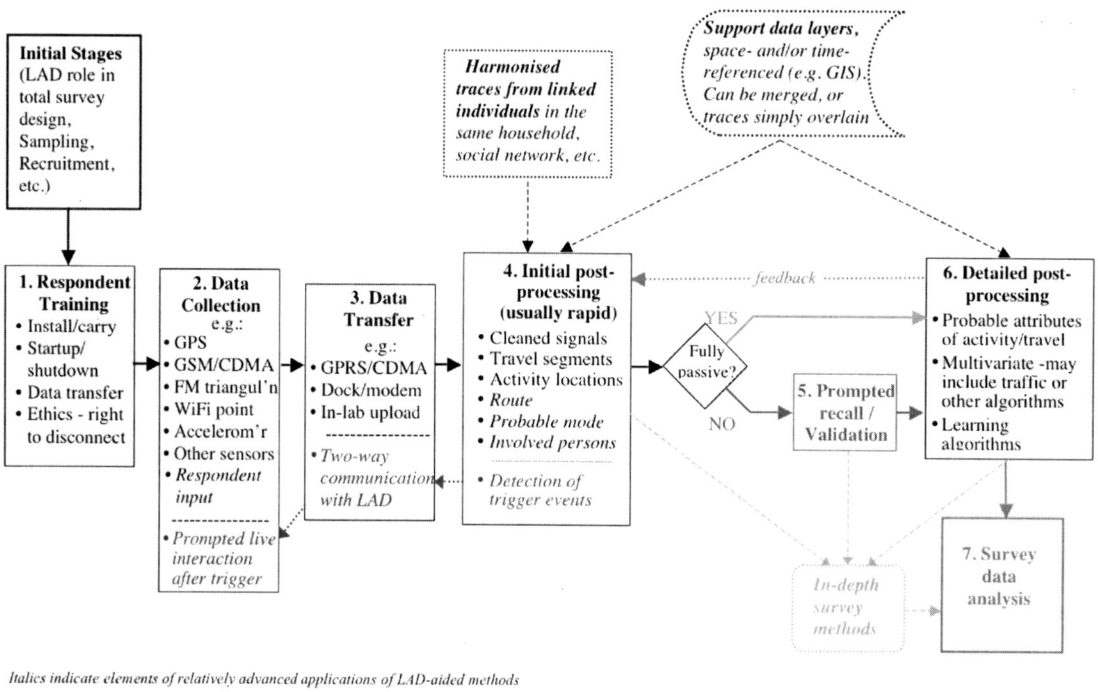

ficiently precise for urban travel, especially in the case of North American GSM, which uses quite large cells, but may give sufficient precision for long-distance travel patterns. Similar considerations pertain to the potential use of the currently increasing numbers of WiFi access points to locate individuals with mobile Internet devices. The choice also depends on the level of automation in post-processing of data: we see in the succeeding section that multi-sensing devices such as GPS with an accelerometer, are of considerable promise here. In addition, data collection may also include the parallel use of physiological or other sensors. Further, a few devices request modest amounts of input from respondents in real time. This may be regularly requested, for example devices that are programmed to query respondents at the beginning and/or end of an episode of travel. Recently, as shown in italics at the bottom of box 2, there is also the possibility of a prompted live interaction with a human operator or researcher following the detection of predetermined events or situations, such as an unusual physiological state.

The chosen data transfer method (3) can be: (a) at the end of the observation period, such as by direct connect to a computer in a laboratory; (b) by periodic store-and-forward via a home-based modem, the Internet or a mobile phone data network, usually GPRS or CDMA; or (c) in real time using various types of wireless data network. With the exception of the event-triggered interaction with a respondent, generally by mobile phone (see italics), this is one-way uploading of data.

The boxes (4 to 7) to the right of these three stages describe, with some simplification, the differences between passive and active LAD-aided survey methods. There are substantial differences in the nature and phasing of data processing, depending on whether the survey is intended to be fully passive, that is without seeking post-validation of monitoring data or supplementary variables obtained retrospectively from respondents. Particularly in the case of GPS, researchers quickly came to appreciate the number and complexity of post-processing algorithms that are needed for a valid and efficient survey package.

The upper part of Figure 1 shows two sources of supplementary data that are of growing importance to the passive and active streams. First, above box 4 is the possibility that the survey could involve the harmonised collection of data from respondents who are known to be connected, such as members of the same household or who belong to a particular social network. Such a possibility allows the analysis of the joint execution of activities and of interpersonal dependencies, but it also recognises that the spatio-temporal behaviour of social groups is also open to study using ICTs. Second, layers of spatially and/or temporally referenced support data are shown feeding into boxes 4 and 6, normally using Geographic Information Systems (GIS) or their derivatives. Data from these layers may be fully merged with mobility traces (data fusion), or simply overlain to aid respondents and analysts in their interpretation.

THE "PASSIVE" STREAM

The passive stream, simply put, represents a "turn it on and forget it" concept, although in principle a predominantly passive approach is also possible for data that includes some non-observable characteristics, such as accompanying persons, costs, or specific purposes, collected by asking respondents to interact with the mobile LAD throughout the observation period (box 2 in Figure 1).

Much depends on the performance of the personal tracker employed. In its ideal form, a passive mobile device would function well in a pocket or handbag, achieve a rapid spatial fix after a period of dormancy, run on its own power for days or weeks, and automatically transmit its data at least several times per day to allow remote monitoring of data quality and to identify problems. At the current state of affordable technology, all these characteristics have been passably achieved, except for on-board power to keep everything running for long periods, even with intelligent power management, notably employing an accelerometer, that is not dependent on good reception of satellite signals to determine that the equipment is stationary. Currently, the best we can expect from "pocketable" devices with internal batteries is to record and transmit a full day of mobility between overnight charges, or 2 to 3 days of mobility without data transmission. Nevertheless, to get this far is a significant milestone for feasible LAD-aided surveys. At the time of writing, even better performance characteristics are within grasp, but it must also be said that a number of promising instruments have been designed but not built, or built but not sustained[1].

At the heart of a passive system is the largely automatic interpretation of spatio-temporal monitoring data. No post-validation or supplementary data (such as activity purposes or accompanying persons) is normally sought from respondents. The different procedures for data cleaning and interpretation (boxes 4 and 6 in Figure 1) are combined into a single process. In order to achieve usable data on travel, much more than map-matching is required: at a minimum, the data require segmenting into periods of movement versus periods at activity locations (as opposed to transfer points or traffic delays), and the identification of probable mode. If the purposes served at activity locations are also to be estimated, as shown in Figure 1, information may also be drawn from GIS layers that identify land uses, building types, or addresses useful in predicting purposes. But this is only one possible use of GIS data; in particular, road and transport network variables can contribute to all of the procedures. Considerable effort has been invested in the algorithms required to automate the procedures.

In part of the Canadian inter-university network research programme, Tsui and Shalaby (2006)

Figure 2. Tsui and Shalaby's "Version 1" GIS-independent processing sequence for GPS-aided travel surveys (from Tsui & Shalaby, 2006).

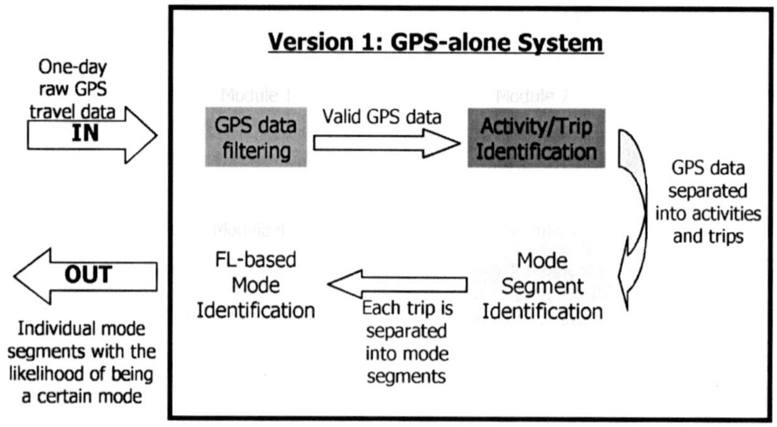

reported on the work of their team to push the limits of automatic processing of multi-modal personal travel without links to GIS data layers. Figure 2 summarises their approach.

Four modules allow cleaned-up GPS data to be sorted into travel and activity segments, and the travel segments to be further segmented in between apparent shifts in travel mode. Within the mode segments, fuzzy logic was used to interpret 95th percentile speed, average speed, median acceleration and data quality.

In "Version 2", a more sophisticated set of algorithms were added downstream from the GPS-alone version, to better deal with the ambiguities of multi-modal travel in central cities. This work, which was calibrated on downtown Toronto, is also described in Tsui and Shalaby (2006); more detail on the link matching algorithm is found in Chung and Shalaby (2005). Figure 3 summarises the modules that were added. A series of procedures first validated the mode identifications from Version 1, using the additional contextual information, such as transit stops and metro routes.

The four modules treat, and help interpret, the circumstances in which the GPS trace is spatially discontinuous, an interval of time in which there is no signal and during which the respondent has moved, or in which travel takes place off the road and transit networks. These circumstances include the use of underground or indoor infrastructure, including subway trains, pedestrian walkways and shopping malls, public spaces such as parks, or simply the delay in obtaining a GPS fix when a respondent emerges from a shielded to a clear sky position.

The results from testing both versions on 58 days of GPS data collected by research personnel were encouraging, although it is possible that a larger, fully "naturalistic", trial would encompass some additional challenges to automatic identification. No activity locations were missed, although there were a few false positives resulting from long traffic delays. Mode identification was 91% overall for Version 1 and 94% for Version 2; the detection rates were higher for automobile travel and walking, but lower for travel by bus. Road and transit links were identified at the rate of 94% by Version 2.

The high degree of success in the automatic detection of mode by Version 1 was an important finding of this work. This showed that much useful data can be obtained on travel even if GIS data are not available, or as is often the case, a GIS is available only for graphical overlays. Further refinement of the above data processing system has resulted in the capability of decomposing

Figure 3. Chung, Tsui and Shalaby's "Version 2" Integrated GPS/GIS processing sequence for GPS-aided travel surveys (from Tsui & Shalaby, 2006).

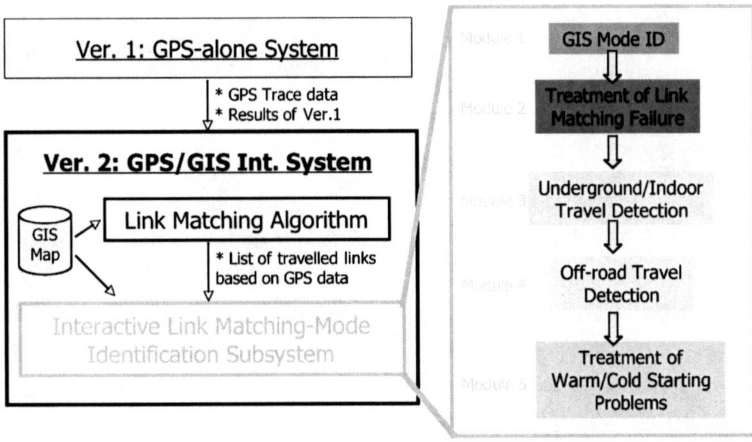

transit travel times into their individual components, namely walk access time, waiting time, in-vehicle time, transfer time (if any) and egress time (Li & Shalaby, 2008). Such data are very useful for both transit service planning and mode choice analysis.

There is, however, considerable room for improvement, in at least three areas. First, more could be done to generate "intelligent guesses" of activity types using data on land use with distance buffers in a GIS, as proposed by Doherty, Noel, Lee-Gosselin, Sirois, and Ueno (2001). Secondly, where the period of observation is more than one week, learning algorithms or Bayesian methods could improve estimations based on an individual respondent's patterns. The resulting feedback (from box 6 to box 4) could also accumulate across respondents and update the initial post-processing algorithms. Thirdly, a logical next step, especially for the more valid treatment of travel in public transport vehicles, where reception problems are common, is to extend the algorithms to the interpretation of multi-channel data. This is likely to happen first by taking advantage of the on-board accelerometers that are increasingly used to detect movement in order to optimise the management of the battery. It is possible to record accelerometer readings up to many times per second for analyses that are synchronised to the often incomplete horizontal speed and acceleration profiles obtained by positioning systems such as GPS, A-GPS[2] or GSM. Other multi-channel technologies are being used by researchers, including the comparison of altitude data from GPS to data from barometric sensors, and the use of recorded sound profiles pre-calibrated to the noise level or "signature" of different travel modes and activity locations. This also opens up other possibilities for data collection from tracking devices, a point to which we return in the section on the active stream.

Supposing a reasonable progression in the automation of processing, the passive stream has enormous potential for household travel surveys. The current context for such surveys is one of increasing response bias, especially in North America where ICT developments have made the representativeness of the predominant telephone-based travel survey methods increasingly problematic. This is not to ignore the significant problems of equipment deployment and recovery (or use of subjects own personal hand-held devices that would introduce significant software compatibility issues), or the privacy concerns, but the overall feasibility of LAD-aided surveys is no more formidable than the alternatives. In addition, the potential to observe personal mobility over periods of one or several weeks without a high respondent burden makes it cost-efficient to invest more field staff resources in each household or individual surveyed.

THE "ACTIVE" STREAM

As shown in Figure 1, the key characteristic of the active stream is the involvement of the respondents in post-validating, and usually augmenting, the interpretation of the LAD data that is fed back to them in summary form, typically every 1-2 days (box 5). Generally labelled "Prompted Recall (PR)", this is also the approach most often used to capture data that cannot be observed with sensors, such as the refinement of activity purposes, the refinement of activity location classifications (e.g. the type of business within a commercial centre), and accompanying persons. In addition, other kinds of data may be collected such as attributes of activity scheduling decisions, or characteristics of physical activity (see the section below on emerging classes of data collection). Prompted Recall can be self-administered, typically via a web site that displays the respondent's tracking data in tabular, graphical and/or map formats, or conducted by an interviewer who has access to similar tools. The sophistication of PR is growing as post-processing algorithms develop, become more efficient and exploit increasing computing power. It is to be expected that algorithms associated with box 6 in Figure 1 will increasingly move to initial processing and so improve the realism of the summaries used in PR: initially, this was not feasible because the run times were too long.

While both the passive and active streams make use of GIS data and capabilities, they have different priorities. The emphasis of the passive stream is smarter automatic detection of activity and travel attributes using contextual information from spatial and temporal data layers. The active stream benefits most from visualisation capabilities, such as maps with landmarks or "fly over" satellite imagery.

The passive and active streams are complementary, both in terms of methodological development, and sometimes as parallel measurement approaches within a single study. One example is the multiple layers of microdata that are needed as input to behaviourally-rich simulations of urban processes, notably those driving accessibility and land-use. Another is that some classes of in-depth survey methods are starting to use LADs, as shown in Figure 1. In-depth interviews, such as are used in stated response surveys or studies of scheduling processes, are usually built around the record of a period of revealed behaviour. A passive approach to this simply echoes back to respondents a skeleton of their own spatial and temporal tracking data, but an active approach would build out from prompted recall and dynamically correct and enhance the display summaries.

SOME EMERGING CLASSES OF ICT-AIDED PERSON-BASED DATA COLLECTION

The work to bring LAD-aided personal-based activity and travel surveys into substantial practice is incomplete at this time. In part, this is a reflection of the technical challenges about which some researchers were over-optimistic in the early 2000s, when vehicle-based GPS-aided survey methods had proven successful. In part, as mentioned in the introduction, it is a problem of "old wine in new bottles": the dominant expectation was that technology would provide a better way to collect the same variables as in conventional household travel surveys in metropolitan areas.

We have now reached a most interesting point in the emergence of ICT-aided surveys. While the development of functional LAD-aided surveys for the conventional needs of the transport research community is accelerated, new classes of measurement are emerging that are shared with other communities. Some of these will come from converging interests of different constituencies: for example, there is an increasing number of public health studies using portable human physical activity monitors, and some of these are starting to link to household travel surveys, or to seek LAD-aided survey methods as part of their own protocols (an example is Rodriguez, Brown, & Troped, 2005).

The public health survey research field is another user of multi-channel instruments. In Canada, a system that is usable for monitoring a variety of physiological states, physical activity and mobility attributes has been developed by Doherty and applied in collaboration with health agencies such as Toronto Rehabilitation Institute and Bloorview Kids Rehab. The first pilot launched in November 2006 involved 50 diabetic patients carry a mobile device (a GPS-equipped Blackberry), that will store and forward over a cellular network, spatio-temporal data, accelerometer readings, heart rate measurements, and digitally recorded hand-written food diary data. A separate continuous monitoring device passively logs blood sugar levels at 5-minute intervals.

Respondents have access to a web/java-based activity and travel diary that is used for prompted recall, building upon earlier work (Doherty, Papinski, & Lee-Gosselin, 2006). A partial image of this is shown in Figure 4. The remainder of person's diary is infinitely scrollable up or down, or by changing dates via right clicks on the left-most column. A person's daily events (activities and trips) are depicted in sequence in a tabular format, along with their attributes. The height of each row is proportional to the event duration. Clicking in the "Event" column brings

Figure 4. Part of a web/java-based activity and travel diary that is used for prompted recall in a travel/ health application (Ontario study of physiological states, physical activity and mobility attributes among diabetics).

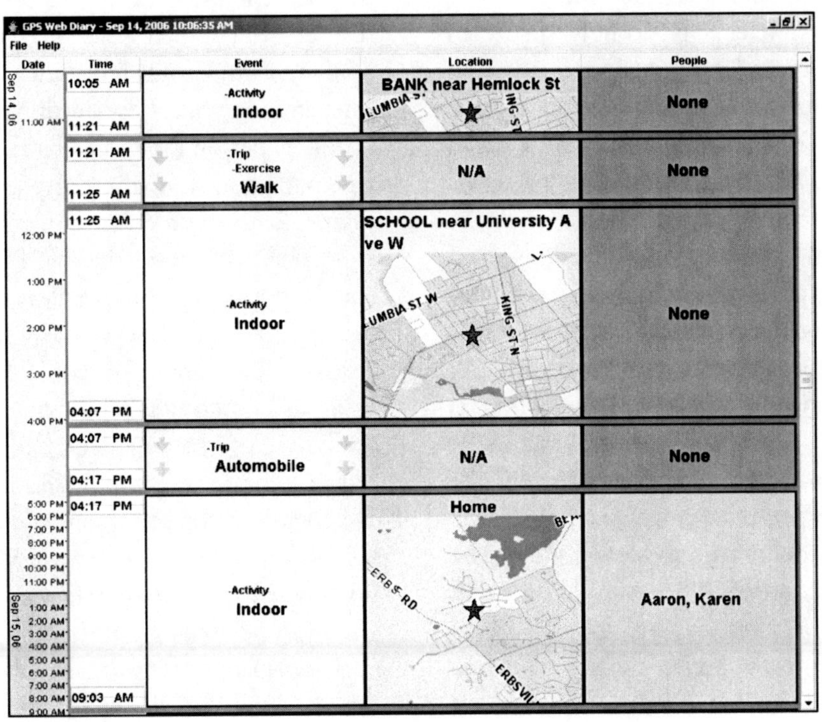

up a hierarchical fly-out style listing of event types, starting with main categories Activity and Trip, and various sub-categories (which can be customized for any given application). Interacting with the "Location" column invites respondents to refine activity locations and their names using re-scalable maps and an address search facility; provisional names are drawn from GIS data in the generic form: <Building name OR land-use name> near <nearest street/intersection>. Accompanying persons are specified from a check-list menu in the "People" column, populated as new person names are defined (their relationship, age, etc. can also be obtained and stored). Colour is used to signify attributes that require user input (red), user confirmation (yellow), or are complete (green). This is an important experiment for both the public health research and transport research communities, and it is backed by a number of private sector concerns whose interests no doubt go beyond epidemiological studies to various real-time services to clients with health challenges.

From a transport research perspective, the possibility to simultaneously measure travel and physiological parameters takes us into new territory. For example, it is strongly suspected but thinly documented that urban air quality is a major concern for the occupants of vehicles, as it is for people living along major roads. ICT-aided measures of the health effects of different levels of motor traffic could provide better evidence than, for example, counting hospital admissions. Similarly, levels of physical activity and outcomes such as obesity are being related to the design of transport facilities and even to urban form.

Another transport research domain that is currently seeing a growing interest in the use of GPS sensors, and where vital synergies could be formulated with personal mobility surveys, is the area of traffic monitoring and management. The motivation for the use of GPS technology in traffic monitoring stems primarily from the limitations of conventional technologies, mainly inductive loop detectors embedded in the pavement along key roads. Although effective and reasonably accurate, loop-based detection systems are expensive to install and maintain, and the financial burden is usually assumed by the government in full. As a result, their coverage of the network is usually limited to major freeways, to the exclusion of less critical freeways and arterial roads.

A growing interest has been developing lately in the promising potential of personal GPS sensors to serve as traffic probes. The appeal of GPS-based probes has also increased with the emergence of GPS-equipped cellular phones, which are anticipated to become the industry norm in the near future, and cellular-assisted GPS (A-GPS) promises to help overcome the limitations of GPS-alone systems. Given the high market penetration of cellular phones, there is a great potential that GPS-equipped cellular phones can provide valuable and relatively inexpensive information on traffic conditions at all segments of the road network, not only along freeways (Kirk, Fagan, & Renner, 2005; Smith, Zhang, Fontaine, & Green, 2006). However, a number of critical processing issues are currently being addressed and resolved by researchers in order to successfully realize the full potential of such systems (Byon, Abdulhai, & Shalaby, 2007). The algorithms required overlap substantially with those described in the section of this paper on the passive stream, notably the automatic detection of activity locations and trip segments, the correct identification of road links along which the device has been travelling along, and the correct detection of travel mode.

There is a great opportunity to advance both personal data collection and traffic monitoring using ICTs, through recognizing and employing the synergies between them. One could foresee a future scenario where the continuous monitoring of network traffic conditions is undertaken using GPS-equipped cellular phones, while a sample of the cell-phone population is tracked periodically to collect personal mobility data.

It is also interesting to speculate on other likely opportunities for ICT-aided data collection in transport research from the creative use of new or newly affordable ICTs. A few examples are: employing geofencing to shift tracking devices between configurations designed for urban mobility to those designed for inter-urban travel; using inexpensive radio frequency ID tags (RFIDs) to map the aggregate activity spaces of population samples over long periods; applying new types of low-cost sensors to study ambient noise or light pollution; or linking software that observes cybermobility to LAD-aided surveys of personal mobility over extended periods of time.

In general, these newer, multi-channel methods of data collection, especially the heavily interactive methods described in the section on the active stream, go beyond a statistical appreciation of spatio-temporal behaviour and lend themselves to the evaluation of effects. These are needed not only for public health and environmental issues. More fundamentally, ICTs promise to enable us to collect the data needed to track the broader trends towards an uncoupling of activities, times and places that ICTs themselves seem to be enabling, such as through linked innovations in mobile telephony, mobile Internet access, real-time travel and location-based service information, and personal scheduling aids.

CONCLUSION: HOW MUCH NEW WINE?

This discussion points to a rethinking of the architecture of human activity measurement to accommodate multiple dimensions and multiple scales. But to what extent has data collection aided by some ICTs put us in a position to understand the effects of ICTs in general on spatio-temporal behaviour? At the current time, the greatest benefit no doubt comes from the rapidly improving potential for person-based (as opposed to vehicle-based or place-based) observation. The combination of micro-behavioural detail and long periods of observation would seem to offer, in the near future, the prospect of holistic analyses of activity space/time at a level of sophistication never before seen, and with unprecedented opportunities to study the spatio-temporal fragmentation of activities. In addition, multi-person streams of data can extend this to households, social networks and groups. There is promise, albeit less imminent, to observe the articulation between personal movement and personal communication, by linking automatically logged communications episodes to automatically logged personal movement using LADs. The interpretation of such articulation, for example in terms of scheduling and micro-coordination, and the observation of other ICT-related phenomena such as multi-tasking, depends for the foreseeable future on what we have called the "active track", in which respondents react to summary representations of their own data and add content to the observations. But the tools of "prompted recall" are also developing rapidly.

From a research perspective, this can be seen as a highly adaptive approach to surveys. But in addition to the increasing difficulty to set priorities among a large number of possibilities, we can expect that the collective effects of a burgeoning number of ICT-aided studies will lead to increased public concern about privacy. Until now, in Canada at least, the properly informed respondents have almost universally welcomed, not feared, technologies such as GPS when used in surveys; indeed, we could reasonably conclude that far fewer respondents have thought in terms of "Big Brother" than was widely anticipated. But there is a legitimate and real concern about the cumulative build up of data, especially those on personal movements. We should expect increasingly strong ethics guidelines for publicly funded research. In the United States, certain types of data must be destroyed when the study is finished, and to the authors' knowledge this has been applied to at least one GPS study: such a rule, widely applied, would be a major loss to any work on the evolution of activity and travel patterns. There are some obvious risks: sloppy data management; the abuse of research data in law suits; or the publication of results that compromise an individual, a household or a neighbourhood. These risks have been around for decades and are largely in the hands of researchers to avoid. In the case of most universities, researchers must seek the counsel and approval of ethics review boards. It is the experience of the authors that such boards help to clarify the available solutions. They will want assurance that respondents are fully informed about the nature of the monitoring, and are well instructed in how to opt out at any stage, before giving their consent, and that data security measures are commensurate with the sensitivity of personal time-space traces. The nature of ICT-aided measurement means that some errors of judgement by researchers could attract sensational publicity. As in other fields, there are some things that we may learn how to do that we choose not to do. There is a need for survey researchers to share ways of respecting reasonable limits on our data collection, and of being proactive in designing our own procedures to protect our respondents.

One key way to continue to offset future privacy concerns is to focus on applications that are seen to "empower" consumers rather than "surveille" them. This implies that the plethora of data that can be obtained automatically from a person, be it GPS, physiological, or otherwise, is transformed into useful information that the *respondent* has complete control over using, sharing, selling, or feeding into other applications for their *own* personal benefit.

On the question of "old wine in new bottles", further advances in wearable tracking devices are sure to open not only *new* bottles, but also those with more *capacity*. In the case of LAD technologies, new bottles/capacity will come in the form of indoor movement tracking (as signals become more usable), longer term observation for the detection of behavioural patterns such as routines, disturbances or triggers (as battery drain is minimized), and more specific behavioural pattern identification such as impaired movements (as coordinates become more accurate and/or precise). New bottles will also surely accrue as GPS is combined with other wearable or local information to produce new synergies (accelerometers, physiological conditions, telecommunications use, weather, ambient), or as larger segments of the population are become location-enabled, allowing detection of group or mass-movement behavioural patterns. In principle, evaluation methods could thereby embrace new classes of real-time experiment, for example of user response to dynamic congestion pricing of roads.

Perhaps the key to *radically* new wine will not be how it is bottled, but rather how it is processed – meaning the transformation of large amounts of personal movement data into more viable, simple and useful information for the analyst and the respondent alike. This is beginning to happen in the context of particular studies, such as the public health examples discussed in this chapter. But there is also tremendous potential from the integration of data over large territories and long periods, provided that the concerns about cumulative archiving are addressed. Some aspects of individual and public decision-making could be transformed by the synoptic mapping of such data, made possible by the flexibility to freely aggregate the data at many different temporal and spatial scales. For example, the mapping of energy use or carbon production resulting from spatio-temporal behaviour could integrate patterns that currently are observed only within silos such as vehicle trips or residential fuel demand. In the future, such synopses may have wide implications for decision-support in public policy areas such as land-use and transport planning, resource allocation or energy efficiency. It may also turn out that some of the same tools will help individuals make more sense of the environmental and financial consequences of the choices they make as consumers, home-buyers or job-seekers.

ACKNOWLEDGMENT

The Canadian Network of Centres of Excellence in geomatics, GEOIDE, supported various stages of the research described in this paper. The patient monitoring system being conducted by Wilfrid Laurier University and the Toronto Rehabilitation Institute is primarily funded by the Canadian Institutes of Health Research, The Health Technology Exchange and University of Toronto Innovations, in collaboration with the industrial partners Blackberry (RIM), Telus, Expedata and Life:WIRE. The authors would like to thank the anonymous reviewers for their constructive suggestions, including one that we have paraphrased in our conclusions about the longer-term potential of archived data.

REFERENCES

Byon, Y., Abdulhai, B., & Shalaby, A. (2007). Impact of Sampling Rate of GPS-enabled Cell Phones on Mode Detection and GIS Map Matching Performance. In *Proceedings of the 86th Annual Transportation Research Board Meeting*, Washington D.C.

Chung, E., & Shalaby, A. (2005). A Trip Reconstruction Tool for GPS-Based Personal Travel Surveys. *Journal of Transportation Planning and Technology*, *28*(5), 381–401. doi:10.1080/03081060500322599

Dijst, M. (2004). ICT and accessibility: An action space perspective on the impact of new information and communication technologies . In Beuthe, M., Reggiani, A., & Zamarini, L. (Eds.), *Transport Developments and Innovation in an Evolving World* (pp. 27–46). Berlin: Springer Verlag.

Doherty, S. T., Noel, N., Lee-Gosselin, M., Sirois, C., & Ueno, M. (2001). Moving Beyond Observed Outcomes: Integrating Global Positioning Systems and Interactive Computer-Based Travel Behaviour Surveys. In Transportation Research Circular Number E-C026: Personal Travel: The Long and Short of It, (pp. 449-446). Washington, DC: Transportation Research Board, National Research Council, Washington, D.C.

Doherty, S. T., Papinski, D., & Lee-Gosselin, M. (2006). An Internet-based Prompted Recall Diary with Automated GPS Activity-trip Detection: System Design. In *85th Annual Meeting of the Transportation Research Board*, January 22-26.

Kalfs, N., & Saris, W. E. (1997). New data collection methods in travel surveys . In Ettema, D. F., & Timmermans, H. J. P. (Eds.), *Activity-Based Approaches to Travel Analysis*. New York: Pergamon Press.

Kirk, B., Fagan, K., & Renner, R. (2005). Development and Demonstration of a System for Using Cell Phones as Traffic Probes. In *ITS Society of Canada Annual Conference and General Meeting*.

Lee-Gosselin, M. E. H., & Harvey, A. S. (2006). Non-web technologies . In Stopher, P. R. (Ed.), *Travel Survey Methods: Quality and Future Directions* (pp. 561–568). Hoboken, NJ: Elsevier.

Li, Z., & Shalaby, A. (2008). Web-based GIS System for Prompted Recall of GPS-assisted Personal Travel Surveys: System Development and Experimental Study. In *Proceedings of the 87th Annual Transportation Research Board Meeting*, Washington D.C.

Matthew, C. E. (2005). Calibration of Accelerometer Output for Adults. *Medicine and Science in Sports and Exercise*, *37*(11), S512–S522. doi:10.1249/01.mss.0000185659.11982.3d

Murakami, E., Wagner, D. P., & Neumeister, D. M. (1999) Using Global Positioning Systems and Personal Digital Assistants for Personal Travel Surveys in the United States. In TRB Transportation Research Circular E-C008: Transport Surveys: Raising the Standard.

Rodriguez, D. A., Brown, A. L., & Troped, P. J. (2005). Portable Global Positioning Units to Complement Accelerometry-Based Physical Activity Monitors. *Medicine and Science in Sports and Exercise*, *37*(11), S572–S581. doi:10.1249/01.mss.0000185297.72328.ce

Smith, B. L., Zhang, H., Fontaine, M., & Green, M. (2006). Cellphone Probes as an ATMS Tool. In *Research Project Report for the National ITS Implementation Research Center and U.S. DOT University Transportation Center*. Retrieved March 3rd 2006 from http://ntl.bts.gov/lib/23000/23400/23431/CellPhoneProbes-final.pdf

Tsui, S. Y. A., & Shalaby, A. S. (2006). (forthcoming). An Enhanced System for Link and Mode Identifications for GPS-Based Personal Travel Surveys. *Transportation Research Record*. doi:10.3141/1972-07

Ueno, M., Noël, N., Doherty, S. T., Lee-Gosselin, M. E. H., Théberge, F., & Sirois, C. (1999). Extending the scope of travel surveys using differential GPS. In ION-GPS 1999, Nashville, Tennessee (CDROM).

ENDNOTES

1 A detailed discussion of mobile equipment is beyond the scope of this paper. However, it is useful to note that highly robust, mass-market consumer devices such as GPS-equipped telephones and handheld computing devices, or personal navigation aids, would seem to best meet the needs of LAD-aided surveys, but usually these are not designed to be powered up continuously for long periods, although improvements are constantly emerging. Custom-designed miniaturised survey devices, on the other hand, have not represented a large enough market to attract the capitalisation necessary to make them as reliable and robust as is required.

2 The combination of cellular positioning with GPS (commonly known as Assisted GPS or A-GPS).

Chapter 2
Challenges of Sensor Discovery

Simon Jirka
Westfälische Wilhelms-Universität, Germany

ABSTRACT

This chapter addresses the challenges of sensor discovery on the background of the Sensor Web Enablement architecture of the Open Geospatial Consortium. In the first part of the chapter two kinds of sensor discovery are introduced: sensor instance discovery and sensor service discovery. After this, four key challenges are presented: the dynamic structure of sensor networks, the integration of sensor status information, the semantics of the observed phenomena and sensor metadata models as well as metadata harvesting mechanisms. Additionally, a practical solution for sensor discovery is introduced. This solution is integrated into the Sensor Web Enablement architecture and aims at addressing the challenges of sensor discovery.

INTRODUCTION

Sensor networks have become a very important means for gathering data about a broad range of physical, chemical, and biological characteristics and phenomena. The scenarios in which geosensor networks are used comprise use cases like weather station networks, environmental monitoring or crisis management systems (Stefanidis, 2006).

For achieving the aim of increased sustainability for mobility, sensor networks can play a significant role. For example sensor networks are used for measuring the traffic density, detecting traffic jams, monitoring real-time information about public transportation networks (e.g. train delays) or for measuring pollution caused by traffic. In addition mobile sensors (e.g. mounted on cars) may be able to deliver direct or indirect traffic data (e.g. deriving the traffic density from the speed of cars). This input delivered by sensor networks will make it possible to optimize traffic planning, avoid traffic congestion or to develop new applications for traffic management taking into account real time data.

However, for being able to use and flexibly integrate the information collected by sensors into

DOI: 10.4018/978-1-61520-769-5.ch002

applications, the availability of discovery mechanisms for sensors and sensor data is essential. Especially due to the usually distributed character of access points to sensors and sensor networks the user must be able to determine which source provides the data that is needed for a specific task or which components allow submitting certain tasks to specific sensors. This means that a user must be able to find the sensors that are available (e.g. searching for all traffic density sensors in an area), all sensor data that is available (e.g. all traffic density measurements performed on a certain street segment at a given time) and how the data can be accessed.

Compared to conventional data sources (e.g. maps, geometry information), sensor networks possess special characteristics that need to be taken into account when creating a catalogue or registry for sensors and sensor data. Especially the often highly dynamic structure of sensor networks must be considered as new sensors may be deployed, older sensors may disappear or mobile sensors may move to new areas (Dantu et al., 2005). Furthermore, additional information sources like sensor status data and specific sensor metadata formats have to be considered.

This chapter starts with a short overview of related work in the field of sensor discovery. After this, an introduction into the Sensor Web Enablement (SWE) architecture of the Open Geospatial Consortium (OGC) will be given. The SWE framework is currently the most important solution for integrating sensors into distributed applications and especially spatial data infrastructures by using the concepts of service oriented architectures. As this chapter will address the topic of sensor discovery on an application layer level, SWE will be an important technological foundation for illustrating how sensor discovery can be addressed in practice.

After this, challenges that have to be considered when enabling the discovery of sensors will be presented:

- Dynamic network structure.
- Integration of sensor status information.
- Exploitation of semantic relationships for describing the observed phenomena.
- Metadata models for sensor networks.

In addition, this chapter will describe an approach how the previously presented challenges can be handled in practice. For this section the SWE architecture will serve as a background. A SWE based approach for a sensor catalogue/ registry framework will be shown that aims at tackling the challenges of sensor discovery. Thus, the practical impact of these challenges will be illustrated.

RELATED WORK

Although this chapter will focus on the OGC SWE framework, in this section several further activities dealing with sensor discovery outside the SWE context are shortly presented.

Whereas the SWE architecture concentrates on the sensor integration on an application layer level, other standards deal with sensors in a way that is closer to the sensor hardware. An important example for such a suite of standards is IEEE 1451 (http://ieee1451.nist.gov/). This set of standards allows describing sensors (and actuators) as well as their inputs and outputs. Although first approaches for mapping the IEEE 1451 world to the SWE framework are available (Hu, Robinson, & Indulska, 2007), future research will be necessary to integrate IEEE 1451 sensor networks into (SWE) discovery solutions.

Within this chapter, the SWE standard SensorML will be used as a metadata format as it is already applied in a broad range of applications. However, for future work also additional metadata formats will be considered. One example for such an additional metadata format is described by Nath, Lie, and Zhao (2006). The Sensor Description Markup Language (SDML) possesses a reduced complexity compared to SensorML as it does not address the description of the internal sensor structure. In addition the authors describe an approach for harvesting sensor metadata from web sites through which sensors are made available.

Finally, the topic of using peer-to-peer organization models for sensor networks shall be mentioned. Currently several efforts are made to investigate on using the peer-to-peer technology for developing sensor networks with a higher scalability and improved performance compared to conventional sensor networks that are organized in a centralized manner. With regards to the discovery approach discussed within this chapter, sensor discovery within peer-to-peer networks will require new solutions as central search indexes will not be available. A solution like the peer-to-peer GIS architecture presented in Liang (2008), seems to be very promising. Future research will have to address the integration of peer-to-peer organization models into the SWE architecture and thus also the combination with SWE based sensor discovery solutions.

OGC SENSOR WEB ENABLEMENT

The term Sensor Web Enablement goes back to a working group with the same name within the OGC. Consisting of more than 380 members, the OGC is an international organization that aims at the development of standards that form the foundation of geospatial data infrastructures. The standards, which address service interfaces as well as data formats, are developed within an open consensus process (http://www.opengeospatial.org/).

Within the OGC, the SWE working group is focusing on developing a standardized framework that enables the integration of sensors and sensor networks into spatial data infrastructures and thus also into any kind of application. In order to achieve this goal the SWE framework defines a set of standards for service interfaces and data formats. The following aims have been of special importance during the development of the SWE architecture (Botts, Percivall, Reed, & Davidson, 2007b):

- Discovery of sensors and sensor data.
- Description of sensors and their measurements.
- Real-time access to measurements and provision of time series data.
- Tasking of sensors.
- Alerting based on measured values and user defined alert conditions.

The SWE framework can be dived into two main parts: the information model dealing with data encodings and the service model comprising standards specifying the interfaces of the SWE services. The following standards form the information model:

- **Sensor Model Language (SensorML):** This encoding is intended to allow the standardized description of sensor metadata. Thus, the SensorML specification is a core element for realizing any kind of sensor discovery service (Botts & Robin, 2007a).
- **Observations & Measurements (O&M):** The O&M standard provides means for encoding the measurements of a sensor (Cox, 2007).
- **Transducer Markup Language (TML):** This format combines the capabilities of SensorML and O&M in a manner that is optimized for streaming data. However as TML is very seldom used, it will not be considered within the presented discovery system (Havens, 2007).

Within the SWE service model the interfaces of web services, providing different types of sensor related functionality are defined. This comprises the following standards:

- **Sensor Observation Service (SOS):** The SOS provides a pull based access to sensor data which are usually returned using the O&M format (Na & Priest, 2006).
- **Sensor Alert Service (SAS):** The core functionality of the SAS is the push based delivery of sensor data based on user defined criteria (e.g. deliver all measurements that exceed a certain threshold) (Simonis, 2006).
- **Sensor Planning Service (SPS):** The SPS has been defined for tasking sensors. This comprises also functionality for managing the tasks of a sensor or the ability to check if it is feasible to execute a requested task (Simonis & Dibner, 2007).
- **Web Notification Service (WNS):** The WNS is a helper service for realizing asynchronous communication between SWE services and/or clients (Simonis & Echterhoff, 2006).

Like all OGC web service interfaces also the SWE services provide an operation for requesting self-descriptions of service instances. Furthermore it is possible to retrieve (SensorML encoded) metadata about the sensors encapsulated by a SWE service, although the way how to access the metadata varies from service specification to service specification. These two elements form the basis for realizing harvesting mechanisms in order to collect the necessary metadata. Currently the SWE architecture offers no specialized service for enabling sensor discovery. However in the OGC Context a catalogue service is defined (Nebert, Whiteside, & Vretanos, 2007). Later an approach will be described how this existing catalogue can be related to SWE discovery. In summary, the OGC SWE framework offers standardized web service interfaces and data formats that allow the interoperable access to sensor data, the alerting based on measured values and means for controlling sensors.

WHAT IS "SENSOR DISCOVERY"

This chapter aims at discussing the challenges of sensor discovery and potential solutions on the background of the SWE architecture. The focus is put on ways to enable sensor discovery from the perspective of service oriented architectures, whereas discovery mechanisms which work on a lower level of the network organization model will not be addressed. Within this chapter two different viewpoints of sensor discovery are considered: sensor instance discovery and sensor service discovery.

A common aim of both discovery types is to search for matching catalogue or registry entries that provide further information about the requested sensors or sensor service instances. This information provides clients the necessary knowledge to access or use the sensors or sensor services he needs. Whereas the first of the two viewpoints concerns the search for individual sensors or sensor systems the second viewpoint deals with finding the access points (in case of service oriented architectures: the (web) services) which provide access to the sensors. The next two sub sections will discuss these two types of sensor discovery.

Sensor Instance Discovery

The term sensor instance discovery is defined as the discovery of sensors based on criteria like:

- **Thematic aspects:** Finding sensors based on the phenomenon they are measuring or the information they are delivering.
- **Location:** Finding all sensors within a certain area or at a certain location.
- **Time dependent data availability:** Finding all sensors that provide data for the time period the user is interested in.
- **Sensor (status) properties:** Finding all sensors that possess certain sensor specific properties.

- **Sensor identification:** finding a specific sensor or a set of sensors based on their id(s).

It is important to note that these criteria are usually not used in an isolated way. Instead, in many cases these criteria are combined in order to form more complex search requests. Besides these very common criteria, further properties have to be taken into account. Additional search criteria range from information about the reliability and accuracy of the measurements, usage restrictions, sensor model or sampling rate to any full text search within the sensor metadata descriptions. In summary sensor instance discovery addresses the search for individual sensors within a sensor network and the retrieval of according metadata descriptions of these sensors.

Sensor Service Discovery

In the previous section the idea of sensor instance discovery was introduced. However, there is usually a need for integrating sensor data into higher level applications (e.g. as input for simulation models that predict the traffic density). In order to feed these models with the necessary sensor data, the access points of the according sensors must be discovered.

For integrating sensor data into any kind of application, service oriented architectures are a common approach (i.e. the SWE architecture). Within an architecture consisting of web services, there is a need for finding those web services which provide the desired functionality (i.e. data access, alerting, tasking). Furthermore it is necessary to determine which sensors are encapsulated by which web service instances. This leads to search requests based on one or more of the following criteria:

- **Thematic aspects:** Finding web services based on the phenomena and observed properties they are dealing with.
- **Location:** Finding web services based on the area they are providing data for.
- **Time dependent data availability:** Finding all web services that provide data for the time period the user is interested in.
- **Usage restrictions:** Often sensor data is underlying legal restrictions. This aspect has to be taken into account when searching for an access point to sensor data.
- **Functionality:** The access points to sensors or sensor networks can offer different kinds of functionality (e.g. data access, alerting and sensor control). For a user it is important to state in a search request which of these functionalities he needs.

Also in the case of sensor service discovery the search criteria are usually combined. Although the search criteria in case of sensor service discovery are quite similar to the ones of sensor instance discovery it is a different approach. In this case, finding individual sensors is not in the focus. Instead, this approach is based on a data centric viewpoint. This means that a user queries for the sensor data or sensor related functionality he needs. The question which specific sensors are measuring the data is usually not of direct interest.

In summary sensor service discovery allows to find information about the sensor services that either encapsulate specific sensor instances or that provide a requested type of information or functionality. As a conclusion it can be stated that the term sensor discovery as used in this chapter connotes the process of searching within catalogues or registries for entries that describe and point to either sensor instances or sensor services.

CHALLENGES OF SENSOR DISCOVERY

In order to enable the two sensor discovery types described in the previous section, several challenges and problems have to be solved. In the next subsections four key challenges that are specific to sensor discovery are introduced.

- Dynamic network structure.
- Integration of sensor status information.
- Semantics.
- Metadata models and harvesting.

These challenges will show the special character of sensor discovery compared to other kinds of discovery mechanisms (e.g. discovery of web services within spatial data infrastructures). For each of these aspects, examples will be given and their implications on the design of sensor discovery mechanisms will be discussed.

Dynamic Network Structure

Compared to conventional data sources (e.g. static maps or vector based geodata) sensor networks are characterized by a much more dynamic structure. Whereas maps are often updated in very large time intervals (sometimes up to years), sensor networks are usually underlying much quicker changes. This is obvious for mobile sensors which move around and provide measurements for changing locations. However, stationary (non mobile) sensor networks are changing their configuration, too: sensors may change their sampling rate or measurement parameters, new sensors may be deployed or defective sensors may be removed (Heinzelman, Murphy, Carvalho, & Perillo, 2004). Another aspect is that sensors may be even deployed on demand (e.g. during road construction work additional traffic density sensors for an optimized traffic management might be deployed). Hence, the availability of sensors is quickly changing. As a consequence, any discovery solution must face the challenge of a potentially rapidly changing availability of sensors and sensor data.

One of the most common queries in the field of sensor networks is the search for sensors that deliver measurements regarding a certain phenomenon in a region of interest for a specified time period. However, as described before, the structure of a sensor network may continuously change. Thus, depending on the time period there may be different sensors that have measured the phenomena the user is interested in. This leads to the requirement that a sensor catalogue or registry must keep track of the sensors, the phenomena they are measuring and their time dependent locations. Obviously this is relevant in case of mobile sensors. But also static sensor networks, especially in case of sensors with a short life time, are affected by this issue. In order to solve this challenge two aspects have to be considered: On the one hand there is a need for a data model that is capable of handling such dynamics. For example, the concepts of moving objects databases are one possible approach which can be used for realizing such a data model (Güting & Schneider, 2005). On the other hand a flexible update mechanism is needed that allows refreshing the sensor geometry information within sensor catalogues and registries. This means that there must be an efficient way to notify a sensor catalogue or registry of changed sensor positions, configuration changes of sensor constellations or even the deployment or removal of sensors. However, as sensor networks can often comprise thousands of sensors this update mechanism needs to support a level of granularity that avoids excessive amounts of data transmission but at the same time ensures a sufficiently accurate knowledge of the sensor network structure within the catalogue or registry.

However, when looking beyond the area of sensor instance discovery another challenge arises with regard to the sensor service discovery. As explained before, in a service oriented framework like the OGC SWE architecture the data measured by sensors or the tasking of sensors is made available through according web service instances. Often, the links between a sensor and the SWE services change during the course of time or the same sensor can even be available through multiple services. For example, the raw data collected by a sensor can be offered by multiple re-sellers which provide their own sensor services. Another good example illustrates the situation that a sensor is linked to several sensor services at the same time: In case of a controllable pan-tilt-zoom camera (e.g. for monitoring the traffic at critical locations) there might be one service for tasking the camera (e.g. setting the zoom) and one service for accessing the recorded video data. When a user wants to retrieve the data that was recorded by the camera he has tasked, he needs to find also the data access service encapsulating the same sensor that was controlled through the tasking service. Thus, there is not only a need for handling the dynamic structure of the sensor network itself, but also the potentially varying link to the sensor web services. Another aspect that has to be mentioned when dealing with the links between sensors and sensor services is the identity of sensors. Often, the usual approach is that each service instance assigns a service specific sensor ID to a sensor. Thus, in one service a sensor might have the id "sensor1" whereas in another service the id might be "sensorA". But for being able to recognize that the data provided by two sensor services is basically the same or was delivered by the same sensor, a service independent id is needed. At the moment, an approach based on globally unique identifiers like Uniform Resource Names (URN) seems to be promising. However, the problem of assigning and ensuring the uniqueness is still not solved.

To sum it up, the dynamic structure of sensor networks creates specific challenges for the design of discovery mechanisms. This includes especially the frequently changing availability of sensors, changing sensor configurations and positions and finally changing links between sensors and sensor services.

Integration of Sensor Status Information

For several discovery tasks also the sensor status plays an important role. Within this chapter the status of a sensor is understood as all kind of data that describes (dynamic) properties of a sensor that are not directly related to the measured phenomenon (e.g. active/inactive, sensor health, battery level). A common use case concerns the administration of geosensor networks. Whereas users are usually interested in retrieving data measured by sensors, the operators of the networks also need to be able to retrieve sensor status information. An important example is the monitoring of sensor properties like the battery state or the general health of a sensor. In order to replace sensors which are defective or need to be serviced (e.g. changing the battery), a sensor catalogue or registry must provide search functionality using the sensor status. A typical search request in this context would be: "Search for all sensors located in the city of Münster with a battery level lower than 10%".

Although the sensor network maintenance is one of the most important reasons for integrating sensor status information into search mechanisms, also discovery requests made by users can significantly benefit from taking into account the sensor status information. A typical example is the tasking of sensors. If a user wants to perform a change of sensor parameters or settings, it has to be ensured that the sensor is not blocked by other tasks and that it is able to execute such a request. For example a pan-tilt-zoom camera is only able to look into one single direction during one instant of time and cannot execute multiple user requests in parallel. Thus, the integration of sensor status information into sensor catalogues and registries provides significant advantages when finding those sensors that can be tasked by a user or that are able to execute a certain request.

In addition, the integration of sensor status information can be used for addressing the previously mentioned challenge created by the dynamic sensor network structure. By integrating the sensor status property "location" into the search indexes of a registry, it is possible to execute spatial search requests for sensors taking into account the latest sensor position information.

It can be concluded that sensor status information offers the possibility to enhance the sensor discovery process (e.g. taking into account changing sensor positions) and that it is an important factor in case sensor discovery functionality shall be used for sensor network management purposes.

Semantics

When searching for sensor data, semantics often play an important role (Lutz & Klien, 2006, McFerren, Roos, & Terhorst, 2006). One of the most interesting aspects within this context concerns the descriptions of the phenomenon or property that is observed by a sensor: How can it be described what a sensor is observing? If sensor data shall be used across different domains, it is important to describe the meaning of the terms that are used. Thus, for sensor discovery not just a purely string based approach for searching sensors measuring certain phenomena would be used. Instead the discovery would rely on pointers to the concepts that are of interest to the user.

In two different domains the same phenomenon might be identified by two different names although is means the same. For example in one context the phenomenon "carbon dioxide concentration" might have the name "carbon_dioxide_concentration" whereas it is named "CO2_concentration" somewhere else. Although both names mean exactly the same it is not possible to find all "carbon dioxide concentration" sensors by just executing a string based search. Although a basic solution for addressing this issue would be to rely on controlled vocabularies, it might become difficult to agree on such a controlled set of terms across multiple domains. Thus, relying on ontologies and on mappings between ontologies seems to be a more promising solution.

Another aspect is that users sometimes do not have an exact idea for which specific phenomena they need information. For example in the scenario of an accident of a chemical transport truck a cloud consisting of unknown pollutants is released. In order to find out which pollutants are contained within the cloud the user needs to find all sensors which measure some kind of air pollutant concentration. Also in this case a purely string based search would not help to fulfill the needs of the user. Instead, by using references to the semantics of the observed phenomena, it can be discovered which phenomena can be considered as an air pollutant (i.e. by applying reasoning techniques to ontologies). By exploiting the meaning of phenomena, the user gets the ability to search for sensors that observe phenomena which are similar (e.g. equivalent, generalization or specialization) to the one he is interested in.

In summary the consideration of semantics makes it possible to develop more sophisticated discovery mechanisms that go beyond a purely text/string based search and take into account the meaning of observed phenomena.

Metadata Models and Harvesting

Finally, for implementing a sensor and sensor service catalogue or registry, it is necessary to collect metadata about sensors and sensor services. Only the indexing of such metadata will make it possible to answer search requests for sensors and sensor services. Two aspects of metadata collection must be considered:

- Metadata models for describing the information needed for discovering sensors.
- Harvesting mechanisms for (automatically) retrieving the necessary metadata.

In order to provide the metadata about sensors and sensor systems in an interoperable way, it is necessary to rely on standardized metadata models and formats. This makes it possible for catalogues to interpret the metadata about any sensor as long as it follows the according metadata standards. However, just defining the metadata format is not sufficient. There must also be some kind of rules which content a metadata document must provide. The necessary information ranges from the sensor position, the observed phenomenon or the time for which the metadata document is valid to legal constraints regarding the data delivered by the sensor. One good example for such a metadata standard will be provided later within this chapter: the OGC Sensor Model Language (SensorML). A special focus will be put on the creation of metadata profiles for sensors that describe a minimum set of information necessary for enabling sensor discovery.

If the metadata descriptions of a sensor are available in a standardized format, there are two ways to integrate them into a registry: manually entering the metadata or automatic harvesting mechanisms. As the first of these two approaches results in a quite big amount of work, especially in case of a large number of sensors, any kind of automatic solution is advantageous. On the one hand, there is a need for collecting data about the sensors on a lower level directly within the sensor network. An example is the search within peer-to-peer networks which is a common organization model for sensor networks. If a sensor joins such a sensor network, registration mechanisms are needed in order to make it discoverable. On the other hand, collecting metadata on the application layer level is crucial for allowing sensor service discovery. Here, it is necessary to collect information about the available sensor services as well as the sensors they encapsulate. Thus, when developing a discovery solution for sensors and sensor services, an approach must be provided that relies on the service interface standards by using those operations that provide the access to standardized metadata.

As a conclusion it can be stated that standardized metadata models for sensors and sensor services form the basis for automatic metadata harvesting mechanisms. This allows handling large sensor networks which would be difficult or even impossible if the database of a catalogue has to be managed manually.

HOW TO HANDLE SENSOR DISCOVERY IN PRACTICE

In this section an approach will be described how the challenges of sensor discovery can be addressed in practice. The SWE architecture of the OGC will serve as a background framework for which the design and realization of a discovery solution will be described.

The Sensor Instance Registry (SIR) as well as the Sensor Observable Registry (SOR) described below was developed within the EC project OSIRIS which aimed at building risk monitoring and crisis management applications based on the SWE architecture. The OSIRIS components, and thus also the presented discovery solution, were successfully tested within four practical scenarios: forest fires, water pollution, fires in industrial buildings and air pollution (OSIRIS Consortium, 2008a).

In the first of the following two sections the SIR will be described which provides functionality for sensor instance discovery as well as sensor service discovery. The issue of linking sensor discovery to the meaning of the observed phenomena will be discussed in the section about the SOR. Furthermore, an approach for a SWE metadata model as well as the integration into existing Catalogue services will be treated.

The Sensor Instance Registry

Within OSIRIS project mentioned before there was a need for a registry managing the metadata of sensor instances as well as their links to SWE services.

The interface of the SIR (OSIRIS Consortium, 2008b) comprises operations for handling sensor descriptions (based on SensorML), sensor status data and the information how a sensor can be accessed or controlled. Basically the SIR interface can be split into two parts: sensor discovery and sensor status handling.

With regard to sensor discovery the SIR is capable of discovering requested sensors, SWE services which provide a requested type of information and SWE services which encapsulate a given sensor. In order to expose this functionality the following operations for searching sensors and entering metadata are specified:

- SearchSensor for searching sensors based on user defined criteria (observed phenomenon, observed area, temporal criteria, text fragments, unit of measurement, sensor id, SWE service instances).
- DescribeSensor for requesting a SensorML description of a specific sensor contained in the SIR.
- HarvestService for automatically harvesting the metadata about all sensors that are contained in a SWE service.
- InsertSensorInfo for manually entering or changing sensor metadata within the SIR.

In addition, the SIR offers several operations that support the handling of sensor status information. The following operations are offered in order to make the sensor status handling functionality available:

- GetSensorStatus for retrieving the status of a specific sensor or a set of sensors that match user defined criteria.
- SubscribeSensorStatus for subscribing to status information of a specific sensor or a set of sensors that match user defined criteria. This includes also status based filter criteria.
- InsertSensorStatus for inserting and updating sensor status information. This is especially important in order to ensure that new sensor status information can be integrated into the SIR as soon as it is available.

For fulfilling the different kinds of search requests the SIR needs to deal with text based data, spatial data and temporal data. In order to handle these data types the SIR relies on three different search index types. For the text based search a text indexing library is used whereas a quadtree based solution is applied for the spatial indexing. Finally, a temporal index is integrated for an optimized support of time dependent metadata.

Although the SIR was mainly developed for supporting sensor instance discovery it also supports sensor service discovery. For each sensor it maintains a list of SWE services to which the sensor is connected. Thus, it is able to return for each sensor a list of SWE services. The SIR as described in this section was developed and successfully deployed within the OSIRIS project. However, the work on the SIR will continue. Especially the optimization of the search indexes and harvesting mechanisms will be one of the most important goals for the future.

The Sensor Observable Registry

As indicated before, the SOR can be seen as an approach for handling in a simple way the semantics of the phenomena observed by sensors (OSIRIS Consortium, 2008b). This means that the SOR is capable of handling the references to definitions of phenomena and observables as well as linking these definitions to ontologies where the according concepts are defined in a formal way. Additionally, the SOR offers functionality for exploiting links between textual phenomenon definitions and formal ontologies: by using reasoning mechanisms the SOR is able to determine those phenomena which are similar to a given one (i.e. specializations and generalizations of a given phenomenon). This results in a set of operations which are defined in the SOR interface:

- GetDefinition for retrieving the definition of the phenomenon which is identified by a given URN; this definition includes a textual description as well as a pointer to an ontology providing a formal definition.
- GetMatchingDefinitions for retrieving a list of phenomena that are similar to a given phenomenon; this operation returns a list of URNs identifying the similar phenomena. Currently only generalization-specialization relationships are taken into account. However, in the future more complex relationships will be used for this operation.

For each phenomenon contained in the SOR an according an entry in an internal dictionary is available. The entries are identified by the phenomenon URN and contain the textual description of the phenomenon as well as a reference to an ontology where a formal definition can be found. Experiments performed within the OSIRIS project which relied on the SWEET ontology (NASA, 2008) lead to satisfying results, although for future solutions more complex and powerful semantic models may be desirable.

In order to benefit from the capabilities of the SOR it must be linked to the search functionality provided by the SIR. This can be achieved as follows (see figure 1): In the SIR specification it is foreseen that the user must identify the phenomena he is interested in by using URNs. These phenomenon URNs can then be used for making the link to the SOR. If the user wants to benefit from the SOR functionality, the SIR allows setting a flag in search requests so that the SOR will be used for finding sensors for similar phenomena. This means that the SIR receives the URN of a certain phenomenon in a search request. Then the GetMatchingDefinitions operation of the SOR is invoked and the SIR receives a list of URNs identifying the phenomena that are similar to the given one. Subsequently the SIR can query its search indexes for all phenomena that are identified by one of the returned URNs.

This leads for example to the following cases in which the SOR is able to allow a more sophisticated phenomenon based discovery:

- The URNs "urn:ogc:def:phenomenon:co2_concentration" and "urn:anystandardizationbody:observables:carbon_dionoxide:_concentration" may identify the same concept of the carbon dioxide concentration as it is defined in an ontology. Thus both URNs, although completely different, refer to the same observable. By asking the SOR for similar phenomena the SIR will return all sensors that measure a phenomenon identified by one of the two URNs.
- The URNs urn:ogc:def:phenomenon:temperature, urn:ogc:def:phenomenon:airTemperature and urn:ogc:def:phenomenon:waterTemperature point to different definitions of temperature. In an ontology it can be formally defined that air and water temperature are some kind of temperature. If a user just wants to search for sensors that measure any type of temperature the SIR

Figure 1. Interactions between SIR and SOR

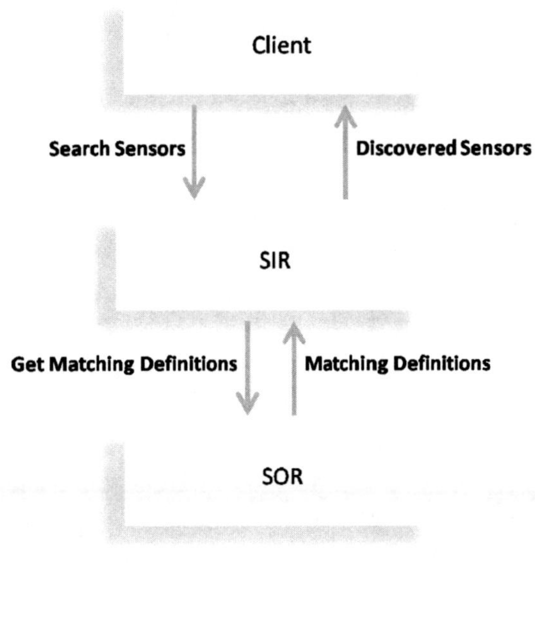

can request the SOR for finding all phenomena that are some kind of temperature. A list of URNs identifying different types of temperature is returned by the SOR and the SIR can extend the search also to the sub types of temperature and return the according sensors.

This functionality of the SOR has been successfully tested in the OSIRIS project. However, in the future it is desirable to use more powerful approaches for expressing the relationships and similarity between different phenomena.

Metadata for Sensor Discovery

The SIR was already introduced within this chapter. However, it was only described which kinds of requests and queries can be supported by the SIR and how the queries can be internally handled. In addition a solution is needed for filling the SIR with the metadata needed in its search indexes. Besides describing a SensorML profile which serves as a format for providing the necessary metadata also a harvesting mechanism will be sketched that performs an automatic analysis of SWE service instances and gathers the required information.

In the SWE introduction earlier in this chapter, SensorML was explained as a metadata format for sensors and sensor systems. Consequently SensorML is well suited for providing the metadata needed for populating the search indexes of a sensor registry. However, in its current state SensorML has one significant deficit. When looking at the SensorML specification it becomes quickly apparent, that most of the elements in the SensorML XML schema are optional and that there is no mandatory way how SensorML has to be structured for describing a specific sensor. The reason behind this is that SensorML is designed to support nearly any kind of sensor. However this leads to difficulties when interpreting SensorML documents and may lead to sensor descriptions which do not contain all information that is necessary for discovery purposes. A solution for this problem is to create a SensorML profile that specifies which information must be given if a SensorML document shall be usable for providing metadata to a sensor registry.

Based on the experiences of the OSIRIS project the following information has been identified as necessary (OSIRIS Consortium, 2008b):

- Identification of the sensor in order to uniquely identify it within a SWE service instance (the identification across different SWE services is a currently ongoing research topic).
- Keywords describing terms that help to discovery a sensor.
- Sensor type.
- Information about the time span in which the sensor description is valid (important due to the high dynamicity of sensor networks).
- Geometry data about the area or point that is observed by the sensor.
- Data about the sensor operator.
- Information which phenomenon is observed (ideally linked to the approach described in the SOR section).
- Description of the output of the sensor (which information in which unit of measurement is delivered).

These required items of sensor metadata have been used as an input for the formal definition of a SensorML profile using the Schematron language. Schematron allows to define additional constrains and rules for validating XML documents. Thus, by validating a SensorML description against the Schematron schema, it can be checked if all required information is given in the correct structure.

Besides the presented sensor metadata format, it is necessary to develop a mechanism to harvest metadata. A common characteristic of all OGC services is the so called GetCapabilities operation for retrieving a description of service instances and their contents. Thus, this operation makes it possible to receive a list of all sensors that are encapsulated by a SWE service. Furthermore, all of these services provide SensorML based metadata about the sensors they contain. However, the way these services make the SensorML documents available is varying: Whereas the SOS provides a dedicated operation for accessing sensor metadata (DescribeSensor), the SPS for example just returns links to SensorML files in its responses to a GetCapabilities request. This can be exploited by a harvesting metadata mechanism. In a first step a SWE service is examined and the sensors it contains are retrieved. Based on this, the SIR is able to link sensor instances and the services which encapsulate them. In a next step the SIR retrieves detailed metadata about the different sensors contained in the service. For example in case of a SOS this would lead to a series of DescribeSensor requests so that the SIR receives for each sensor a SensorML description.

Integration Into Existing Catalogues

When discussing discovery in the context of spatial data infrastructures, the OGC Catalogue Service (CS-W) has to be mentioned (Nebert et al., 2007). The CS-W provides functionality for searching data and web services within spatial data infrastructures. Thus, when integrating sensor networks into spatial data infrastructures it is desirable that they can also be searched using a CS-W.

In the sections before the SIR, which is highly specialized with regard to the characteristics of sensor networks, was presented. This results in a very powerful framework which, however, is not congruent with the functionality and data models of the CS-W. Thus, there is a need for linking the specialized sensor network discovery functionality to the CS-W so that on the one hand the CS-W is enabled to search sensor networks and on the other hand the extensive functionality of the presented approach is not lost.

Currently, the link between the CS-W and the presented sensor discovery framework is still a research topic. In a first step it has to be determined to which extend the specific sensor network discovery requirements can be covered by the CS-W. Depending on the outcome an architecture consisting of two layers seems to be promising. The SIR could be used as a highly specialized and powerful sensor network discovery tool which is used by those who need very sophisticated sensor discovery functionality. At the same time there should be a mechanism for generalizing and feeding the complex metadata managed within the SIR into CS-W instances. This would make it possible that those users, who do not need the complex SIR functionality, can send their queries to a normal CS-W instance like they would do for any other search request. This ensures a seamless integration of sensor discovery into the concepts of existing spatial data infrastructures while at the same time making the complex SIR functionality available on a different level.

Another aspect that is currently being worked on is the mapping of the specific sensor network metadata formats (i.e. SensorML) to the data models that are used by the CS-W. The two most important data models that are in the focus of the ongoing work are the ebRIM model as well as the standardized ISO 19115 metadata model. This will lead to a first integration of sensor metadata into existing catalogues so that basic searches for sensors and sensor data using a CS-W become possible. This mapping of SensorML to the existing catalogue data models will furthermore form the basis for the realization of feeding mechanisms from SIR to CS-W instances.

FUTURE RESEARCH DIRECTIONS AND CONCLUSION

In summary, this chapter has provided an insight into the requirements of sensor discovery and the special characteristics of sensor networks that have to be taken into account when designing a sensor catalogue or registry. Besides an introduction what can be understood by the term "sensor discovery", several challenges like the dynamic network structure, semantics, the integration of sensor status information and specific sensor metadata models have been presented. Their implications on the design of sensor discovery were discussed and potential approaches for dealing with them have been shown. Finally, it should be stated that the subject of sensor discovery is an ongoing research topic and that this chapter shall be understood as a description on the current state of the art with a special focus on the SWE architecture.

In the future, several ideas presented in this chapter may be enhanced in order to improve the SWE discovery framework. One important aspect will be the improvement of SWE metadata formats, and especially the definition of more comprehensive SensorML profiles. Furthermore, it is desirable to explore the possibility of integrating more powerful means for exploiting the semantic relationships between the observed phenomena. Especially the integration of similarity measures into the GetMatchingDefinitions operation of the SOR seems to be very promising. A third idea which has to be investigated further is the linking of the presented SWE discovery framework to existing catalogues and especially to the CS-W specification. Finally it is planned to connect the SWE discovery framework to lower level sensor network discovery mechanisms (e.g. to investigate how peer-to-peer sensor networks with their own discovery solutions can be integrated into the SIR).

However, already now the presented discovery approach enables the discovery within the SWE architecture and provides an effective solution that addresses the challenges of sensor discovery.

REFERENCES

Botts, M., Percivall, G., Reed, C., & Davidson, J. (Eds.). (2007b). *OGC White Paper - OGC Sensor Web Enablement: Overview and High Level Architecture (OGC Document Number: 07-165)*. Wayland, MA: OGC.

Botts, M., & Robin, A. (Eds.). (2007a). OpenGIS Sensor Model Language (SensorML) Implementation Specification 1.0.0 (OGC Document Number: 07-000). Wayland, MA: OGC.

Cox, S. (Ed.). (2007). Observations and Measurements - Part 1 - Observation schema 1.0 (OGC Document Number: 07-022r1). Wayland, MA: OGC.

Dantu, K., Rahimi, M., Shah, H., Babel, S., Dhariwal, A., & Sukhatme, G. S. (2005). Robomote: Enabling Mobility in Sensor Networks. In *IEEE/ACM Fourth International Conference on Information Processing in Sensor Networks (IPSNSPOTS)* (pp. 404 – 409), New York.

Güting, R. H., & Schneider, M. (2005). Moving Objects Databases, (Morgan Kaufmann Series in Data Management Systems). San Francisco, CA: Morgan Kaufmann.

Havens, S. (Ed.). (2007). OpenGIS Transducer Markup Language (TML) Implementation Specification 1.0.0 (OGC Document Number: 06-010r6). Wayland, MA: OGC.

Heinzelman, W.B., Murphy, A.L., Carvalho, H.S., & Perillo, M.A.. (2004, January). Middleware to support sensor network applications. *IEEE Network Magazine Special Issue*.

Hu, P., Robinson, R., & Indulska, J. (2007). Sensor Standards: Overview and Experiences. In *Proceedings of the 3rd International Conference on Intelligent Sensors, Sensor Networks and Information Processing ISSNIP'07*, Melbourne, Australia, December 3-6.

Liang, S. H. L. (2008). A New Fully Decentralized Scalable Peer-to-Peer GIS Architecture. *Int. Arch. Photogram Rem. Sens. Spatial Inform. Sci.*, *37*, 687–692.

Lutz, M., & Klien, E. (2006). Ontology-based retrieval of geographic information. *International Journal of Information Science*, *20*(3), 233–260.

McFerren, G., Roos, S., & Terhorst, A. (2006). *Fire alerts on the geospatial semantic web*. Paper presented at the Workshop Terra Cognita 2006, Athens, GA.

Na, A., & Priest, M. (Eds.). (2006). Sensor Observation Service 1.0 (OGC Document Number: 06-009r6). Wayland, MA: Open Geospatial Consortium Inc.

NASA. (2008). *Semantic Web for Earth and Environmental Terminology (SWEET)*. Retrieved February 20, 2009, from http://sweet.jpl.nasa.gov/ontology/

Nath, S., Lie, J., & Zhao, F. (2006). *Challenges in Building a Portal for Sensors World-Wide*. Microsoft Technical Report MSR-TR-2006-133. Redmond, WA: Microsoft Research.

Nebert, D., Whiteside, A., & Vretanos, P. (Eds.). (2007). OpenGIS Catalogue Services Specification 2.0.2 (OGC Document Number: 07-006r1). Wayland, MA: OGC.

OSIRIS Consortium. (2008a). *Deliverable 2400 - Scenario Definition. Paris: OSIRIS Consortium.* Retrieved from http://osiris-fp6.eu/doc/OSIRIS-WP2400-DEL-0017-scenario%20definition.pdf

OSIRIS Consortium. (2008b). *Deliverable 6001 Revision A - OSIRIS Architecture Specification and Justification. Paris: OSIRIS Consortium.* Retrieved from http://osiris-fp6.eu/doc/OSIRIS-WP6000-DEL-0037-Archi%20Spec%20and%20 Justification.pdf

Simonis, I. (Ed.). (2006). OGC Sensor Alert Service Candidate Implementation Specification 0.9 (OGC Document Number: 06-028r5). Wayland, MA: OGC.

Simonis, I., & Dibner, P. C. (Eds.). (2007). OpenGIS Sensor Planning Service Implementation Specification 1.0 (OGC Document Number: 07-014r3). Wayland, MA: OGC.

Simonis, I., & Echterhoff, J. (2006). *Draft OpenGIS Web Notification Service Implementation Specification 0.0.9 (OGC Document Number: 06-095).* Wayland, MA: OGC.

Stefanidis, A. (2006). The Emergence of GeoSensor Networks. *Directions Magazine.* Retrieved February 20, 2009, from http://www.directionsmag.com/article.php?article_id=2100&trv=1

Chapter 3
Ubiquitous Travel Environments and Travel Control Strategies:
Prospects and Challenges

Caspar Chorus
Delft University of Technology, The Netherlands

Harry Timmermans
Eindhoven University of Technology, The Netherlands

ABSTRACT

In the near future, ubiquitous travel environments in which travelers have direct access to information and communication technology will become widely available. In principle, it will allow various organizations and companies not only to provide real-time dynamic information but also to induce travelers to behave in ways that contribute with some underlying personal or system-wide objective (e.g. avoid risk, environmental concerns etc). The latter application assumes a sufficiently valid and reliable model of traveler response to information provision or recommendation. As these two fields have mainly developed in isolation, the purpose of this chapter is to provide an overview of the state of the art of these two fields.

INTRODUCTION

In recent years, transportation research has witnessed a tremendous growth of interest into the use of modern Information and Communication Technology (ICT) for data collection. Because of rapidly declining response rates of traditional travel surveys, modern mobile technology is viewed as an important alternative for these traditional surveys in collecting data about individual travel patterns. Mobiles devices range from laptop computers, Personal Digital Assistants (PDAs) and mobile phones to mobile game computers. In the near future, new technologies such as Augmented Reality may become widely available. Also changes in wireless networking will offer new opportunities: cellular telephony is moving from low-bandwidth to higher bandwidth (UMTS-Universal Mobile Telecommunications System) to support more advanced services (e.g. graphics). Wireless networks provide a higher bandwidth but are primarily used for laptops and PDA's. Hotspots support access to the internet. Bluetooth provides a low-bandwidth, short range protocol for communication between devices (e.g.

DOI: 10.4018/978-1-61520-769-5.ch003

mobile phone and head set). An upcoming technology is WIMAX (Worldwide Interoperability for Microwave Access) that provides a long range, fast connection and potentially competes with the UMTS standard.

Another line of transportation research that has gained substantially in academic interest concerns the provision of travel information in an attempt to trigger particular travel behaviors that would contribute to sustainable mobility. Originally, travel information was descriptive and distributed to a group of travelers. The tendency however is to move from static, descriptive information to dynamic, prescriptive information. Applications include tourist guides (e.g., Abowd et al., 1997; Arikawa, Tsuruoka, Hideyuki & Ome, 2007; Baus, Chevherst & Kray, 2005), leisure and health (e.g., Mc. Creadie et al., 2006) and travel assistants (e.g., Chorus & Timmermans, In Press; Hurtig, 2006; Rehl, Bruntsch & Mentz, 2007; Winter & Nittal, 2006). These developments lead to ubiquitous travel environments: the idea that information can be shared in a network environment from some geo-sensors providing users with information readily available anywhere, for any person and at anytime. This information can be descriptive, but can also relate to recommendations.

These two research frontiers in transportation research have largely emerged as separate lines of development, and researchers have predominantly made a contribution to either one or the other line of development. In this chapter, however, we argue that these separate literatures can potentially be combined in interesting and innovative ways. One of the key factors influencing the potential of using ICT to influence individual behavior in a particular way is to detect with the least possible error the current and future behavior of individual travelers. This requires accurate information in time and space about current and perhaps past behavior of individual travelers. When used in isolation, tracking technology such as GPS is insufficient in this regard as it only provides information about route choice behavior and the current position of the person tracked. However, in principles such data can be enhanced with other spatial and non-spatial data. Moreover, GPS (Global Positioning System) traces can be interpreted using particular algorithms to derive information about other facets of activity-travel patterns. This information can then be used to provide personalized travel recommendations. The problem here is that little is known about how travelers use such information, whether they act strategically and how they adjust their behavior. If however models of response behavior can be developed successfully, information can also be provided taking into account strategic behavior to better meet the objectives of the underlying control strategies.

In this chapter, using the prospect of merging these lines of development as a framework, we will discuss recent progress in both fields of research. In terms of data acquisition, we will discuss the development of the use of Geographic Information Systems (GIS), radio frequency identification (RFID) and other technologies to collect data on space-time behavior of individuals. This is followed by a discussion of models that have been suggested to predict traveler response to information and advice. In that context, it is relevant to distinguish between neutral personal information and control strategies that provide advice to individual travelers in an attempt to persuade them to adopt particular behaviors such as to optimize some underlying system-wide objective.

DATA COLLECTION STRATEGIES

Collecting data on people's space-time behavior has a long history in disciplines such as transportation, geography, tourism and urban planning. Such data have been predominantly been collected using travel surveys. Respondents were typically asked for one or more days to report the activities and travel they conducted during that time, plus some trip characteristics such as duration, motive, destination and transport mode. Because response rates have been declining, researchers and survey agencies have explored new data collection strategies, which are less demanding for participants. The use of GPS technology has been dominant in this regard.

Table 1 provides an overview of studies that examined the use of GPS to collect data on travel information. It shows that the late 1990's evidenced the first pilot studies on using GPS data collection. Most of these were small scale pilot studies and were based on in-car GPS systems. Later these were complemented with wearable devices. In the beginning, the devices were quite heavy, but the last decade has witnessed a rapid increase in light GPS devices with longer lasting batteries, PDA's and cellular phones with a GPS device. The availability of such systems and the fact that they are easier to use increased significantly the potential use of GPS data collection. GPS has been therefore been introduced into official data collections, predominantly in the U.S. (Wolf, Oliviera & Thompson, 2003). Moreover, Table 1 shows that in addition to transportation, it has been used in retailing, tourism and related disciplines.

The overview also shows that early studies have focused on the general acceptance of GPS, their accuracy and on the comparison of travel data collected using traditional means and by using GPS. These studies suggest that GPS outperforms traditional data collection approaches in terms of collecting more accurate spatial and temporal data on travel of vehicles and people. For example, Battelle (1997) concluded for the Lexington Area Travel Data Collection that respondents take more trips of shorter distances than past national estimates would suggest. Similar results were also found by Murakami and Wagner (1999).

In particular, GPS traces provide information about the route taken, distance covered, timing and duration of travel, and speed, while additional data can be derived from these data. For example, stops and the number of trips are often derived by searching the traces for some arbitrarily set period of no movement (usually from 30 seconds to 2 minutes; see e.g. Bohte & Maat, 2008; Chung & Shalaby, 2005; Forrest & Pearson, 2005; Li & Shalaby, 2008; Schönfelder, Axhausen, Antille & Bierlaire, 2002; Wolf, Schönfelder, Samaga, Oliveira, & Axhausen, 2004). In some cases, this may be too crude and additional information may help to identify trip ends and activity locations.

It should also be acknowledged however that although GPS traces provide a lot of information, they are not perfectly reliable. For example, Ohmori, Nakazato, Harata, Sasaki, and Nishii (2006) found that the averaged error in distance was about 50-150 meters. Moreover, data may be missing. This may be due to weak signals and other technology issues, respondent burden or inherent characteristics of the technology. For example, in-vehicle GPS systems not necessarily provide information about trip destination as the car may be parked and the last leg of the trip is likely to be performed by foot. Hence, GPS traces need to be filtered and some imputation will be necessary as errors may occur due to weak signals from a network of satellites, urban canyons and tunnels, limitations of the technology itself, such as the time required for the GPS device to locate the satellites (cold starts, see Stopher, Bullock & Horst, 2003), limited battery life, and errors due to human behavior (e.g., GPS not turned on).

If we wish to provide travelers with context-sensitive information or recommendations, in addition to these basic data, we need to interpret and enrich these data to induce what activity they are doing, and we may also need to capture information of their past performance and identify any routines and responses to particular information/advice. Table 1 shows that this kind of work is still in its infancy. Existing research relies on accurate and detailed geo-spatial information systems, including transport networks with stop locations. In addition, geo-coded land uses data (educational establishments, shopping centers, hospitals, parks, etc.) are used jointly

Table 1. summary of studies

Authors	Year	Country	Technology	Sample	Length	Facets					Learning	Remarks
						Trips	Route	Destination	Mode	Activity		
Auld et al	2009	US	Portable GPS device + web-based prompted recall	5	2 weeks	x	x	x	x	p	No	
Bellemans et al	2008	Belgium	GPS-enabled PDA	816	1 week	x	x	x	x	x	No	
Schuessller & Ax-hausen	2008	Switzerland	Portable GPS device	4882	6 days	x	x	x	x		No	Fuzzy logic
Bohte & Maat	2008	Netherlands	Portable GPS device + web-based prompted recall	1104	1 week	x	x	x	x		Limited	
Li & Shalaby	2008	Canada	Portable GPS device + web-based prompted recall	15	35 days							
deVries et al	2008	Eindhoven	Cell phone	8	1 day		x	x			No	Pictures, etc
Spek van der	2008	Europe	Portable GPS device	150-250	hours		x				No	pedestrians
Krijgsman et al	2008	South Africa	Cell phone	83-129	2 days	x	x	x	x		No	
Zou & Golledge	2007	USA	GPS+pocket PC	20	1 week	x	x	x	x	x	No	speech
Shoval & Isaacson	2007	Israel	Portable GPs device	1	hours		x				No	tourism
Asakura & Iryo	2007	Japan	Wearable GPS	56	1 day		x	x			No	tourism
Ahas et al.	2007	Estonia	Cell phone	Roaming 1 year			x	x			No	tourism
Hato et al.	2006	Japan	Cell phone + web-based prompted recall	100	1 month	x	x	x	x	x	No	
Ohmori et al.	2006	Japan	Cell phone + PDA	49	2 days	x	x	x	x		No	
Hato	2006	Japan	Multiple sensor	1	trip		x	x	x	x	No	Sound, pres-sure
Itsubo & Hato	2006	Japan	Cell phone + web-based prompted recall	31	5 days	x	x	x	x		No	No
Tsui & Shalaby	2006	Toronto Canada	GPS	9	1 day	x	x	x	x		No	Hot/cold stops Fuzzy logic
Doherty et al	2006	Canada	Portable GPS device + web-based prompted recall	1	1 day	x	x	x	x	x	No	

continued on the following page

Table 1. continued

Shoval & Isaacson	2006	Israel	Portable GPs device	1	2+ hours		x				No	pedestrians
Ohmori et al	2005	Japan	Cell phone + PDA	13-38	1 week	x	x	x			No	
Li et al.	2005	Atlanta	In-vehicler GPS	182	10 days	x	x	x			No	Commuter trips
Stopher et al	2005	Australia	Wearable GPS	32-51	6 days	x	x	x			No	
Stopher & Collins	2005	Australia	GPS device+ web-based prompted recall	29	1 day	x	x	x			No	
Chung & Shalaby	2005	Toronto	Wearable GPS	1	---		x		x		No	
Forrest & Pearson	2005	Laredo	In-vehicle GPS	150	1 day	x						
Axhausen et al.	2004	Sweden	In-vehicle GPS	186	30 days	x	x	x		x	No	Ad hoc processing
Wolf et al.	2004	Sweden	In-vehicle GPS	186	30 days	x	x	x		x	No	Ad hoc processing
Stopher et al.	2003	Sydney	In-vehicle GPS + prompted recall	52	5 days	x	x	x			No	
Pierce et al.	2003	USA	In-vehicle GPS + PDA	20	4 m	x	x	x			No	
Zmud & Wolf	2003	USA	Wearable GPS	292	20weeks	x	x	x			No	
Marca et al.	2002	USA	In-vehicle GPS	4	?	x	x	x		x	No	Kriging surfaces
Schonfelder et al	2002	Sweden	In-vehicle GPS	310 v	14 months	x	x	x		x	No	
Wolf et al	2001	USA	In-vehicle GPS	24	3 days	x	x	x		Partially	No	
Bachu et al	2001	USA	In-vehicle GPS + Prompted recall	18	2-3days	x	x	x			No	
Hato & Asakura	2001	Japan	Cellular phone + prompted recall	10	10 days	x	x	x	x	x	No	
Wolf et al.	2001	Atlanta, US	GPS	30		x	x	x	x	x		
Draijer et al.	2000	Netherlands	Wearable GPS	121	1 day	x	x	x			No	
Murakami & Wagner	1999	Kentucky, US	GPS + onboard computer	100	6 days	x	x	Cars only			No	
Guensler & Wolf	1999	Atlanta, US	GPS + handheld diary	---	---	x	x	x	x	x	No	

with ad hoc heuristics to induce activities (e.g., Wolf, Guensler, Frank & Ogly, 2001). As argued by Tsui and Shalaby (2006), and Schuessler and Axhausen (2008) fuzzy logic seems to provide new possibilities in extracting data on activity type from GPS traces. Unfortunately, no empirical results are presented yet in these studies. However, ongoing work based on Bayesian belief networks rather than fuzzy logic indicates that activities can be induced with sufficient accuracy (Moisteeva, Arentze & Timmermans, 2009).

Because GPS traces and the data imputation algorithms are not perfect, and because researchers may require additional data such as number of people traveling and costs associated with travel, so-called prompted recall surveys have become increasingly popular. It means that respondents receive the interpreted GPS traces and are invited to check these results, modify incorrect data and provide details for the additional questions. These so-called prompted recall instruments vary in terms of administration and the use of different technology to collect the information. As discussed by Doherty, Lee-Gosselin and Papinski (2006), Sequential methods involve systematic querying of subjects for missing or supplemental activity-trip pattern attributes either later by phone or internet. Alternatively, the data can be collected simultaneously to location tracking. Temporal/tabular prompted recall methods involve a time-ordered display of GPS or cellular phone-detected activities and trips, including start and end times, travel times/modes and activity locations. Spatial prompted recall methods involve use of a GIS or cellular phone to generate a map showing a person's routes, activity stops/location, and an array of text boxes or other map attributes depicting such items as mode, speed, location name, start/end times, trip and activity sequence (in the day), overlaid on the road and land-use network for context.

Prompted recall surveys have been typically used to provide check interpreted GPS traces and provide supplementary data for a single day. One aspect of GPS data that does not seem to have been taken into account is that multiple observations of essentially the same trip may allow a better identification of travel information, and that observations of other respondents may be used to better estimate missing information of a particular respondent. Moisteeva, Arentze and Timmermans (2009) argue that if learning and data imputation algorithms are applied, respondent burden should decrease over time and hence it may become increasingly feasible to collect multi-day travel information across longer periods of time, with acceptable respondent burden. A pilot study will start soon. Such multi-week information is critical in better understanding the dynamics underlying behavior and in assessing responses to information/advice (e.g. Axhausen, Zimmermann, Schonfelder, Rindsfuser & Haupt, 2002). Geo-information is required to identify activity types. Even if such data are not available immediately, however, prompted recall data can be used to collect data over time, using intelligent learning algorithms. Similarly, temporal information can be used.

INFORMATION CONTROL STRATEGIES

Based on these technologies and the use of advanced models and algorithms, it is feasible to derive with a high level of accuracy situational information and the behavior of the individual traveler, and the induced state of the transportation system at large in real time. This personal information in turn can be the input to personalized advice and to specific control strategies, trying to persuade travelers to adopt particular behaviors. The latter is a risky challenge. If the information provided does not take into account likely traveler response, the effects of the control strategies may be detrimental to the performance of the (transport) system. Hence, a successful application requires knowledge and models of traveler acceptance of and compliance with personalized information technology, theories and models of their strategic behavior and their adaptation of activity-travel patterns. In this part of the chapter, we will discuss newly emerging concepts and models and discuss their findings.

NON-STRATEGIC BEHAVIOR

In case of non-strategic behavior, individual choice behavior under information provision/advice is based only on the traveler's own (expected) utility and not on the behavior of other travelers. The dominant approach to modeling such behavior under uncertainty is that of expected utility maximization (Von Neumann & Morgenstern, 1947), although other preference measures are also applicable (see for example Chorus, Arentze, & Timmermans, 2008a, for an application of expected regret minimization). Under expected utility maximization, the individual is assumed to explore each choice alternative's relevant attributes, trade-off the utilities derived from the attributes for each possible state of the world and then choose the alternative which will maximize his/her expected utility.

Note that this expected utility maximization framework only deals with the assessment of already available or specified alternatives and not with alternative generation. In other words, it is simply assumed that the decision-maker knows all the alternatives he can choose from. Microeconomic search theory, however, applies the principles of utility maximization to the situation where an individual performs a sequential search for alternatives as well as an assessment of the alternatives generated, and can provide us with insights regarding this alternative generation process (e.g. Weibull, 1978). In such a sequential search process, apart from transaction costs, the cost of rejecting the most recently searched-for alternative can also be included. Such costs may be traded off against the expected utility to be derived from the next found alternative. The alternatives found are mostly assumed to be subsequently assessed following the principles of utility maximization. Several notions of microeconomic search theory can be found in travel demand studies. For examples see Williams and Ortuzar (1982), Lerman and Mahmassani (1985), Swait and Ben-Akiva (1987), Polak and Jones (1993), Ben-Akiva and Boccara (1995), Arentze & Timmermans (2005a,b) and Chorus, Walker and Ben-Akiva (In Press). There is one important difference between the application of utility maximization principles for alternative assessment and for alternative generation: utility maximization for alternative assessment deals with choosing from alternatives, while its application on alternative generation in addition to this also deals with choosing from decision strategies (should one proceed or stop searching?). Choosing a search strategy by applying utility maximization principles, the individual may well end up with an alternative having sub-optimal utility because the costs of searching are also taken into account in his decision strategy.

This principle of utility maximization behavior has not escaped criticism, and alternative principles have been suggested. A dominant critique on the validity of utility maximization principles as a base for individual decision-making was formulated by Herbert Simon, introducing the principle of bounded rationality (Simon, 1955). According to Simon, human beings cannot be assumed to have either the wish or the capability to perform extensive search processes and thoroughly assess the alternatives found. A perspective on decision strategies that does not make such strict assumptions is that of satisficing behavior (Simon, 1955); the individual is assumed to be searching for the first alternative that is good enough, i.e. he has certain aspiration levels for relevant attributes of alternatives – these levels may change over time - and searches for an alternative that meets these standards (Olander, 1973). Bounded rationality is often called procedural rationality, representing the idea that most decision-making is performed using simple 'rules of thumb'. An essential consequence of this perspective is that choice between alternatives, and with this choice also the generation and assessment of alternatives, are generally not driven by a determination of the traveler to use information in order to maximize some form of pay-off. Instead, a predecisional information search is performed to end up with an alternative of which relevant attributes meet the aspiration level set for that attribute. Notions of bounded rationality and satisficing behavior can be found in several travel demand studies (e.g. Mahmassani & Chang, 1987; Mahmassani & Jayakrishnan, 1991; Schofer, Khattak & Koppelman, 1993; Van Berkum & Van der Mede, 1993; Emmerink, Axhausen, Nijkamp & Rietveld, 1995; Emmerink, Nijkamp, Rietveld & Van Ommeren, 1996; Gärling et al., 2002; Chorus et al., 2006).

In addition to these principle of (bounded) rational behavior, which implicitly or explicitly assume that decision-making is more or less conscious process, it has been argued that many of the choices individuals repeatedly make are a consequence of the execution of a habit (e.g. Triandis, 1977, Hodgson, 2004), being defined as a repetition of past behavior without deliberating or forming an intention. Since no actual decision is made in the sense of generating and assessing alternatives, habitual behavior is often not regarded as decision-making. Several studies have pointed at the role of habitual travel behavior in the making of travel choices, especially mode choices (e.g. Aarts, Verplanken & Knippenberg 1997, 1998; Verplanken, Aarts & Knippenberg, 1997; Aarts & Dijksterhuis, 2000; Fujii, Gärling, & Kitamura, 2001; Fujii & Kitamura, 2003; Fujii & Gärling, 2003; Schlich & Axhausen, 2003). Although the past behavior on which habitual behavior is based was an actual decision, and could well have been optimal when it was first performed, a sub-optimal situation may arise as changes concerning alternatives and situations are not observed by the individual, because he does not consciously makes his decisions. Han, Arentze, Timmermans, Janssens, and Wets (2008) have formulated a more general, dynamic approach showing that habitual behavior may also result from learning: in the beginning individual will explore different alternatives until their aspiration are met. After that they will exhibit habitual behavior, until changes in the environment and changes in their aspirations result in payoff lower than their aspirations. Information provision or advice may be a trigger to lead travelers to reconsider their options.

Travel decisions are based under uncertainty: the environment is always in a state of flux, while particular characteristics of the system, such as travel times, are the accumulated results of the decisions of multiple actors. When provided with information or advice, one source of uncertainty is added: the credibility of the information. The provision of (travel) information, to the extent that it is reliable, implies a potential reduction of uncertainty about the state of the parts of the system. When receiving information, the question arises how the traveler will combine this information with already existing beliefs to create new beliefs. The extent to which he decides to use available information forms an upper limit of the degree of updating that may possibly be achieved. It is an upper limit since his mental processing ability and his perception of information unreliability may induce a further decrease in the degree to which perceptions are updated through information. There are two paths along which perceptions can be updated (Chorus et al., 2006): changing awareness of choice options and/or changing the perception of attributes of alternatives. Conceptually, the latter is more challenging as the literature has provided ample evidence that individual perceptions of characteristics of alternatives differ widely from their actual values (e.g. Mc Fadden, 1999, 2001; Zhao & Harata, 2001; Lyons, 2001; Li, 2003; Avineri & Prashker, 2003a, b; Bonsall, Firmin & Beale, 2004). It is well known that mostly, individuals overestimate positive features of chosen alternatives compared to non-chosen alternatives. Perception updating is thus an important issue and different approaches have been suggested. Most studies have focused on the updating of perceived travel times, but the basic ideas are applicable to other characteristics. Early studies concerned with the updating of travelers' perceptions have treated them as deterministic variables, consistent with information integration theory (e.g. Anderson, 1981; Einhorn & Hogarth, 1981). For example, Horowitz (1984) considers perception updating as a process of weighed averaging of initial perceptions and a new experience/the uptake of travel information. Other approaches have treated perceptions as stochastic variables and have used Bayesian principles of updating (Raiffa & Schlaifer, 1961; Edwards, Lindman & Savage, 1963): an individual has a prior perception of a characteristic, represented by a probability distribution. Using Bayes' rule, new information is used to update this prior distribution into a posterior distribution that becomes the updated perception. The degree to which the prior perception is updated depends on the individual's perceived unreliability of his initial estimates and the perceived reliability of the new information. Kaysi (1991) suggested using this approach in travel analysis. See Jha, Madanat and Peeta (1998), Chen and Mahmassani (2004), Chorus et al.(2006) and Chorus, Arentze and Timmermans, (2008b) for recent applications and developments. The updating of the attributes of travel destinations using a Bayesian Belief Network was introduced by Arentze and Timmermans (2005a, b). As the attributes of a destination represent the possible activities that can be performed at that destination, this latter application can also be seen as the updating or extension of choice-sets.

It should be mentioned here that the assumption of the individual being a Bayesian updater has been found to have limited behavioral validity in some situations, due to the assumed high degree of information processing capability, in particular an individual's way of dealing with conditional probabilities (e.g. Tversky & Kahneman, 1974; El-Gamal & Grether, 1995; Avineri & Prashker, 2003b). On the other hand, another substantial body of researchers in psychology and behavioral economics have found that while performing a faultless process of Bayesian learning may be too much to ask from most human beings, behavioral outcomes (such as choices) are often fairly in line with Bayesian predictions (e.g. Birnbaum, 1983; Viscusi & Magat, 1992; Gigerenzer & Hoffrage, 1995; Hakes & Viscusi, 1997; Gopnik et al., 2004; Aadland, Caplan & Phillips, 2007; Xu & Tenenbaum, 2007). Furthermore, it is generally felt that Bayes Theorem's theoretical elegance provides a very useful and elegant simplification of alternative, more complex learning models, which provides an additional explanation for its popularity in travel behavior research.

A recent study (Chorus, Arentze & Timmermans, 2009) shows how the paradigms of Expected Utility maximation and Bayesian learning can be combined to generate predictions of traveler compliance with personalized and non-personalized route-guidance. However, whereas virtually all studies mentioned above have examined updating as a function of one source of information, considering the single-stage decisions, in reality decisions may have ramifications for choices later during the day and information may be provided for different links in the network and different times of the day. Hence, especially in an activity-travel framework, a more general approach is required. Sun, Arentze and Timmermans (2005a, 2005b) addressed this problem. The decision problem is modeled as a decision tree representing all paths of possible decisions and outcomes of uncertain events. To avoid the exponential grow of the branches, individuals are assumed to apply heuristics to reduce the number of combinations. To deal with multiple sources of uncertainty, travelers are assumed to use mental scenarios. Each scenario represents a possible outcome of the combinations of possible decisions and outcomes of uncertain events. The traveler assigns probabilities to each outcome indicating how likely this outcome he perceives to be. Under the assumption of rational behavior, possible schedule decisions are evaluated under each possible outcome and the decision that maximizes the expected utility is chosen. A probability distribution across the outcomes of an event represents the uncertainty. It is further conditioned upon context variables. Assume Y is the information received (before the event) and X is the outcome experienced (after the event) towards the same uncertain event, P(Y | X) then represents the individuals beliefs regarding the probability of a message Y (i.e., 'the travel time will be Y minutes') given the true state X (i.e., the true travel time is X minutes). Hence, it reflects the degree of credibility the individual attaches to the information source.

The probabilities representing an individual's beliefs are updated each time an event is experienced during the implementation of the schedule. Beliefs are updated according to Bayesian principles:

$$P^{t+1}(y_i \mid x_k) = \frac{P^t(y_i \mid x_k)W^t(x_k) + w^t}{W^t(x_k) + w^t} \quad (1)$$

$$P^{t+1}(y_j \mid x_k) = \frac{P^t(y_j \mid x_k)W^t(x_k)}{W^t(x_k) + w^t} \quad (2)$$

$$W^{t+1}(x_k) = W^t(x_k) + w^t \quad (3)$$

$$P^{t+1}(y_j \mid x_l) = P^t(y_j \mid x_l) \quad \forall \text{j,l} \neq \text{k} \quad (4)$$

$$W^{t+1}(x_l) = W^t(x_l) \quad \forall \text{j,l} \neq \text{k} \quad (5)$$

where y_i denotes the observed state of Y, x_k is the observed state of X, w_t is the weight assigned to the *t-th* case and $P^{t+1}(y \mid x)$ is the probability representing the individual's updated belief in $Y = y$ given that $X = x$. In applications, Y represents some uncertain event and X represents the contextual variables on which the beliefs depend. W t denotes a running total which can be seen as a measure of the amount of experience accumulated until time t. It implies the impact of the past experience on belief updating. In non-stationary systems it is rational to assign more weight to more recent cases, namely, w^t should decay as a function of time. The perceived credibility of an information source is updated based on the same mechanism.

Updated beliefs are then input to a utility function for the activity-travel schedule. In addition, it is assumed that individuals adopt a myopic view when they generate or revise a schedule under conditions of uncertainty for a given day. This heuristic is mainly meant to avoid the combinatory problem that arises when facing multiple uncertain events. This heuristic assumes that travelers consider one uncertain event, with all its possible outcomes, at a time rather than considering the combinatory space of all possible outcomes of all uncertain events in a day.

STRATEGIC BEHAVIOR

In case of non-strategic behavior, individuals do not take into account the fact that other individuals may receive similar information/advice and that their response/decision should take the decisions of other into account. For example, when travelers arrive at a junction and receive advice which route to take, the experienced travel times after selecting a route will depend on the shares of travelers complying with this advice. Under particular circumstances, individuals may be better off by not complying with this advice. This type of decision problem in general is relevant for congested systems.

Han and Timmermans (2007) provide an overview of models of strategic behavior that may be relevant for different situations. They assume a set of $N = \{1, 2, ..., n\}$ individuals, each with a strategy set $S_i = s_{i1}, s_{i2,} ..., s_{iJ}$ of J_i pure strategic choices. $s = (s_1, ..., s_n) \in S$ is a strategy combination, and it consists of n strategies, one for each individual. $s_{-i} = (s_1, ..., s_{i-1}, s_{i+1}, ..., s_n)$ is a strategy combination of all individuals except i. Each individual has a payoff function, which represents the utility that an individual derives from selecting a particular strategic response in reaction to the strategies that other individuals might choose. An individual i's payoff for strategy j consists of three components: (1) a choice alternative specific component, $U^{exogenous}$, which expressing the exogenous attractiveness for a given alternative; (2) an interaction component, $U^{endogenous}$, which capturing the expected impact of other individuals' choice behavior; (3) an idiosyncratic error term, ε, so-called individual "type", treated as an individual and alternative specific random variable whose distribution is common knowledge among all individuals, but whose exact value is private information to the individual. Thus,

$$U_{ij} = U_j^{exogenous} + U_j^{endogenous} + \varepsilon_{ij} \quad (6)$$

$$U^{endogenous} = h(f_j(n), n) \quad (7)$$

$f_j(n)$ allows a flexible specification of the effect on utilities due to inter-individual interaction, which depends on the characteristics of the context. The choice of an identical linear utility function would imply that the marginal impact of other individuals choosing the same alternative is additively constant across different alternatives. Alternatively, nonlinear functional forms would allow the impact to increase or decrease at a varying rate.

In equilibrium, we expect that individuals take into account the expected decisions of all other individuals when choosing their strategies. That is in equilibrium, individuals choose strategies $(s_1^*(\varepsilon_1), ... s_i^*(\varepsilon_i), ..., s_n^*(\varepsilon_n))$ such that, for each individual i and each of its possible types ε_i, strategy $s_i^*(\varepsilon_i)$ maximizes its expected payoff across alternatives. Following traditional game theory (Fudenberg & Tirole 1991), the Bayesian Nash equilibrium constitutes the optimal response that maximizes an individual's expected payoff, given its conjecture about other individuals' strategies. Each individual's pure-strategy Bayesian Nash equilibrium choice is formulated as:

$$s_i^*(\varepsilon_i) \in \arg\max_{S_i} \int_{\varepsilon_{-i}} U_i(s_i(\varepsilon_i), s_{-i}(\varepsilon_{-i}), (\varepsilon_i, \varepsilon_{-i})) f(\varepsilon_{-i}) d\varepsilon_{-i} \quad (8)$$

$U_i(s_i(\varepsilon_i), s_{-i}(\varepsilon_{-i})$ denotes the utility to individual i of choosing choice strategy $s_i(.)$ conditional on the individual's specific term ε_i, when the other individuals follow strategies $s_{-i}(.)$ across the set of specific preferences ε_{-i} for all individuals, excluding individual i. This function expresses the fact that the expected payoffs are calculated from the equilibrium distribution of joint strategies. The choice probabilities of any one individual will affect the beliefs and hence the expected payoffs of others in the same way, since each of them has its own expected utilities over alternatives with the same payoff structure and is engaging in the same utility-maximizing behavior.

$p(\varepsilon_{-i} | \varepsilon_i)$ denotes individual i's conditional probability about other individual's types $\varepsilon_{-i} = (\varepsilon_1, ..., \varepsilon_{i-1}, \varepsilon_{i+1}, ..., \varepsilon_n)$ given its type ε_i. Based on the common knowledge that the strategies of others are a function of their types, individual i could use its conditional beliefs $p(\varepsilon_{-i} | \varepsilon_i)$ to compute the expected utility to each choice and thus find its optimal responses. This probability is calculated by integrating across the distribution of other individuals' error distributions. Due to the symmetry of individuals' error terms, which are draws from the same distribution, each individual will have the same conjecture about any one other individual k's optimal strategies. That is,

$$\begin{aligned} p_{ij} &= \Pr(s_{kj}(.) = 1) = \Pr(E_k[U_{kj}(\varepsilon_{kj})] \geq E_k[U_{kj'}(\varepsilon_{kj'})]) \\ &= \Pr(E_k[\overline{U}_j(s_{-k})] + \varepsilon_{kj} \geq E_k[\overline{U}_{j'}(s_{-k})] + \varepsilon_{kj'}, \qquad \forall i \neq k, \forall j \neq j') \end{aligned} \tag{9}$$

Consequently, no individual's particular preferences affect the equilibrium. It implies furthermore that in equilibrium the probabilities for all individuals are identical (i.e., $p_{ij}^* = p_{kj}^* = p_j^*$) as are the non-random utilities for each alternative (i.e., $\overline{U}_{ij}^* = \overline{U}_{kj}^* = \overline{U}_j^*$). An individual's vector of conjectures over all alternatives j is consequently defined by the set of J equations. This implies that, in the equilibrium, the belief probabilities that determine the expected payoffs on the right side of (14) match the decision probabilities on the left that result from applying a probabilistic choice rule to those expected payoffs.

The particular characteristics of the model depend on the distribution of the random component. Assuming individuals' types follow exponential distributions (IID), the choice probability of individual i choosing alternative j is a proportional exponential function of expected payoffs:

$$p_j^* = \frac{\exp(E[\overline{U}_j])}{\sum_{j'=1}^{J} \exp(E[\overline{U}_{j'}])} \tag{10}$$

This model can represent different scenarios depending on the structure of the interaction between individuals and the degree of uncertainty. Let μ be a positive "uncertainty" parameter that reflects the degree of the uncertainty of individuals' beliefs about other individuals. The model can then be expressed as:

$$p_j^* = \frac{\exp(E[\overline{U}_j] / \mu)}{\sum_{j'=1}^{J} \exp(E[\overline{U}_{j'}]\mu)} \tag{11}$$

Equation (11) suggests that higher values of the uncertainty parameter make choice probabilities less sensitive to expected payoffs. If μ is not explicitly identified and normalized to one, the model evolves into the traditional multinomial logit model. In the extreme case, when μ tends to zero, the variance of the utility disturbances approaches zero. In this case, there is no uncertainty about other individuals' utility, and a deterministic choice model is obtained because all the information about individual preferences is included in the systematic utilities. This means there is no error, and the model approaches the Nash equilibrium under complete information. In this case, the alternative with the highest expected payoff is selected with certainty, reflecting perfect rational behavior. As μ goes to infinity, the variance of the disturbances approaches infinity. Therefore, the probabilities become equal for all alternatives, irrespective of the expected payoffs. The choice model then provides no information, and consequently the alternatives are equally likely to be chosen. This means that decisions are based on error only (i.e. there is a unique equilibrium with equal probabilities for each alternative), reflecting perfectly random behavior. In reality, individuals are neither perfectly rational ($\mu = 0$) nor perfectly noisy ($\mu = \infty$), and the intermediate values of μ are most relevant for explaining human behavior. In that sense, the "uncertainty" parameter can be interpreted as a measure of bounded rationality of individuals with limited ability of anticipation.

Strategic choice will depend on an individual's past experiences with similar situations. When an individual has to make a strategic choice for the first time, there is a lot of uncertainty about what others think, what others think they think, and so on in a process of iterated reasoning. As they experience a particular situation more often and repeatedly observe the actual payoffs from different choices, they can be expected to make more precise value judgments of the expected payoffs of different strategies. As a consequence, the notion of equilibrium should be viewed as the limiting outcome of a learning process that clarifies over time through experience. Specific models can be derived for each of these situations.

When individual have no previous experiences (one shot situation), decision can not rely on experienced payoffs, but rather individuals can only rely on introspection. Assuming that individuals are increasingly uncertain about what others think, about what others think they think, etc., Goeree and Holt (2004) suggested to model this increased uncertainty by injecting some noise into the model as follows:

$$p_{ij} = \Phi_{ij}^{\mu_0}(B^1) = \frac{e^{\bar{U}_{ij}(B^1)/\mu_0}}{\sum_{j'=1}^{J_1} e^{\bar{U}_{ij'}(B^1)/\mu_0}} \quad (12)$$

- p_{ij} represents an individual's choice probability (B^0),
- B^1 represents an individual's (first-order) belief about the other's choice,
- B^2 represents an individual's (second-order) belief about the others' belief about the individual's own choice.
- μ_0 represents the uncertainty parameter in iteration reasoning with ordinal number.

In this approach, an individual's first-order belief is modeled as the others' logit best response, $B^1 = \Phi_{ij}^{\mu_1}(B^2)$, given the individual's second-order belief, B^2. In other words, the process that produces an individual's first-order belief about what the others will do is modeled as the others' noisy best response given what the individual thinks the others think the individual will do. Thus, the higher-order beliefs follow logit best responses for (n > k) is:

$$\Phi_n^k = \Phi^{\mu_k} \circ \Phi^{\mu_{k+1}} \circ \ldots \circ \Phi^{\mu_n} \quad (13)$$

which maps an individual's n-th order belief into its k-th order belief: $B^k = \Phi_n^k(B^n)$. It is likely that the higher the order of beliefs, the more imprecise it becomes (i.e., $\mu_0 \le \mu_1 \le \mu_2 \le \cdots \le \mu_\infty$). Noise in this successive iteration increases to infinity as the number of iteration increases to infinity (i.e., $\mu_\infty = \infty$). The set of noisy rational responses is obtained by recursively applying the Φ^μ-mapping to P as the limit, using a higher error rate at every step:

$$B^0 = \lim_{n\to\infty} \Phi_n^0(p_0) = \lim_{n\to\infty} \Phi^{\mu_0} \circ \Phi^{\mu_1} \circ \ldots \circ \Phi^{\mu_k} \circ \Phi^{\mu_{k+1}} \circ \ldots \circ \Phi^{\mu_n}(p_0) \quad (14)$$

The starting probability of the iterative thinking process, p_0, can be chosen arbitrarily because when $\mu = \infty$ the logit best response Φ^μ-mapping projects any belief probability to a uniform probability (representing perfect randomness behavior). That is, individuals' higher-order beliefs about what others think about what others think about ... etc, become increasingly more diffuse. It is more plausible that these (infinite) thinking processes start out with uniform beliefs as it is the degree of uniformity that reflects the degree of uncertainty. The outcomes of these thinking processes lead to individuals' decision choice probabilities, which are not uniformly distributed because the exogenous attractiveness of alternatives are different.

Assuming the error rates to increase with iteration order according to a two-parameter function (Goeree & Holt, 2004), it follows that

$$\mu_k = t^k \mu_0 \quad (15)$$

where k is the iteration reasoning order.

When strategic decisions can be based on past experiences, learning should be taken into consideration. In that case, it is realistic to assume that individuals tend to choose strategies which have higher expected payoffs given beliefs derived from past experiences. Two types of learning are relevant in this context: (1) reinforcement learning, which means that successful strategies will be repeated, (2) belief learning, which means that individuals keep track of what other individuals have done to figure out what those individuals will do in the future. Based on such learning processes, individuals will choose strategies that will give the highest payoff given their beliefs. A possible specification of such a learning model, in which individuals begin with arbitrary beliefs and then update their beliefs according to experiences, can be expressed as follows:

$$E_{ij}(t) = \frac{\lambda \cdot N(t-1) \cdot E_{ij}(t-1) + U(s_{ij}, s_{-i}(t))}{\lambda \cdot N(t-1) + 1} \tag{16}$$

$$N(t) = \lambda \cdot N(t-1) + 1 \tag{17}$$

where $U(s_{ij}, s_{-i}(t))$ denote individual i's payoff in a period t by choosing strategy j, and others chosen their corresponding strategies. $N(t)$ is the indicator of past experience relative to current experience. The initial experience weight $N(0)$ reflects the strength of initial expectations relative to incremental changes in expectations due to actual experienced payoffs. If $N(0)$ is small, the effect of the initial beliefs are quickly displaced by observations. If $N(0)$ is large, then the effect of the initial beliefs persist during the learning process. The parameter λ captures the declining weight placed on payoffs from more distant periods of actual experience, compared to more recent periods. This is either because of forgetting or because individuals are aware that the environment is changing. In the extreme case, when $\lambda = 0$, the observations prior to the most recent one are ignored, only the most recent observation counts. At the other extreme, when $\lambda = 1$, all observations count equally, regardless of the number of periods that have passed. Using λ as the learning parameter to indicate the dependence of probabilities and expected payoffs, the expected payoffs based on the updated beliefs in any period determine each individual's decision probabilities of that period via the logit rule in equation (6). Han and Timmermans (2007) suggested the following general form:

$$P_{ij}(\lambda) = \frac{\exp(E_{ij}(\lambda) / \mu)}{\sum_{j'=1}^{J} \exp(E_{ij'}(\lambda) / \mu)} \tag{18}$$

This learning model is a special extension of standard reinforcement learning, which assumes that the only effect on subsequent choice is produced by rewards for actual choices (Timmermans, Arentze & Ettema, 2003). The model is experience-adaptive in the sense that forward-looking expected payoffs are based on backward-looking (in)direct observations: individuals not only respond to their own previous payoffs given the chosen strategy, but also to the knowledge about what others did with respect to the non-chosen strategies. This situation is likely to occur in high information environments in which individuals have real-time information updates. When repeated experience with a similar situation is available, beliefs will evolve as individuals learn. Learning slows down as observed decisions look more and more like prior beliefs, as error and surprises are reduced. In a steady state, learning ceases when beliefs are consistent with observed decisions, reaching equilibrium – the steady-state expectations are equal to steady-state average payoffs.

Now, in reality not only the traveler may exhibit strategic behavior. Information providers may also have different objectives and act strategically. This process can be modeled in terms of a Stackleberg game. Interestingly, experimental work (Han and Timmermans, 2006) has suggested that individual become sensitive of underlying objectives in repeated circumstances and show different strategic behavior, dependent on underlying objective. In general, compliance levels tend to be lower if the underlying objective is less consistent with personal beliefs and utilities.

CONCLUSION

The main idea underlying this chapter has been that recent advances in information and communication together and algorithms to rapidly and automatically impute context-sensitive information from ubiquitous travel environment, together with advances in various kind of predictive models of decision making under uncertainty provides the necessary building blocks for commercial and planning applications. Context-sensitive, real time information can be used to offer individual information and advice in their personal interest or in the interest of location-based service providers. Tourist information, advertising and commercial applications are obvious examples. For example, when pedestrian enter in the vicinity of a particular store, they can be identified and be given a personalized add to trigger them into the store. Classical models of non-strategic behavior are relevant in this context to predict likely responses. Models of strategic behavior are relevant in congested situations and can serve as a personal assistant to support individuals making smarter decisions by more accurately predicting what others will do.

Alternatively, the advice may be given such as to trigger individuals into a behavioral response that would be relevant for society at large or that would improve system's performance. The potential value of strategic choice models in this context seems apparent. These predictions can be used to compute expected payoffs of strategies and recommend a choice with the highest expected value. At the planner's level, the proposed models can be used as a proxy of the population to assess the feasibility of planning actions, predict external effects on existing destinations through the distribution of flows, especially from an urban management perspective. It can be extended to support more complex decision problems, for example a tourist office may wish to influence the behavior of individual tourists to optimize or at least improve some system-level objective. Other problems of location-based services can be envisioned. In this case, any model should also take into account the extent to which individuals comply with the provided recommendations and with anticipation of the extent to which others will comply.

Although empirical results both with regard to advances learning algorithms and sophisticated data imputation and with respect to models of strategic choice behavior are still very limited, it is to be expected that rapid progress can be made. It opens up the possible of several interesting new applications in a future of ubiquitous environments.

REFERENCES

Aadland, D. M., Caplan, A. J., & Phillips, O. R. (2007). A Bayesian examination of information and uncertainty in contingent valuation. *Journal of Risk and Uncertainty*, *35*, 149–178.

Aarts, H., & Dijksterhuis, A. (2000). The automatic activation of goal-directed behavior: the case of travel habit. *Journal of Environmental Psychology*, *20*, 75–82.

Aarts, H., Verplanken, B., & van Knippenberg, A. (1997). Habit and information use in travel mode choices. *Acta Psychologica*, *96*, 1–14.

Aarts, H., Verplanken, B., & van Knippenberg, A. (1998). Predicting behavior from actions in the past: repeated decision making or a matter of habit? *Journal of Applied Social Psychology*, *28*, 1355–1374.

Abowd, G., Atkeson, C.G., Hong, J., Long, A., Kooper, R. & Pinkertin, M. (1997). Cyberguide: A mobile context-aware tour guide. *Wireless Networks*, 421-433.

Ahas, R., Aasa, A., Mark, Ü., Pae, T., & Kull, T. (2007). Seasonal tourism spaces in Estonia: case study with mobile positioning data. *Tourism Management, 28*, 898–910.

Ahas, R., Aasa, A., Silm, S., Aunap, R., Kalle, H., & Mark, Ü. (2007). Mobile positioning in space-time behavior studies: Social Positioning Method experiments in Estonia. *CaGIS, 34*, 259–273.

Anderson, N. H. (1981). Foundations of information integration theory. New York: Academic Press.

Arentze, T. A., & Timmermans, H. J. P. (2005a). Representing mental maps and cognitive learning in micro-simulation models of activity-travel choice dynamics. *Transportation, 32*, 321–341.

Arentze, T. A., & Timmermans, H. J. P. (2005b). Information gain, novelty seeking and travel: a model of dynamic activity-travel behavior under conditions of uncertainty. *Transportation Research, 39A*, 125–145.

Arikawa, M., Konomi, S., Ohnishi, K., & Navitime, S. (2007). Supporting pedestrian navigation in the real world. *IEEE Pervasive Computing / IEEE Computer Society [and] IEEE Communications Society, 6*, 21–29.

Arikawa, M., Tsuruoka, K. I., Hideyuki, F., & Ome, A. (2007). Place-tagged podcasts with synchronised maps on mobile media players. *Cartography and Geographic Information Science, 34*, 293–304.

Asakura, Y., & Iryo, T. (2007). Analysis of tourist behavior based on the tracking data collected using a mobile communication. *Transportation Research Part A, Policy and Practice, 41*, 684–690.

Auld, J., Williams, C., Mohammadian, A. & Nelson, P. (2009). An automated GPS-based prompted recall survey with learning algorithms. *Transportation letters, 1*, 58-79.

Avineri, E., & Prashker, J. N. (2003a). *Sensitivity to uncertainty: the need for a paradigm shift.* Paper presented at the 82nd meeting of the Transportation Research Board, Washington, D.C.

Avineri, E., & Prashker, J. N. (2003b). *The impact of travel time information on travelers' learning under uncertainty.* Paper presented at the 10th International Conference of Travel Behavior Research, Lucerne.

Axhausen, K. W., Zimmermann, A., Schonfelder, S., Rindsfuser, G., & Haupt, Th. (2002). Observing the rhythms of daily life: A six-week travel diary. *Transportation, 29*, 95–124.

Axhausen, K. W., Schönfelder, S., Wolf, J., Oliveira, M., & Samaga, U. (2004). Eighty weeks of GPS traces: approaches to enriching trip information. In *Proceedings of the 83rd Annual Meeting of the Transportation Research Board,* Washington, D.C., USA.

Bachu, P. K., Dudala, T., & Kothuri, S. M. (2001). Prompted recall in Global Positioning System survey – proof-of-concept study. *Transportation Research Record, 1768*, 106–113.

Battelle (1997). *Lexington area travel data collection test.* Final Report. Prepared for Federal Highway Administration, September, 1997.

Baus, J. R., Chevherst, K., & Kray, C. (2005). A survey of map-based mobile guides. In *Map-Based Mobile Services*, (pp. 197-216).

Bellemans, T., Kochan, B., Janssens, D., Wets, G., & Timmermans, H. J. P. (2008). *In the field evaluation of the impact of a GPS-enabled personal digital assistant on activity-travel diary data quality.* Presented at 87th Annual Meeting of the Transportation Research Board, Washington, DC.

Ben-Akiva, M., & Boccara, B. (1995). Discrete choice models with latent choice-sets. *International Journal of Marketing, 12*, 9–24.

Birnbaum, M. H. (1983). Base rates in Bayesian inference: signal detection analysis for the cab problem. *The American Journal of Psychology, 96*, 85–94.

Bohte, W., & Maat, K. (2008)., Deriving and validating trip destinations and modes for multi-day GPS-based travel surveys: An application in the Netherlands. In *Proceedings 87th Annual Meeting of the Transportation Research Board*, Washington DC.

Bonsall, P., Firmin, P., & Beale, J. (2004). *Perception of modal attributes: how accurate and how susceptible to change?* Paper presented at the 83rd meeting of the Transportation Research Board, Washington, DC.

Chen, R. B., & Mahmassani, H. S. (2004). *Travel time perception and learning mechanisms in traffic networks*. Paper presented at the 83rd meeting of the Transportation Research Board, Washington, DC.

Chorus, C.G., Arentze, T.A., Molin, E.J.E. & Timmermans, H.J.P. (2006). The value of travel information: Decision-strategy specific conceptualizations and numerical examples. *Transportation Research, 40*(Part B), 504-519.

Chorus, C.G., Arentze, T.A. & Timmermans, H.J.P. (2008a). A random regret minimization model of travel choice. *Transportation Research, 42*(Part B), 1-18

Chorus, C. G., Arentze, T. A., & Timmermans, H. J. P. (2008b). *A general model of Bayesian travel time learning*. Paper presented at the 87th annual meeting of the *Transportation Research Board*, Washington, D.C.

Chorus, C. G., Arentze, T. A., & Timmermans, H. J. P. (2009). Traveler compliance with advice: a Bayesian utilitarian perspective. *Transportation Research Part E, Logistics and Transportation Review, 36*, 538–551.

Chorus, C. G., & Timmermans, H. J. P. (In Press). Personal Intelligent Travel Assistants. In A. de Palma, R. Lindsey, E. Quinet, & R. Vickerman (Eds.), Handbook in Transport Economics. London: Edward Elgar.

Chorus, C. G., Walker, J. L., & Ben-Akiva, M. E. (in press). The value of travel information: A search-theoretic approach. *Journal of Intelligent Transportation Systems.*

Chung, E., & Shalaby, A. (2005). Development of a trip reconstruction tool for GPS-based personal travel surveys. *Journal of Transportation Planning and Technology, 28*, 381–401.

(2006). Internet-based prompted recall diary with automated GPS activity-trip detection: system design. In *Proceedings of the 85Th Annual Conference of the Transportation Research Board*, Washington, DC.Doherty, S.T.Lee-Gosselin, M.E.H.Papinski, D.

Draijer, G., Kalfs, N., & Perdok, J. (2000). Global positioning system as data collection method for travel research. *Transportation Research Record, 1719*, 147–153.

Du, J., & Aultman-Hall, L. (2007)... *Transportation Research Part A, Policy and Practice, 41*, 220–232.Increasing the accuracy of trip rate information from passive multi-day GPS travel datasets: Automatic trip end identification issues

Edwards, W., Lindman, H., & Savage, L. J. (1963). Bayesian statistical research for psychological research. *Psychological Review, 70*, 193–242.

Einhorn, H. J., & Hogarth, R. M. (1981). Behavioral decision theory: processes of judgment and choice. *Annual Review of Psychology, 32*, 53–88.

El-Gamal, M. A., & Grether, D. M. (1995). Are people Bayesian? Uncovering behavioral strategies. *Journal of the American Statistical Association, 90*, 1137–1145.

Emmerink, R. H. M., Axhausen, K. W., Nijkamp, P., & Rietveld, P. (1995). Effects of information in road transport networks with recurrent congestion. *Transportation, 22*, 21–53.

Emmerink, R. H. M., Nijkamp, P., Rietveld, P., & Van Ommeren, J. N. (1996). Variable message signs and radio traffic information: an integrated empirical analysis of drivers' route choice behavior. *Transportation Research, 30A*, 135–153.

Forrest, T. L., & Pearson, D. F. (2005). *A comparison of trip determination methods in GPS enhanced household travel surveys*. Presented at 84th Annual Meeting of the Transportation Research Board, Washington, D.C., 2005.

Fudenberg, D., & Tirole, J. (1991). Game theory. Cambridge, MA: MIT Press.

Fujii, S., & Gärling, T. (2003). Development of script-based travel mode choice after forced change. *Transportation Research, 6F*, 117–124.

Fujii, S., Gärling, T., & Kitamura, R. (2001). Changes in drivers' perceptions and use of public transport during a freeway closure. *Environment and Behavior, 33*, 796–808.

Fujii, S., & Kitamura, R. (2003). What does a one-moth free bus ticket do to habitual drivers? *Transportation, 30*, 81–95.

Gärling, T., Eek, D., Loukopoulos, P., Fujii, S., Johansson-Stenman, O., Kitamura, R., et al. (2002). A conceptual analysis of the impact of travel demand management on private car-use. *Transport Policy, 9*, 59–70.

Gigerenzer, G., & Hoffrage, U. (1995). How to improve Bayesian reasoning without instruction: Frequency formats. *Psychological Review, 102*, 684–704.

Goeree, J. K., & Holt, C. A. (2004). A theory of noisy introspection. *Games and Economic Behavior, 46*, 365–382.

Gopnik, A., Glymour, C., Sobel, D. M., Schulz, L. E., Kushnir, T., & Danks, D. (2004). A theory of causal learning in children: Causal maps and Bayes nets. *Psychological Review, 111*, 3–32.

Guensler, R., & Wolf, J. (1999). Development of a handheld electronic travel diary for monitoring individual trip making behavior. Washington, DC: TRB, National Research Council.

Hakes, J., & Viscusi, W. K. (1997). Mortality risk assessments: A Bayesian reassessment. *Journal of Risk and Uncertainty, 15*, 135–150.

Han, Q., Arentze, T. A., Timmermans, H. J. P., Janssens, D., & Wets, G. (2008). *Modeling context-sensitive dynamic activity-travel behavior under conditions of uncertainty incorporating reinforcement learning, habit formation, and behavioral and cognitive adaptation strategies*. Paper presented at the 87th Annual Meeting of the Transportation Research Board, Washington D.C.

Han, Q., & Timmermans, H. J. P. (2006). Interactive learning in transportation networks with uncertainty, bounded rationality, and strategic choice behavior: quantal response model. *Transportation Research Record, 1964*, 27–34.

Han, Q., & Timmermans, H. J. P. (2007). Towards models of strategic spatial choice behavior: theory and application issues. *GeoJournal, 67*, 195–206.

Hato. (2006). Development of behavioral context addressable loggers in the shell for travel-activity analysis. In *Proceedings IATRB Conference*, Kyoto, Japan.

Hato, E., Itsubo, S., & Mitani, T. (2006). Development of MoALs (Mobile Activity Loggers supported by GPS-phones) for travel behavior analysis. In *Proceedings 85th Annual Meeting of the Transportation Research Board*, Washington DC.

Hato, E., & Asakura, Y. (2001). New approach for collection of activity diary using mobile communication systems. In *Proceedings 80th Annual Meeting of the Transportation Research Board*, Washington DC.

Hodgson, G. M. (2004). Reclaiming habit for institutional economics. *Journal of Economic Psychology*, *25*, 651–660.

Horowitz, J. L. (1984). The stability of stochastic equilibrium in a two-link transportation network. *Transportation Research, 18B*, 13–28.

Hurtig, T. (2006). A mobile multimodal dialogue system for public transportation navigation evaluated. In *Proceedings of the 8th conference on Human-computer interaction with mobile devices and services*, September 12-15, Helsinki, Finland.

Itsubo, S., & Hato, E. (2006). A study of the effectiveness of a household travel survey using GPS-equipped cell phones and a WEB diary through a comparative study with a paper based travel survey. In *Proceedings 85th Annual Meeting of the Transportation Research Board*, Washington DC.

Jha, M., Madanat, S., & Peeta, S. (1998). Perception updating and day-to-day travel choice dynamics in traffic networks with information provision. *Transportation Research Part C, Emerging Technologies*, *6*, 189–212.

Kaysi, I. (1991). *Framework and models for provision of driver information system*. PhD-thesis. Cambridge, MA: Massachusetts Institute of Technology.

Krijgsman, S., Nel, J., & De Jong, T. G. (2008). *Deriving transport data with cellphones: Methodological lessons from South Africa*. Paper presented at the *8th International Conference on Survey Methods in Transport*, Annecy, France.

Lerman, S. R., & Mahmassani, H. S. (1985). The econometrics of search. *Environment and Planning, 17A*, 1009–1024.

Li, H., Guensler, R., Ongle, J., & Wang, J. (2005). Using GPS data to understand the day-to-day dynamics of the morning commute behavior. In *Proceedings 84th Annual Meeting of the Transportation Research Board*, Washington, DC.

Li, Y. (2003). Evaluating the urban commute experience: a time perception approach. *Journal of Public Transportation*, *6*, 41–66.

Li, Z., & Shalaby, A. (2008). Web-based GIS system for prompted recall of GPS-assisted personal travel surveys: system development and experimental study. In *Proceedings 87th Annual Meeting of the Transportation Research Board*, Washington, DC.

Lyons, G. D. (2001). Towards integrated traveler information. *Transport Reviews*, *21*, 217–235.

Mahmassani, H. S., & Chang, G. (1987). On boundedly rational user equilibrium in transportation systems. *Transportation Science*, *22*, 89–99.

Mahmassani, H. S., & Jayakrishnan, R. (1991). System performance and user response under real-time information in a congested traffic corridor. *Transportation Research, 25A*, 293–307.

Marca, J., Rindt, C. R., & McNally, M. (2002). Collecting activity data from GPS readings. In *Proceedings 81st Annual Meeting of the Transportation Research Board*, Washington DC.

Marca, J., Rindt, C. R., & McNally, M. (2002). The Tracer data collection system: Implementation and operational experience. In *Proceedings 81st Annual Meeting of the Transportation Research Board*, Washington, DC.

McCreadie, C., Raper, J., Gunesh, A., Wood, J., Carey, K., Petrie, H., et al. (2006). Persuasive technology for leisure and health: development of a personal navigation tool. In *Persuasive Technology*, (pp. 187-190).

McFadden, D. (1999). Rationality for economists? *Journal of Risk and Uncertainty, 19*, 73–105.

McFadden, D. (2001). Disaggregate behavioral travel demand's RUM side: a 30 year retrospective. In D. Hensher (Ed.), Travel behavior research: the leading edge. Amsterdam: Pergamon.

Moiseeva, A., Arentze, T. A., & Timmermans, H. J. P. (2009). *Imputing relevant information from multi-day GPS tracers for retail planning and management using fusion and context-sensitive learning.* To be presented at the 16th RARSS conference, Niagara Falls, Canada, July 6 – 9, 2009.

Murakami, E., & Wagner, D. P. (1999). Can using Global Positioning System (GPS) improve trip reporting? *Transportation Research Part C, Emerging Technologies, 7*, 149–165.

Ohmori, N., Nakazato, M., & Harata, N. (2005). GPS mobile phone-based activity diary survey. In. *Proceedings of the Eastern Asia Society of Transportation Studies, 5*, 1104–1115.

Ohmori, N., Nakazato, M., Harata, N., Sasaki, K., & Nishii, K. (2006). Activity diary surveys using GPS mobile phones and PDA. In *Proceedings 85th Annual Meeting of the Transportation Research Board*, Washington DC.

Olander, F. (1973). Search behavior in non-simultaneous choice-situations: satisficing or maximizing? In *Selected proceedings of an interdisciplinary research conference*, Rome.

Pierce, B., Casas, J., & Giaimo, G. (2003). Estimating trip rate under-reporting: preliminary results from the Ohio household travel survey. In *Proceedings 83th Annual Meeting of the Transportation Research Board*, Washington DC.

Polak, J., & Jones, P. (1993). The acquisition of pre-trip information: a stated preference approach. *Transportation, 20*, 179–198.

Raiffa, H., & Schlaifer, R. (1961). Applied statistical decision theory. Boston: Harvard University Press.

Rehl, K., Bruntsch, S., & Mentz, H.-J. (2007). Assisting multimodal travelers: design and prototypical implementation of a personal travel companion. *IEEE Transactions on Intelligent Transportation Systems, 8*, 31–42.

Schlich, R., & Axhausen, K. W. (2003). Habitual travel behavior: evidence from a six-week travel diary. *Transportation, 30*, 13–36.

Schofer, J. L., Khattak, A., & Koppelman, F. S. (1993). Behavioral issues in the design and evaluation of advanced traveler information systems. *Transportation Research Part C, Emerging Technologies*, 107–117.

Schönfelder, S., Axhausen, K. W., Antille, N., & Bierlaire, M. (2002). Exploring the potentials of automatically collected GPS data for travel behavior analysis - a Swedish data source. In J. Möltgen & A. Wytzisk (Eds.), *GI-Technologien für Verkehr und Logis-tik, IfGIprints*, (132, 20, pp.155-179). Münster, Germany: Institut für Geoinformatik, Universität Münster.

Schuessler, N., & Axhausen, K. (2008). *Identifying trips and activities and their characteristics from GPS raw data without further information.* Paper presented at the *8th International Conference on Survey Methods in Transport*, Annecy, France.

Shoval, N., & Isaacson, M. (2006). The application of tracking technologies to the study of pedestrian spatial behavior. *The Professional Geographer*, *58*, 172–183.

Shoval, N., & Isaacson, M. (2007)... *Annals of Tourism Research*, *34*, 141–159.Tracking tourists in the digital age

Simon, H. A. (1955). A behavioral model of rational choice. *The Quarterly Journal of Economics*, *69*, 99–118.

van der Spek, S. (2008). Measuring and observing pedestrian activity: Tracking pedestrians in Norwich, Rouen and Koblenz. In *Proceedings Walk 21*.

Stopher, P., Bullock, P. J., & Horst, F. N. H. (2003). Conducting a GPS survey with a time-use diary. In *Proceedings 83thAnnual Meeting of the Transportation Research Board*, Washington, DC.

Stopher, P., Greaves, S., & Fitzgerald, C. (2005). Developing and deploying a new wearable gps device for transport applications. In *Proceedings 18th Australasian Transport Research Forum*, Sydney, Australia.

Stopher, P., & Collins, A. (2005). Conducting a GPS prompted recall survey over the internet. *TRB Annual Meeting, CD-ROM,* Transportation Research Board, National Research Council, Washington, DC.

Sun, Z., Arentze, T. A., & Timmermans, H. J. P. (2005a). Modeling the impact of travel information on activity-travel rescheduling decisions under conditions of travel time uncertainty. *Transportation Research Record*, *1926*, 79–87.

Sun, Z., Arentze, T. A., & Timmermans, H. J. P. (2005b). Modeling the impact of travel information on activity-travel rescheduling decisions under multiple uncertain events: distributed myopic decision heuristics. In *the Ninth International Conference on Computers in Urban Planning and Urban Management*, London, UK.

Swait, J., & Ben-Akiva, M. (1987). Incorporating random constraints in discrete models of choice set generation. *Transportation Research*, *21B*, 91–102.

Timmermans, H. J. P., Arentze, T. A., & Ettema, D. F. (2003). Learning and adaptation behavior: Empirical evidence and modelling issues. In *Proceedings ITS Conference*, April 1-4, Eindhoven, The Netherlands.

Triandis, H. C. (1977). Interpersonal behavior. Monterey, CA: Brooks / Cole Publishing Company.

Tsui, S. Y. A., & Shalaby, A. S. (2006). An enhanced system for link and mode identification for GPS-based personal travel Surveys. In *Proceedings 85th Annual Meeting of the Transportation Research Board*, Washington, DC.

Tversky, A., & Kahneman, D. (1974). Judgment under uncertainty: heuristics and biases. *Science*, *85*, 1124–1131.

Van Berkum, E. C., & Van der Mede, P. H. J. (1993). *The impact of traffic information*. PhD-thesis. Delft: Delft University of Technology.

Verplanken, B., Aarts, H., & van Knippenberg, A. (1997). Habit, information acquisition and the process of making modal choices. *European Journal of Social Psychology*, *27*, 539–560.

Viscusi, W. K., & Magat, W. A. (1992). Bayesian decisions with ambiguous belief aversion. *Journal of Risk and Uncertainty*, *5*, 371–387.

Von Neumann, J., & Morgenstern, O. (1947). Theory of games and economic behavior. Princeton, NJ: Princeton University Press.

de Vries, B., Lin, Y., & Jessurun, J. (2008). Sense of the City. In *Proceedings DDSS Conference*, Leende, The Netherlands.

Weibull, J. W. (1978). A search model for microeconomic analysis –with spatial applications. In A. Karlqvist (Ed.), Spatial interaction theory and planning models. Amsterdam: North-Holland Publishing Company.

Williams, H. C. W. L., & Ortuzar, J. D. (1982). Behavioral theories of dispersion and the misspecification of travel demand models. *Transportation Research, 16B*, 167–219.

Winter, S., & Nittel, S. (2006). Ad-hoc shared-ride trip planning by mobile geosensor networks. *International Journal of Geographical Information Science, 20*, 899–916.

Wolf, J., Guensler, R., & Bachman, W. (2001). Elimination of the travel survey diary – experiment to derive trip purpose from Global Positioning System travel data. *Transportation Research Record, 1768*, 125–134.

Wolf, J., Guensler, R., Frank, L., & Ogly, J. (2001). The use of electronic travel diaries and vehicle instrumentation packages in the year 2000 Atlanta regional household travel survey: test results, package configurations, and deployment plans. In *Proceedings 9th International Association of Travel Behavior Research Conference*, Gold Coast, Queensland, Australia.

Wolf, J., Loechl, M., Myers, J., & Arce, C. (2001). Trip rate analysis in GPS-enhanced personal travel surveys. In *International Conference on Transport Survey Quality and Innovation Kruger Park, South Africa*, August 2001.

Wolf, J., Oliviera, M., & Thompson, M. (2003). Impact of underreporting on mileage and travel time estimates: results from Global Positioning System-enhanced household travel survey. In *Transportation Research Record: Journal of the Transportation Research Board*, No. 1854 (pp. 189–198). Washington, DC: TRB, National Research Council.

Wolf, J., Schönfelder, S., Samaga, U., Oliveira, M., & Axhausen, K. W. (2004). Eighty weeks of GPS traces: approaches to enriching trip information. In *Proceedings 83th annual meeting of the transportation research board*, Washington DC.

Xu, F., & Tenenbaum, J. B. (2007). Word learning as Bayesian inference. *Psychological Review, 114*, 245–272.

Zhao, S., & Harata, N. (2001). *Travel information, perceived travel time, and route diversion behavior*. Paper presented at the 9th World Conference on Transport Research, Seoul.

Zhou, J., & Golledge, R. (2007). Real-time tracking of activity scheduling/schedule execution within unified data collection framework. *Transportation Research Part A, Policy and Practice, 41*, 444–463.

Zmud, J., & Wolf, J. (2003). Identifying the correlates of trip misreporting - results from the California statewide household travel survey GPS study. In *10th International conference on Travel Behavior Research*.

Chapter 4
Setting the Stage for the Integration of Demand Responsive Transport and Location-Based Services

Frank van der Hoeven
Delft University of Technology, The Netherlands

ABSTRACT

Demand responsive transport systems such as paratransit could deliver services that collective transport simply cannot provide. Location-based services may be capable of bridging the divide between transport services without fixed routes, stops or schedules and their potential users. This chapter outlines how the integration of demand responsive transport and location-based services may help to deliver a flexible transport system that is sensitive to the needs of individual users in urban and rural areas. Such a system would have the potential to liberate urbanism from the need to orient spatial development on rigid transit lines.

INTRODUCTION

Policy makers in Europe have tried for decades to curb the growing use of private vehicles for reasons such as congestion and climate change (Commission of the European Communities, 2001). To that end, national and regional authorities have invested heavily in public transport infrastructure and services. Nonetheless, the share of private vehicles in the modal-split (the percentage share of each mode of transport in total inland transport) is still rising in most European countries (Eurostat, 2008). The efforts to promote public transit have not seemed to successfully provide an alternative means of transportation at a mass scale.

A number of experts blame this fact on the essence of the 'collective transport' concept. With its fixed time schedule, predefined routing and a limited number of possible pick-up and drop-off points, collective transport is fairly insensitive to the actual behaviour patterns of the public. Collective transportation seems to flourish only in dense urban areas, where a collective travel demand can be catered to without the need to facilitate the travel needs of individuals. As such, it seems questionable whether collective transport can provide an adequate

DOI: 10.4018/978-1-61520-769-5.ch004

service level for those that live outside 'classic' metropolitan areas like London, Paris, Madrid, Berlin or Moscow.

A much talked about strategy tries to increase the urban density for the sake of supporting public transport: Transit-Oriented Development, or TOD (Calthorpe, 1993). TOD has had local success stories, such as Portland; Virginia (the Rosslyn-Ballston corridor); Dallas (Mockingbird Station and Addison Circle); Atlanta (Lindbergh Center and BellSouth); San Jose (Ohlone-Chynoweth); and San Diego (Barrio Logan) (Dittmar & Ohland, 2003). However, the TOD approach defies common sense. Transport should facilitate the needs of people. People should not be used to support public transport.

Demand responsive transport (DRT) is often cited as the answer to such issues (Cervero, 1997). DRT is defined as an advanced, user-oriented form of public transport characterised by flexible routing and the scheduling of small or medium-sized vehicles (Mastrogiannidou & Kozanidis & Ziliaskopoulos, 2006). In theory, these demand responsive systems (paratransit) seem very promising, but the first successful implementation of a DRT-system that captures a substantial market share (say: 10 or 20%) has yet to take place.

The Delft University of Technology (TU Delft), Department of Urbanism, participated recently in two large European funded projects: Connected Cities and Spatial Metro. Connected Cities advocated a link between sustainable mobility and urban or regional development (Hoeven, 2008). Spatial Metro allowed the use of techniques such as positioning, tracking and location-based services (LBS) to observe visitors in city centres (Hoeven & Smit & Spek, 2008). In light of these projects, location-based services seem to be a promising method of tackling many of the obstacles faced by the developers and providers of paratransit schemes. We have also observed that the relevant research fields do not intersect in shared projects or joint research groups. Logistics optimisation and location-based services still have to realise the opportunities for collaboration on public transit issues.

This chapter sketches the societal relevance of such mutual collaboration, including the particular areas in which paratransit still seems to struggle and some LBS applications that can address these shortcomings. In the end, it outlines a simple framework that illustrates a possible outcome of such a process. We acknowledge that this picture is speculative and hypothetical by nature. We are nevertheless convinced that it should set the stage for a successful integration of demand responsive transport and location-based services.

BACKGROUND

In 2004, TU Delft developed an initial proposal for a so-called INTERREG IIIC network called Connected Cities. This activity focused on the idea that urban and regional development should reflect public (rail) transport connections. The assumption was that a planning practice that locates housing, employment and services in close proximity to the stops, stations and hubs in the public transport networks would provide the best incentive for people to use public transport (Hoeven, 2005). Such concepts (Transport Development Areas, Transit-Oriented Development) had also been recently developed in the UK (Royal Institution of Chartered Surveyors, 2002) and the US (Belzer & Autler, 2002) in response to issues like congestion, air quality, land use and climate change.

In our talks with potential partners from Greece, we noticed that they were concerned with issues much different than those most commonly discussed in northwest Europe. Air quality, congestion and traffic safety are not big issues outside of Athens, while the lack of transportation in remote rural areas is. Poor accessibility results in social isolation and economic exclusion of large groups of the rural population. In response to this issue, the INTERREG network grew to encompass an additional focus on the social dimension of sustainability (Hoeven, 2008). The Greek University of Thessaly proposed to look into paratransit as a demand responsive transport system that could address the need of people in low population density areas. As such, paratransit became an integral part of the work that would follow in the next three years and introduced us, as urban planners, to the world of systems optimisation.

At the same time, TU Delft was approached by the Norwich City Council with a request to participate in its INTERREG IIIB project Spatial Metro. Spatial Metro aimed to optimise the conditions that would allow people to discover a historical city centre on foot (Hoeven, Smit, & Spek, 2008). The Spatial Metro project prompted TU Delft to tap into a greater European experience that integrates aspects such as urban renaissance, the built heritage, public space, pedestrian mobility, leisure economy and even sustainability. TU Delft examined the question as to how to assess the effectiveness of the investments made in Norwich, Rouen and Koblenz. How can aspects like the accessibility and navigability of public spaces be measured? Much of the effectiveness of a public space naturally depends on the way people use it. We therefore used 'GPS tracking' to analyse in detail the movement patterns of people visiting these three city centres (Schaick & Spek, 2008). Through collaboration in the Spatial Metro project, we, as urban planners, were introduced to the world of tracking, positioning and location-based services. Nevertheless, we had to overcome some initial scepticism on what these novel techniques could actually do in cities like Norwich, Rouen and Koblenz. Staff members representing the local authorities had their doubts, but in the end, our outcomes did convince them all.

To explore this terrain further, we organised in early 2007 an international seminar on how tracking and positioning could contribute to urbanism called Urbanism on Track (Delft University of Technology, 2008). Forty people from the Netherlands, throughout the rest of Europe and the United States came to Delft to exchange knowledge and look for innovative possibilities for applying tracking technologies in the context of urban design and planning. A year later, in 2008, we published a scientific book on the application of tracking technologies in Urbanism: Urbanism on Track (Schaick & Spek, 2008).

USING LOCATION AWARENESS TO SUPPORT PARATRANSIT

Positioning and Tracking

Excited as we were about the prospects of tracking, positioning and location-based services in our own field, we noticed that these concepts were still relatively unknown among planners and policy makers in the fields of travel demand management and public transport (including paratransit). True, GPS is used for fleet management (Schweiger, 1996), but demand responsive services imply that a provider of a service is able to see and respond to actual 'demand' (or customers), not just merely to its own 'supply' (or fleet).

So far, most DRT-services depend on 'static' input through telephone and/or websites. In this way, a DRT-service is only able to see a fragment of the actual demand for travel. Often, that fragment only addresses the needs of people who have no real transportation alternative: the elderly or the mobility impaired (Suleyman, 2008).

Paratransit as a Demand Responsive Transport System

Paratransit is often identified as a 'system' that can provide a valuable demand responsive transport service. Paratransit systems (or dial-a-ride systems, as they are also called) were introduced in the early seventies in the UK and the US as public transport services in low-density areas. Later, they also evolved into a door-to-door service for the mobility impaired (Mastrogiannidou, Kozanidis, & Ziliaskopoulos, 2006).

The basics of a paratransit system are rather simple. Someone makes a phone call to a call centre or uses a website to submit a request by means of an online form. The transport request is registered in a system. The transport provider optimises its operations by combining several requests into one route. Sophisticated algorithms are often deployed to perform such optimisations (Waller & Ziliaskopoulos, 2003). The passengers receive word on when to expect their ride, while at the same time a driver receives his or her schedule. Using a taxi or a minivan, that driver picks up the passengers one by one and delivers them to their destinations, one by one. The driver is required to arrive at the destinations at the time-intervals that were communicated to the customers. Depending on the service, the passengers will pay either when leaving the vehicle or automatically through an account. All in all, a paratransit system aims to be faster and more efficient than a regular bus service and more cost-effective than a taxi service (Mastrogiannidou et al., 2006).

Issues to be Resolved

While the promise of such a demand responsive system is clear, its potential problems are obvious as well. Much depends on whether a transport provider is able to solve the vehicle routing problem effectively. While much has gone wrong at this point in the past, let us assume for a moment that a transport provider works with a high quality algorithm that provides efficient and reliable routes. Then, the first issues that a DRT-provider faces are the following questions: Is there a ready market in our region? Where can we locate potential customers who would benefit from the service? If the provider is able to identify such areas, another question is bound to arise: How can we make it easier for a customer to communicate his or her travel demand? A customer may not find it easy to ask for transport in an area with which he or she is unfamiliar. Next, there is a need to provide a choice of services. A DRT-service tries to balance the trade-off between travel time and travel costs. A provider probably will want to offer several price-time combinations to the customer. After reaching a deal on a price for a trip, the service then has to find its customer. A demand responsive transport system does not use fixed stops or a rigid time schedule. This approach offers flexibility but introduces uncertainty as well. What happens if a passenger or vehicle is later than expected or waits at the wrong spot? Finally, when the customer arrives at the destination, a financial transaction must take place. Most likely, the customer wants to take a second look at the bill. A DRT-system could strengthen its overall trustworthiness if it were able to provide the means to review the trips that are made and give insight on the trade-off between travel time and travel costs.

Location Awareness and Vehicle Tracking

Location-based services and advances in related technologies may change the prospects for demand responsive transport systems dramatically. Based on new location aware technologies and applications, we can sketch a simple but effective framework in which all of these applications can work together to solve the issues illustrated above. The available technologies should be regarded as building blocks for an interface between a location-aware paratransit system and its users. We refer to them as a 'travel demand miner', a 'geotagger', a 'trip negotiator', a 'ride finder' and a 'track viewer'. Jointly, these tools could support paratransit and other demand responsive collective modes of transportation, such as carpooling or employer commute services.

Travel Demand Miner

The first issue that paratransit needs to solve is its inability to identify potential travel demand. The Dutch navigation firm TomTom developed, in collaboration with Vodafone, an application that addresses this problem. As of 2007, TomTom has been tracking congestion on major roads in the Netherlands by positioning, in real time, all four million subscribers to Vodafone's mobile phone services (Cyrus, 2006). Four million subscribers represent one quarter of the entire Dutch population, a sample size that is quite sufficient to thoroughly track congestion on Dutch roads. There is no technical hurdle, however, to increasing this number even further. All that TomTom would need to do would be to include data from other carriers, such as KPN and T-Mobile, in its scheme. With a mobile phone market penetration of over 100%, almost everyone can be tracked. TomTom's ability to track the movements of all mobile phone users, combined with the fact that almost everyone carries a mobile phone, means that we can conduct an accurate and continuous survey of travel demand.

With this tracking ability, we can look at national travel demand. Furthermore, we can examine demand at specific locations by isolating a university campus or a major airport, for instance, and looking at specific time intervals (like 9.00h - 9.30h) to see how many people arrive at the location and determine from where they actually came. With this approach, we can investigate the habits of groups of travellers: Do they share the same origin? Are their origins situated along the same route? Such travel behaviour can be observed over the course of several days in the week and several weeks in the year until we establish clear and recurring patterns. By analysing these travel patterns, it is rather simple to determine how long it takes for specific groups to travel to their destination, whether by car, bike, train or bus. Simulations, using sophisticated algorithms, can determine if paratransit is able to deliver a competitive service. If it does, there is considerably less risk than there would be without this sophisticated analysis in initiating a demand responsive transport service and addressing the users who would benefit from it.

A 'travel demand miner' that explores potential markets for demand response transport systems is a vital first step in the development of such a system. Travel demand mining will involve heavy number crunching beyond the range of a typical computer. It will also require that a transport provider, a large employer or an airport work in close cooperation with a mobile phone provider. As long as privacy is secured, there is no need to inform potential users of the data mining taking place. The requirements for this data mining will therefore not be limited by the specifications of regular mobile phones or home computers.

Geotagger

A first 'requirement' that the paratransit user should meet is using a mobile phone that can work as a 'geotagger'. A paratransit service needs to be flexible so that it can adapt to the needs of its users. The users, in turn, must be able to indicate to the service WHEN they want to travel and FROM where and TO where they to want to travel. Those who have ever tried to use an online trip planner on a mobile phone know how user-unfriendly the experience can be. Most mobile phones are just not fit for typing, scrolling or selecting options from a long list. Worst of all, there is a real chance that one will not really know how precise one's location is when in need transportation. If one is a regular visitor to an area, he or she may know the address by heart. If one is unfamiliar with the place, however, he or she probably does not and may need some assistance.

Location-aware technologies can provide that help. GPS chipsets are increasingly used in mobile phones. GPS chipsets can accurately position the user, although some chipsets will have some problems in densely built urban areas (urban canyons). If the mobile phone does not include GPS, the mobile phone provider can still determine the user's position through techniques such as triangulation. The accuracy of such approaches depends on the density of local GSM-antennas. In rural areas, this requirement can pose a problem. (Schaick & Spek, 2008).

Positioning will make the user's task of selecting his or her location much easier. The mobile phone will assist in disclosing the user's (approximate) location to the paratransit service. While positioning could identify the place FROM which the user wants to travel, he or she must disclose the destination TO which he or she is going. Many of these destinations may be places that are visited regularly: work, day care, the supermarket, a swimming pool or friends. These locations must only be bookmarked once to be stored on the mobile phone, and within a week or so, most of them may have been geotagged. To most potential paratransit users, 'geotagging' (bookmarking one's location) may sound unfamiliar. Geotagging is nevertheless rapidly becoming a widespread component of all sorts of applications. Right now, geotagging is already included in some digital cameras. By 2011, 40 million GPS-enabled digital cameras are expected to have been shipped worldwide, meaning that about twenty percent of the digital cameras will tag each photo with data about where that picture was taken (Shankland, 2008).

In a few years' time, geotagging has become a word in daily usage, like 'internet', 'surfing' or 'bookmarking'. Since photo cameras and GPS chipsets are increasingly built into mobile phones, geotagging has a very real chance of becoming an integral part of one's Nokia, Samsung or iPhone in the near future. There will be potential paratransit users who decide to live without a smart phone, but they can always indicate their frequently used destinations online on a Google map and SMS the corresponding numbers: 1 = HOME, 2 = SCHOOL.

Trip Negotiator

Once the destinations are tagged, a simple application should help to negotiate the transportation needed. In most cases, the user will want to travel from his or her current location. Because the paratransit system is based on location awareness, it will already be aware of where the user is. However, he or she will still need to tell the system WHEN he or she wants to travel. Will that be NOW (as soon as possible), within an HOUR or within a DAY? Second, the user needs to select a geotag to indicate where TO go. After the destination is selected, the paratransit system will report the travel options, including the fare. Based on the likeliness that the trip can be combined with other trips, the user may receive several time/price offers. This feature means that there is a third step involved in the paratransit process: choosing WHICH price/travel time combination suits the user the best. Finally, the user may receive his or her travel details through SMS or any other messaging protocol, and that message can act as a ticket if necessary. In summary, this type of advance paratransit system requires indicating only WHEN one wants to go, TO where one wants to go and WHICH offer one prefers.

Ride Finder

The next application deals with the fact that a paratransit service has no regular schedule and no fixed stops. In case of delay, it is hard for a user to tell if the minibus is late or possibly stopped at a different place than he or she anticipated. In the case of a no-show of one of its passengers, it is equally difficult for a driver to determine whether he or she should wait a while or drive on.

As for the other applications mentioned thus far, we see new services emerging that can be adapted for a paratransit system. In February 2009, Google launched its Latitude service. Google claims that Latitude can show where one's friends are, as long as they possess a smart phone, have Google Maps installed and choose to disclose their locations.

[Gundotra, V. (2009) See where your friends are with Google Latitude, The official Google

blog, (2/04/2009 12:01:00 AM) on-line: http://googleblog.blogspot.com/2009/02/see-where-your-friends-are-with-google.html, retrieved 1 March 2009]

If Latitude can pinpoint one's friends, the same technology can show a driver where his or her passengers are and show the passenger the whereabouts of his or her driver. If a passenger and driver are in close proximity, they can decide to wait a few more minutes without becoming nervous or making a wrong decision by not stopping or otherwise abandoning the pick up.

Track Reviewer

When taking a public transport bus, a rider travels along a predefined route and pays a fixed fare. Taking a cab is a much different experience. The cab driver decides what route to take, and once in a while, the passenger may feel cheated if the cab takes 'the long way home'. This should not happen with the paratransit system that we describe here. A paratransit user should be able to verify the route and the pricing. Today, most of us are accustomed to checking a credit card or mobile phone bill online. Free available services, such as 'InstaMapper', offer easy ways to track someone's trip and visualise it instantly on the web, including details such as speed and travel time.

[InstaMapper http://www.instamapper.com]

With a few additions, we can turn such an application into a virtual taxi meter. As soon as transportation is requested, the paratransit system positions the user and starts tracking him or her. As soon as the destination is reached, it stops tracking. The travel distance can be calculated automatically, with a fare based on that distance. Depending on the overall travel time (response time + trip time), the provider can ask for the full price or offer a reduction. If the driver had to pick up an additional passenger and turned up late, the passenger may deserve a lower fare. The beautiful aspect of such a tracker service is that it allows the passenger to review the trips that he or she has made, using a regular web browser at any time.

Integration in a Closed or Open Shop Model

We have discussed concepts such as the 'travel demand miner', 'geotagger', 'trip negotiator', 'ride finder' and 'track viewer'. The interface to these services needs to be combined in a small app that can be easily installed and used on a mobile phone. Alternatively, the interface could be combined in a small web-application.

The overall system that runs that interface can be developed with one specific transport provider in mind. In that sense, the system would use a 'closed shop' model. The same system can also be developed as an open system, in which different firms, institutions or transport providers can create their own accounts, groups or communities.

FUTURE RESEARCH DIRECTIONS

We have provided a brief overview of sustainable and collective transport issues and explained the potential role of demand responsive transport systems within such a framework. Demand responsive transport systems (DRTs) can provide something that collective transport cannot deliver, by meeting the actual needs of individual clients. In order to function properly, DRTs need insight into the actual mobility patterns of people in cities and regions. We referred to the services currently provided by TomTom to illustrate that it is now possible to track everyone. Developing an instrument that can 'mine' these daily tracks in order to find small groups of similar travel demands seems possible. We have discussed the benefits of geotagging specific destinations, how a suitable combination of travel time and travel costs could be negotiated through a simple procedure, how the user could find the driver and vice versa, and how the user can review the trips that he or she has made. Finally, we have proposed an integrated interface to combine these services: a small app that can be easily installed and used on a mobile phone.

We wish to emphasise that we do not believe that it will ever be possible to develop a successful paratransit system based on desk research alone. The two EU-funded INTERREG projects have taught us that a concept like the one presented here can only be the point of departure of a research project that applies such technologies and services in real situations and in different cultural and geographic areas. Mobility has to do with people, with their willingness to adapt to new technologies and services and their intuitive understanding of the possibilities that such technologies and services offer in a specific city or region. Solutions may only be developed within such a framework by testing and improving them in real life situations. This fact requires a rather unconventional research approach, a living lab at different sites that works with test users and researchers and is able to communicate and exchange their findings and experiences in a collaborative setting.

Such a lab should develop, in conjunction with key stake-holders, additional strategies that can effectively kick off a demand responsive service in a highly regulated public transport market that is, in most countries, monopolised by the state. If that may seem like an impossible task, one should look at the transport revolution (Rote Punkt Aktion) that took place in Hanover (Germany) in June 1969, when citizens organised a bottom-up urban transport system, based on ride sharing, that successfully forced the city's transit operator (Üstra) to reduce its fares and become a publicly owned company (Sander & Hoeven, 2008).

CONCLUSION

Demand responsive transport systems could deliver a service that collective transport cannot provide: tailor-made transportation that meets the needs of individuals. Abandoning the idea that a public transport system needs schedules, dedicated routes and fixed stops require that we change the way transport services interact with their users and users with their providers. Location-based services could bridge the gap that currently exists between vision and reality. Unfortunately, the research fields of tracking/location-based services on the one hand and systems optimisation on the other do not seem to meet in joint projects or programmes. TU Delft did participate in both disciplines and sketched the outlines of what such a joint undertaking may deliver in terms of system-user interactions. Such ventures may deliver a flexible transport system that allows urbanism to liberate itself from a practice in which planners attempt to feed people to public transport systems instead of offering transportation to the people. In order to deliver such a system, we take the position that that a living lab-like environment is required to bring the concept of paratransit to maturity.

REFERENCES

Belzer, D., & Autler, G. (2002). *Transit-Oriented Development, Moving From Rhetoric to Reality*. Berkeley, CA: Strategic Economics. Retrieved July 1, 2009, from http://www.reconnectingamerica.org/public/reports/90

Calthorpe, P. (1993). The Next American Metropolis, Ecology, Community and the American Dream. New York: Princeton Architectural Press.

Cervero, R. (1997). Paratransit in America: Redefining Mass Transportation. Santa Barbara, CA: Praeger Publishers.

Commission of the European Communities. (2001). *European transport policy for 2010: time to decide*. Policy White Paper. Retrieved July 1, 2009, from http://ec.europa.eu/transport/white_paper/documents/doc/lb_com_2001_0370_en.pdf

Cyrus, F. (2006).Vodafone, TomTom partner to create real-time traffic data network. *Engadget* October 2006. Retrieved July 23, 2008: http://www.engadget.com/2006/10/27/vodafone-tomtom-partner-to-create-real-time-traffic-data-networ/

Delft University of Technology. (2008). *Delft - Urbanism on Track*. Retrieved July 1, 2009 from: http://www.bk.tudelft.nl/urbanismontrack

Dittmar, H., & Ohland, G. (2003). The New Transit Town, Best Practices In Transit-Oriented Development. Washington, DC: Island Press.

Eurostat (2008). *Modal split of passenger transport - % in total inland passenger-km: tran_typ > Passenger cars*. Retrieved July 1, 2009, from http://epp.eurostat.ec.europa.eu/tgm/table.do?tab=table&init=1&language=en&pcode=tsdtr210&plugin=1

Gundotra, V. (2009). See where your friends are with Google Latitude. *The official Google blog*. Retrieved March 1, 2009 from: http://googleblog.blogspot.com/2009/02/see-where-your-friends-are-with-google.html

Hoeven, F.D. van der (2005). Connected Cities. *Nova Terra*, December, 3.

Hoeven, F.D. van der (2008). Connected Cities: the Final Cut. *Nova Terra*, April, 3-6.

Mastrogiannidou, C., Kozanidis, G., & Ziliaskopoulos A.K. (2006). Paratransit systems, a case study for the Mount Pelion region in Greece. *Nova Terra*, June, 18-22.

Royal Institution of Chartered Surveyors. Policy Unit (2002). *Transport Development Areas, Guide to good practice*. Retrieved July 1, 2009, from http://connectedcities.eu/downloads/3rdparty/rics_tda_mainreport.pdf

Sander, H., & Hoeven, F.D. van der (2008). Transport as a field for local anarchy. *Nova Terra*, April, 7-10.

Schweiger, C. L. (1996). Deployment of technology for paratransit: Examples of innovation. In *Vehicle Navigation and Information Systems Conference, VNIS '96* (Vol. 7, pp. 197- 204). Retrieved July 1, 2009 from http://ieeexplore.ieee.org/xpl/freeabs_all.jsp?tp=&arnumber=1623750&isnumber=34087

Shankland, S. (2008). Cameras with built-in geotagging on horizon. *CNET*. Retrieved March 1, 2009 from http://news.cnet.com/8301-13580_3-9868159-39.html

Suleyman, K. (2008). A nested decomposition approach for solving the paratransit vehicle scheduling problem. *Transportation Research Part B: Methodological*, *43*(4), 448–465.

van der Hoeven, F. D., Smit, M. G. J., & van der Spek, S. C. (Eds.). (2008). Street Level Desires: Discovering the city on foot, Pedestrian mobility and the regeneration of the European city centre. Charleston, SC: Booksurge Publishing.

van Schaick, J., & van der Spek, S. C. (Eds.). (2008). Urbanism on Track. Application of tracking technologies in Urbanism (Research in Urbanism Series, Vol. 1). Amsterdam: IOS Press.

Waller, S. T., & Ziliaskopoulos, A. K. (2003). On the Online Shortest Path Problem. *Networks*, *40*(4), 182–202.

Section 2
Exploring New Methods on the Analysis of Movement Behaviour

Chapter 5
Developing an Interactions Ontology for Characterising Pedestrian Movement Behaviour

Daniel Orellana
Wageningen University, The Netherlands

Chiara Renso
KDD Lab – ISTI - CNR, Italy

ABSTRACT

Since the introduction of Time Geography, the literature has witness a growing interest in representing and understanding human movement and its relationship with the environment. Although recent technology in personal tracking devices brought new potentialities in collecting and representing individual movements, methods to deal with the complexity and dynamism of collective movements are still lacking. This chapter introduces a spatial knowledge representation for the conceptualisation of pedestrian movement as a complex system based on the interactions. Movement interactions are defined and classified to represented global characteristics of the movement as emergent properties other than as a set of individual properties. The devised approach is exemplified through a case study on characterizing visitor behaviour in the Dwingelderveld National Park in The Netherlands.

INTRODUCTION

The interest in representing the movement of people in order to understand their relationship with the environment dates back to the '70s, when Hägerstrand posed the basis of Time Geography. He studied the space-time path of individuals to identify the spatial and temporal constraints that characterise human movement (Hägerstrand, 1970). This idea has proved very suitable to represent the individual movement as an essential relationship between an individual and the environment, as well as between individuals. Moreover, analysis of pedestrian movement has been widely recognised to be essential in understanding human behaviour (Blythe, Miller & Todd, 1996). The space-time paths used by Hägerstrand (i.e. trajectories) have become the most common representation for human movement, due in part to its intuitive visualisation and interpretation, but also to its feasibility to directly represent data from tracking technologies. As a result, several technologies and tools have

DOI: 10.4018/978-1-61520-769-5.ch005

been developed for the collection, storage and recovery of large trajectory data sets (Güting & Schneider, 2005; Manco et al., 2008; Ortale et al., 2008; Pfoser & Jensen, 2001; Renso et al., 2008), analysis (Andersson, Gudmundsson, Laube, & Wolle, 2008) and visualisation (G. Andrienko, N. Andrienko, Kopanakis, Ligtenberg, & Wrobel, 2008). Although trajectories have proved to be useful in representing the movement of individuals, they seem to be an inadequate representation to deal with the complexity and dynamism of collective movement and the interactions that are the cause and the consequence of it. For example, the spatio-temporal path of an individual would have different meanings depending on the presence or absence of other individuals as well as the relationship of the individual with the environment; the trajectory-based representation is not enough sophisticated to represent these different meanings. This idea will be further developed in the next sections.

Our research assumption is that this limitation arises from an underestimation of the representation's roles and implications. Davis stated in (Davis, Shrobe, & Szolovits, 1993) that a "knowledge representation" has five roles that should be taken into account: a surrogate for the real world, a set of ontological commitments, a fragmentary theory of reasoning, a medium for efficient computation and a medium for human expression. We believe that the broad use of the trajectory-based representation has led to an underestimation of some of these roles. Mainly, the ontological commitments of this representation (i.e. what is represented and what is not) have been undervalued. For example, there is a commitment that the movement of an individual starts at the first point of the trajectory and ends at the last point, and that the whole movement can be represented as an interpolation of the intermediate points, which is not necessarily true. Another commitment is that intermediate points would represent either stops or moves, but this distinction, although intuitive, is not always precise, since individuals are hardly ever still and therefore the conceptualisation of stops depends on the scale. Consider for example, a commuter's trajectory: It would have long stops (i.e. home, office) and short stops (i.e. kiosk, traffic lights), but the people are not completely static in these places and it would even be interesting to represent the movement inside those "stops". As we can see, these commitments limit the possibilities of the representation.

The main contribution of this chapter is the description of a Spatial Knowledge Representation for conceptualisation of pedestrian movement. Indeed, pedestrian movement can be conceptualised as a dynamic complex system based on the interactions that happen between individuals as well as between individuals and the environment in which they move. This conceptualisation becomes useful since it allows for the representation of some global characteristics of the movement as emergent properties rather than as a set of individual properties. Moreover, it emphasises the feedback effects of the movement of a pedestrian on other pedestrians and the environment. These characteristics and effects are the basis of the interactions based representation.

This representation is formalised according to the five previously mentioned roles. However, special emphasis will be placed on the first three roles, which allows us to answer the following questions:

- What are the movement interactions?
- What elements and relations are important to consider?
- How can we interpret them?

On the one hand, we propose a simple taxonomy of pedestrian interactions and we show how this taxonomy can be part of an ontology of interactions, where concepts like Context, Behaviour and Patterns are combined to offer a more general understanding of the essence of pedestrian movement behaviour. On the other hand, we have developed an ontology of pedestrian behaviour based on the proposed interactions. This last aspect benefits from the reasoning capabilities provided by formal ontologies in defining behaviour in terms of combination of interactions, movement patterns and contextual information. The viability of our approach to analyse the collective pedestrian movement will be demonstrated through the case study of visitors to the Dwingelderveld National Park in Netherlands.

The remaining parts of this chapter are organised as follows. Next section presents previous related research on pedestrian movement representation, patterns and behaviour. After, the focus is on interactions for pedestrian movements, where definitions of pedestrian interactions are given. We also propose an ontology for interactions where movement patterns, context, and interactions are combined to express pedestrian behaviour. Then, a case study about inferring the behaviour of visitors in a park is depicted, whereas conclusions are reported in the last section.

RELATED WORK

The tasks of detection, representation and interpretation of movement patterns have attracted the attention of several scientific communities. Important contributions have been made from such diverse areas as artificial intelligence, video surveillance, emergency management, environmental planning, human geography, transportation management, animal and human behaviour, and computational geometry, among others. Exhaustive review of research related to the representation of pedestrian movement exceeds the purpose of this chapter; nevertheless, we would like to mention some of the works that inspired us in the development of our approach.

The proposal of analysing people movement for the interpretation of their behaviour was first glimpsed in Time-Geography carried out by Hägerstrand, who showed how human activity can be determined not only by decisions made by individuals, but mainly by the spatial and temporal constraints they are subject to (Hägerstrand, 1970). Hägesrtrand's proposal of representing movement as spatio-temporal paths (the sequence of spatio-temporal positions of a moving object) still remains the foundation of most current approaches. Other authors use different terms as "trajectories" (Smyth, 2001; N. Andrienko, G. Andrienko, Pelekis & Spaccapietra, 2008) or "geospatial lifeline" (Hornsby & Egenhofer, 2002) for the same concept. Miller (2005) confronted the potential lack of rigour of Hägesrtrand's approach by extending the moving object database techniques to develop more rigorous definitions of basic concepts such as space-time path, prisms, bundles and intersections in the context of a measurement theory for Time Geography.

The most recent revision of Time-Geography was proposed by Yu and Shaw (2007), in which the authors propose integrating the representation of human activity together in both the physical and virtual worlds in order to study diverse types of interaction. Later, the authors recognised that trajectories themselves were not sufficient for the representation of patterns when using large movement data sets, and suggested the use of a generalised representation called Generalised Space-Time Paths (GSTP) (Shaw, Yu, & Bombom, 2008). More recently, they presented the implementation of this approach in a Space-Time GIS, where they demonstrated how the activities and interactions of individuals in the physical and virtual worlds could be suitably represented with Linear Reference Systems (LRS) in space-time paths (Shaw & Yu, 2009).

A different approach was proposed by Alvares et al. (2007) where authors addressed the complexity of analysing large trajectory data sets by enriching trajectories with semantic geographical information. In their work, the authors proposed the transformation of traditional raw trajectories into *semantic* trajectories, through a pre-processing of stops and moves (Spaccapietra et al., 2008). This promising approach is conceptually similar to that presented in this chapter; however we have expanded the formalisation of movement patterns beyond stops and moves.

Daamen and Hoogendoorn (2003a, 2003b) described controlled experiments with a view to studying pedestrian behaviour in various situations. The authors represent movement at two levels: microscopic level (trajectories) and macroscopic level (flows). They also show how the former can be used to understand individual behaviour, while the latter are more useful for characterising collective behaviour.

Moreover, a significant research project has been carried out on context-aware representation and reasoning of spatial knowledge. Location-Based Systems (LBS) use context to retrieve information that is relevant to the user, depending on their location. In an interesting study by Schlieder and Werner (2003), authors suggested that the location itself is not enough to determine the user's intentional behaviour, and proposed a location model that takes into consideration both the user's movement patterns and their context in order to infer behaviour. The main approach is based on the translation of movement patterns into sequences of intentions, assuming that these intentions are specific for each type of space and are characterised by the user's movement pattern.

Helbing et al. (1993) proposed a representation of pedestrian collective movement based on fluid dynamics models. The main assumption of this approach is that some of the fluids' properties could be also found in pedestrian collective movement. Therefore, the authors adapted the classic equations of fluid dynamics (particularly gas-kinetic) to integrate the effects of interactions between individuals and to represent movement as macroscopic flows. In a later work, the authors explored how non-linear interactions between individuals can result in surprisingly predictable emerging properties in collective movement, and this brought a new pedestrian behaviour model based on *social forces* where attraction and repulsion forces between pedestrians are represented (Helbing, Molnár, Farkas, & Bolay, 2001). They have also explored interactions in which pedestrian movement both influences and is influenced by the environment, as happens, for example, in the formation of trails on deformable terrain.

The study of movement as the interaction between people and the environment is hardly new. Hillier, Penn, Hanson, Grajewski and Xu (1993) studied the effect of so-called "attractors", defined as points with a particular interest, or capable of generating flows of people movement between them. The authors determined that these attractors are only the multipliers of a more important effect caused by the configuration of travelable space, thus stressing the importance of representing the environment in order to understand people's movement. Following the same viewpoint Turner, Doxa, O'Sullivan, and Penn (2001) approached the interactions between individuals and built-up environments through the study of visibility graphs, suggesting that some movement patterns, such as way-finding or route-choosing, are closely related to the visual perception of space.

In the field of computational geometry, several researchers have focused on studying techniques for pattern matching and recognition for movement data, proposing formalisations of patterns and developing algorithms for their detection (Gudmundsson, van Kreveld, & Speckmann, 2004; Andersson, Gudmundsson, Laube & Wolle, 2008; Benkert, Gudmundsson, Hübner, & Wolle, 2008). In particular, Laube, Imfeld, and Weibel (2005) proposed a complete approach in which the movement of individuals is represented as an analysis matrix having movement parameters of various individuals that are compared over the course of time. The authors also proposed a framework called REMO (RElative MOtion) for the formalisation of possible movement patterns, as well as data mining algorithms for the detection of such patterns. Although these implementations were mainly carried out in data sets of animals' movement, this approach is perfectly valid for the analysis of human behaviour.

Data mining over trajectories is a new and promising research field aimed at investigating techniques to extract patterns from large datasets (Nanni, Kuijpers, Korner, May, & Pedreschi, 2008). Clustering has been exploited to uncover a variety of global behavioural patterns, such as density-based clustering (Rinzivillo et al., 2008), moving clusters, and identifying groups of objects that move similarly and close to each other for a long time (Kalnis, Mamoulis & Bakiras, 2005). The identification of local patterns in movement data, i.e. of concise representations of interesting local behavioural patterns of moving objects, has been also a fertile area of research. Among them, trajectory pattern mining in (Giannotti, Nanni, Pinelli, & Pedreschi, 2007) pursued to unveil sequences of temporally-annotated spatial regions.

Despite the amount of research carried out on movement patterns, there appears to be no agreement on how they must be organised or represented. Recently Dodge, Weibel, and Lautenschütz (2008) recognised this lack of organisation and suggested a systemic representation of movement patterns. They proposed two main classes, namely *Generic Patterns* and *Behavioural Patterns,* and arranged subclasses of the first according to main movement parameters, whereas subclasses of the second were defined using the complex patterns in a specific domain. Dodge classification constitutes a first attempt to organise the existing definitions of movement patterns, but a more formal and comprehensive classification is still needed. Actually, Wood and Galton (2008) investigated the relationship between collective paths from Dodge classification and some examples of collective phenomena and detected some gaps in their approach; they have also concluded that the classification system could be formalised in First-Order Logic.

Finally, an important contribution for the representation of the dynamics of pedestrian movement has been provided by research on simulation and AI. This type of research seeks to create models that reproduce pedestrian behaviour and use them as the basis for the design and planning of transportation infrastructures, evacuation facilities and leisure areas. Various approaches have been suggested for these models; among them, the most noteworthy ones are based on Agent Based Modelling (Penn & Turner, 2001; Batty, 2003; Batty, DeSyllas, & Duxbury, 2003; Bierlaire, Antonini, & Weber, 2003; Antonini, Bierlaire, & Weber, 2006), and Cellular Automata (Blue & Adler, 2001; Burstedde, Klauck, Schadschneider, & Zittartz, 2001).

A REPRESENTATION OF PEDESTRIAN MOVEMENT BASED ON INTERACTIONS

Let us consider a recreation area with a large number of visitors freely walking around. If the movement of each person is represented using a trajectory, it will be possible to detect certain movement patterns such as the most widely used routes, the places where several people meet or the most frequently visited areas. However, it is worth noting that the higher the number of visitors, the less adequate those trajectories will be for movement representation. Indeed, consequently the whole space will be covered by them, thus making patterns more difficult to understand and the causes of these patterns more variable. Moreover, these patterns do not explain *per se* the behaviour of people, because the interpretation can vary. For example, if people meet by stopping in front of some attractive element or if, on the contrary, they all stop at some uninteresting area. As a result, the movement pattern may have been characterised by an interaction with the environment (e.g. to stop and look at some interesting element) or by an interaction between the pedestrians themselves (e.g. to stop for a chat). Therefore, it is obvious that these movement patterns can be interpreted by looking at the possible interactions that caused them and

Figure 1. Flocking pattern

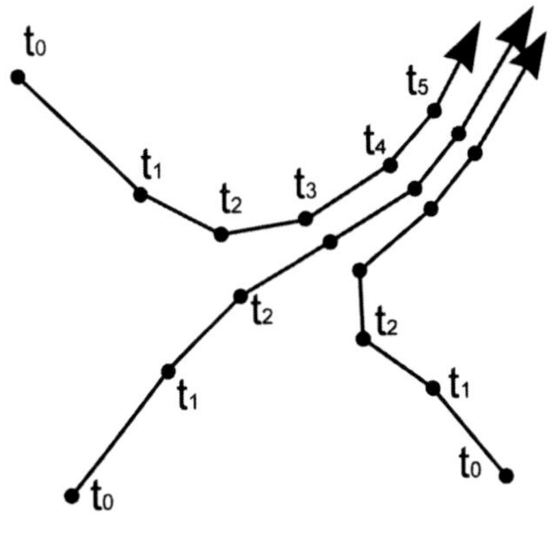

making them explicit through the representation of the context in which movement takes place. Pedestrian behaviour can be ultimately inferred from the analysis of interactions. For example, the "socialising" behaviour in a pedestrian can be inferred from the interaction type they show when encountering other people in certain places.

Our approach suggests a new spatial knowledge representation for pedestrian movement through the exploitation of ontologies for the formalisation of movement interactions and related concepts: patterns, context and behaviour. This interactions-based representation enables us to semantically represent movement patterns computed by different tools and to incorporate the knowledge of domain experts on pedestrian behaviour. Interactions operate therefore as a conceptual bridge between movement patterns and pedestrian behaviour.

As we mentioned in the introduction, the analysis approach in current research is based on the assumption of a univocal relationship between movement patterns and behaviour, or, as some authors have stated, *Movement is behaviour* (Blythe et al., 1996, p. 13; Dodge et al., 2008, p. 245). However, given that interpretation relies on context, we require a conceptual bridge in order to explicitly integrate that context. Our view is that *Movement is behaviour, but patterns are not.* Patterns are the evidence of the interactions that take place during pedestrian movement. For example, the pattern shown in Figure 1 can be interpreted as a group of individuals flocking at time t_3 and moving together (Gudmundsson et al., 2004).

Figure 2 shows different behaviours that can be inferred from the same movement pattern depending on the information we have about the context. Consider the first pattern on the left; as we saw, this could be interpreted as a flocking behaviour in which the individuals meet and move together intentionally. The second picture shows the context from which we can interpret the pattern as people following fixed pathways. Analogously, the third pattern discloses a typical type of behaviour of people seeking a goal (e.g. going to take the bus). Finally, the last pattern can be interpreted as a leadership behaviour in which the central individual guides the other ones. This contextual information allows us to understand the primary elements that influence the movement and produce the pattern. Therefore, the existence of a movement pattern is of course a necessary, but not sufficient condition for the interpretation of pedestrian behaviour.

Our approach introduces the notion of interactions as the primary element of analysis and movement representation. Movement interactions can be defined as the result of active and reciprocal relationships between pedestrians themselves, and between pedestrians and their environment, interpreted in a given context. This definition is of considerable usefulness, as it allows us to integrate concepts from previous research and expands the possibilities of analysis and representation. Therefore, interactions act as a conceptual bridge between patterns and behaviour.

Thus, the traditional approach of direct mapping from patterns to behaviour is enriched with a formal context representation employed for the reasoning and interpretation of observed patterns (Figure 3). The behaviour can be ultimately disclosed by the analysis of a set of interactions and the relationships between them. In our opinion, this representation allows more formal differen-

Figure 2. Different behaviours interpreted from the same movement pattern

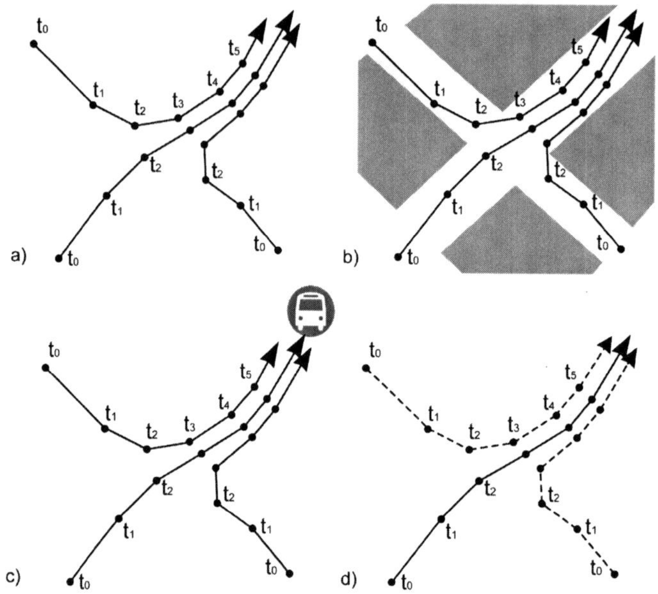

tiation between the various levels of movement interpretation, which in addition provides a higher expressivity and more independence from the application domain.

Elements of Representation

Next we will study the elements that form our proposal and the way they are represented.

Patterns

The most elementary unit of representation in our approach is the pedestrian movement pattern. As we have previously mentioned, a pattern is the evidence of the interaction between pedestrians and their environment. Patterns would be conceptualised as the building blocks of movement, each one with specific and observable characteristics in the spatial, temporal or spatio-temporal dimensions. These observable characteristics are used to detect the patterns in movement data sets. Besides the diversity of patterns, there is an increasingly wide range of tools and methods to detect these patterns in movement datasets. In our approach, we deal with this variability by establishing a set of properties for each pattern that defines its characteristics and computing methods. Once

Figure 3. Interactions based representation

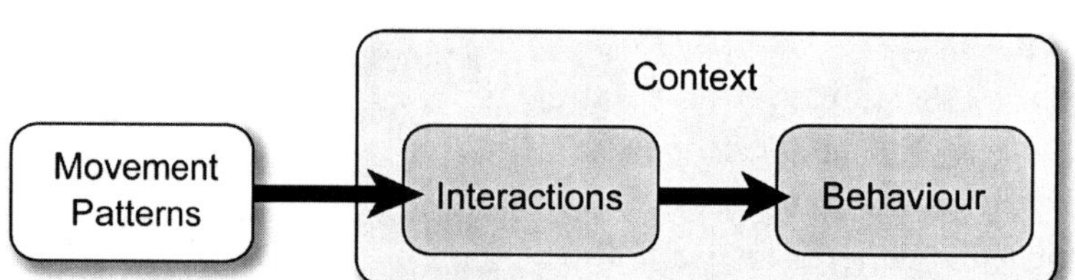

Figure 4. Contexts for movement representation. Each context is described by a set of parameters [P] and values [V]

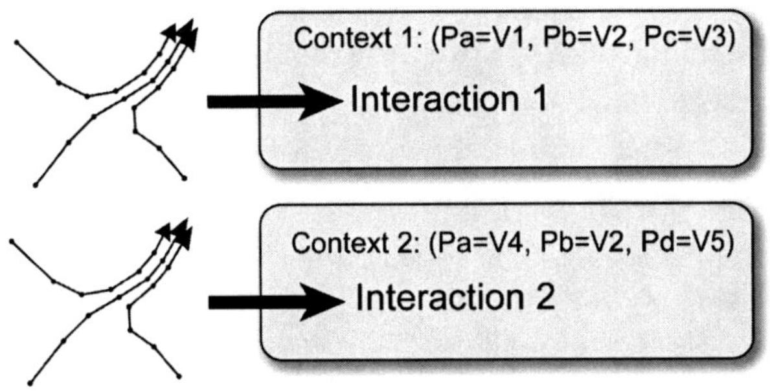

computed (using a specific tool or method), the pattern can be formalised with its properties and values as an instance of the ontology.

Context

The context is formed by a partial set of parameters and values that is never complete, precise nor objective, but it is useful for movement representation and for reasoning on it. Employing the metaphor suggested by Giunchiglia and Bouquet (1997), this can be illustrated as a box, inside which we find the expressions that establish or explain the domain or the phenomena to be studied, whereas outside the box we find a set of parameters and their values, which determine, at least partially, the interpretation of the expressions inside the box. Although it is obvious that any type of representation (and reasoning) will depend on the context, we have seen that most approaches either assume that the context is implicit, or rather exclude it. Thus, the interpretation of patterns is carried out only on the movement's own features.

By contrast, our approach seeks to explicitly and formally represent a context in order to sustain the reasoning and the interpretation of movement patterns, and to expand the possibilities of such interpretation. Moreover, representation of the context is enriched with the results of this interpretation, and can be used again for subsequent interpretations in an iterative process. This approach paves the way for a more dynamic form of representation and contextual reasoning about pedestrian interactions.

Considering again the example of a recreation area with a considerable amount of visitors: collective movement can show patterns of high pedestrian concentration in certain places, which will be interpreted as attraction areas. The location and extension of these attraction areas are used to enrich the representation of the context, allowing us to establish new interpretations of movement patterns: for example, visitors who stop by in these attraction areas can be interpreted as a visitor type dissimilar from those who do not.

Although context is usually linked to a geographic environment, there may be other parameters that can describe it, such as the purposes and intentions of visitors, temporal context, the degree of constriction in a given space, and relationships with other patterns. Interactions can become a part of the context themselves, thus interpreting new interactions (Figure 4).

Behaviour

Hoogendoorn, Bovy and Daamen (2002) suggested that pedestrian behaviour is determined by a hierarchic structure of decision-making at three levels. (i) A strategic level in which the pedestrian decides about his/her destination, the activities to be carried out or his/her aims (ii) a tactical level in which the pedestrian decides about the route to follow, the response to unexpected events or the avoidance of "unwanted" areas or spots; and finally, (iii) an operational level in which the pedestrian decides in which direction the next step will be taken, which means that they intuitively choose a direction and speed, depending on the immediate environment.

We must point out that each level carries an implicit conceptualisation of the corresponding space: at the strategic level, space is conceptualised in terms of aims, needs and desires; at the tactical level, space is conceptualised in terms of assigned symbolic values of utility, opportunity, benefit, etc.; whereas at the operational level, conceptualisation is only observational and based on perceived features. In our approach, operational and tactical behaviours can be represented by the simplest interactions, while representation of the strategic behaviour requires more complex sets of interactions. Consider for example a visitor who takes a detour (operational level) in order to avoid an overcrowded area (tactical level), and then discovers a new interesting spot that leads him/her to reprogram his/her activities (strategic level).

Analysts or domain experts can define diverse behaviour categories depending on the movement interactions expected in different levels. These definitions can be represented in an ontological formalisation, with reasoning tasks being assigned in order to find examples of pedestrians performing every type of behaviour from a set of movement patterns.

Interactions

Although the notion of movement interactions is not new, in our approach we develop the idea of interaction as the key element for the representation of pedestrian movement. This idea is based on the assumption that there are different elements that play an active role, and which influence and are influenced by movement. These elements can be conceptualised as "agents" because of the active effect they have on movement interactions, and they would be "individual", "collective" and "environmental" agents.

Movement interactions create specific dynamics of action and reaction between the agents in any given context. Consequently, pedestrian interactions must be represented bearing in mind the kind of agent participating, the context in which they occur, and the specific movement patterns that demonstrate them. The wide variety of possible interactions requires for a type of classification that allows us to organise them in an ontological representation. In our approach, we suggest a natural type of classification based on the kinds of agents involved.

- *Pair-Wise Interactions*: They occur when the movement is determined by interactions between individuals. Examples include encounters, guiding-following and joint-walking. Although pair-wise interactions would involve more than two individuals, the conceptualisation remains as a relationship between pairs of pedestrians: the elements that determine the movement are individuals.
- *Environmental Interactions*: They occur as a consequence of the relationship between pedestrians and features of the environment. These features have physical, social or psychological functions that have an influence on the movement attracting, repelling or restricting pedestrians when they are walking. Examples of environmental interactions include attraction, route following or trail formation. These interactions imply the conceptualisation of environment as an active agent that influences and is influenced by pedestrian movement.
- *Collective Interactions*: They appear as a consequence of interactions between several individuals. Collective interactions differ from pair-wise interactions not

(only) in number but in the appearance of observable systemic properties of the collective as a whole. These properties usually include a progressive synchronisation of one or more movement parameters, resembling a moving herd. Although collectives would normally be defined by their member's characteristics, in our approach they are defined by the global properties of their movement, that is, collectives are conceptualised as the cause and consequence of movement. Examples of collective interactions include lane formation, incremental concentration and flocking, or circulating flows at intersections (Helbing et al., 2001).

This simple but intuitive classification offers several advantages: firstly, it focuses on the active role played by the environment, which influences and is influenced by pedestrian movement; secondly, it allows us to formally represent the systemic properties of collective movement reported by various authors (Helbing, Buzna, Johansson, & Werner, 2005; Wood & Galton, 2008); thirdly, it allows for the representation of more than one type of interaction operating at the same time, and finally, there is the practical advantage in defining interaction types according to combinations of patterns and contexts for a wide variety of applications.

Moreover, this conceptualisation represents pedestrian movement as an expressive collection of relationships between the different agents involved, relationships that are demonstrated by movement patterns in defined contexts. Finally, it is worth noting that movement actually involves more than one type of interaction at the same time. For example, an encounter at a point of interest would be represented as the combination of pair-wise and environmental interactions.

Examples of Interactions

Since the potential number of possible interactions depends on the context, and therefore on the application, next we will describe some of the most common interactions that can be defined from a relatively small number of patterns, which in turn can be easily extracted with the available tools and methods. Therefore, it is not our intention to describe all the possible interactions, but to use some significant examples to how the interactions are used to represent pedestrian movement.

For the representation of the context we have used the "constriction level" as an example of a contextualising parameter. The possible values to be assigned to that parameter could be: *constrained space*, *semi-constrained space*, and *open space*. Other contextual parameters are employed in order to evaluate the concepts of spatial and temporal proximity and threshold. Here, a number of spatial and temporal values are represented as threshold to indicate a short, medium or long value. These values make it possible to formalise the concepts related to distance and time. For example, when two people stop close to each other for a long period, the concepts "close to" and "long" are formalised as semantic concepts more than quantitative measurements of distance and time. An additional parameter is used to evaluate the concept of attractiveness of geographical features. The level of attractiveness indicates how much specific points or elements of environment tend to attract pedestrians. This attractiveness is represented as low, medium or high levels and an additional "neglected" level to indicate a negative attraction (i.e. repulsion).

Encounter

Main interaction type: Pair-wise interaction.
Associated movement pattern: Spatio-temporal coincidence.
Contextual parameters: Constriction level, spatio-temporal threshold.

Definition: The coincidence in space and time of two or more individuals who remain together for a long time-span. During the encounter, pedestrians may stop moving or may keep on walking together. Spatial context is used in order to evaluate the influence of spatial constriction or attraction elements as well as the spatio-temporal threshold conceptualisation: If the duration of the pattern is long, then the restriction level does not have any influence. However, if the coincidence happens in an open space, a medium duration and spatial threshold are enough, since there is no influence from the environment that encourages the individuals to approach each other. If the interaction takes place in a location with high attractiveness, it cannot be inferred as an encounter since the pattern would be due the environment.

Approaching

Main interaction type: Pair-wise interaction.
Associated movement pattern: Spatio-temporal coincidence with relaxed temporal threshold.
Contextual parameters: Constriction level, spatio-temporal threshold.

Definition: The space-time coincidence of two or more individuals who do not remain together. Spatial context is used in order to evaluate the influence of spatial constriction or attraction elements. Contextual parameters also define the relaxation value of temporal proximity in order to differentiate crossings from encounters.

Guiding - Following

Main interaction type: Pair-wise interaction.
Associated movement pattern: Spatial coincidence with temporal delay.
Contextual parameters: Constriction level, spatio-temporal threshold.

Definition: A similar sequence of positions with a certain delay between one individual and the other. Spatial context is used in order to evaluate the influence of spatial constriction or attraction elements, the spatio-temporal threshold conceptualisation and the temporal delay. This delay must be shorter and the duration longer in constrained spaces than in open spaces to be inferred as a guiding-following interaction.

Visiting

Main interaction type: Environmental interaction.
Associated movement pattern: Stopping.
Contextual parameters: Attractiveness level, spatial proximity.

Definition: A medium or long stop of an individual or a group of individuals in a geographic location with medium or high attractiveness level.

Route Choosing

Main interaction type: Environmental interaction.
Associated movement pattern: Short stops.
Contextual parameters: Constriction level, temporal duration.

Definition: A sequence of motion-stop-motion located in semi-constrained points in which pedestrians must choose a route from a limited number of options. Contextual parameters define the constriction level and the duration of stops.

Attraction

Main interaction type: Environmental interaction.
Associated movement pattern: High density / Movement suspension.
Contextual parameters: Attractiveness level.

Definition: The suspension of movement in locations containing interesting features. Contextual parameters define the attractiveness level.

Flocking

Main interaction type: Collective interaction.
Associated movement pattern: Collective coordination of relative motion parameters.
Contextual parameters: Constriction level, temporal threshold.

Definition: The formation of groups that move together in a similar way for a given time-span in an open space. Contextual parameters define the constriction level and the duration of the interaction.

Aggregation

Main interaction type: Collective interaction.
Associated movement pattern: Concentration (High-density pattern).
Contextual parameters: Constriction level.

Definition: The grouping effect produced by the concentration of pedestrians in defined locations, causing an increase in the density in a given time-span. Contextual parameters determine the constriction and attractiveness level. This interaction is equivalent to the environmental interaction "Attraction" but here, the attractive element is the collective interaction itself.

Trail Formation

Main interaction type: Environmental interaction.
Associated movement pattern: Linear movement clusters.
Contextual parameters: Constriction level.

Definition: The formation of linear clusters in open spaces. The repeated passing of pedestrians leaves trails that, depending on the features of the area, may modify the space forming a trail that consequently acts as a linear attractor. Contextual parameters define the spatial constriction level and the time-span that has been calculated for the trail formation.

In the following section, we introduce the formalisation of the above mentioned concepts in a taxonomy of interactions, namely the Interactions ontology. We give the conceptual view of the ontology in the next section, followed by a simple case study to disclose pedestrian behaviour in a park.

THE INTERACTIONS ONTOLOGY

Following the Davis roles for a knowledge representation as mentioned in the introduction (Davis et al., 1993), this section introduces a formal representation of the above defined classes by means of a *formal ontology*. The definition given by Gruber (2008) is used to define formal ontology as "a technical term denoting an artefact that is *designed* for a purpose, which is to enable the modelling of knowledge about *some* domain, real or imagined." Such ontologies determine what can be represented and what can be inferred about a given domain, using a specific formalism of concepts. An ontology language is a formalism used to express such knowledge.

Web Ontology Language (OWL) is a well known standard that came from the Semantic Web and it is now a W3C recommendation (W3C Consortium, 2004). OWL is based on a family of languages known as Description Logics (DL) that provide a deductive inference system based on formal well founded semantics (Baader, Calvanese, McGuinness, Nardi, & Patel-Schneider, 2003). The basic components of DL are *concepts (classes)*, *properties (roles),* and *instances (individuals)*. Concepts describe the common properties of a collection of instances and properties are binary relations between concepts. The special relation *is_a* represents the *specialization* property between two concepts and describes a taxonomy in the ontology, based on subsumption relation. Furthermore, a number of language constructs, such as intersection, union and role quantification, can be used to define new concepts by means of *axioms*. In other words, concepts may be intentionally defined by axioms, which express the properties that characterize such defined class. The Description Logic formalism comes with a number of primitive reasoning tasks, such as *classification* and *satisfiability*, *subsumption* and *instance checking*. Classification is the computation of a concept hierarchy based on subsumption, satisfiability checks if the ontology defined concepts are consistent, whereas instance checking verifies that an individual is an instance of a concept. In this chapter, we use OWL DL, the OWL sub-language that allows for the maximum expressiveness while retaining computational completeness, corresponding to Description Logics. Other OWL languages are OWL Lite and OWL Full. OWL Lite is a subset of OWL DL which restricts the logical operators allowed, thus resulting in an efficient inference checking system. On the contrary, OWL Full allows for the maximum expressiveness, however, it loses completeness.

The developed interactions ontology defines the four main concepts depicted in Figure 3, namely Movements Patterns, Interactions, Context and Behaviour. Figure 5 shows the top level classes and properties of the interactions ontology. The Interaction concept may be associated to a (pre-computed) movement pattern and may be located in a specific context (spatial and/or temporal). Furthermore, each Pedestrian may participate in an Interaction.

In the following we specify the taxonomy levels of the main concepts: Interaction, Movement Pattern and Context. The three kinds of interactions are specified by the following taxonomy (Figure 6).

Interaction concepts are defined in terms of movement patterns and context parameters. In other words, each interaction concept is a *defined class*, where a new concept is defined in terms of a logical combination of other concepts and/or properties. Consequently, the reasoning engine can automatically infer pedestrian movement interactions based on pre-computed movement patterns. As an example, consider the *Encounter* interaction. Encounter can be specified by the following OWL DL axiom:

Figure 5. Main classes for the Interactions ontology

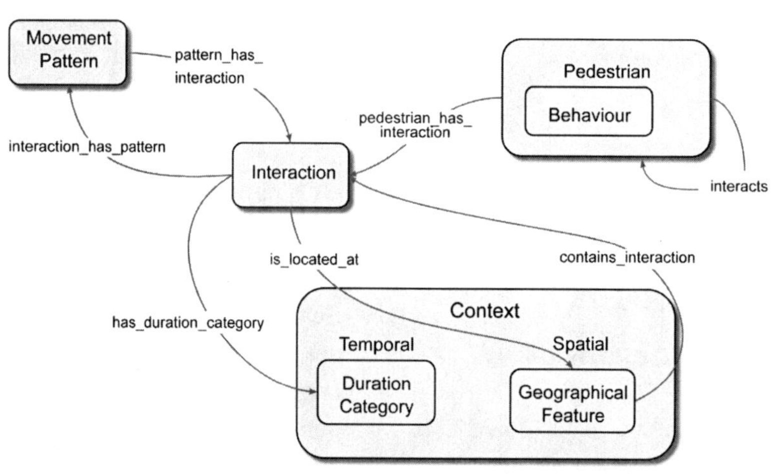

Figure 6. The interactions taxonomy

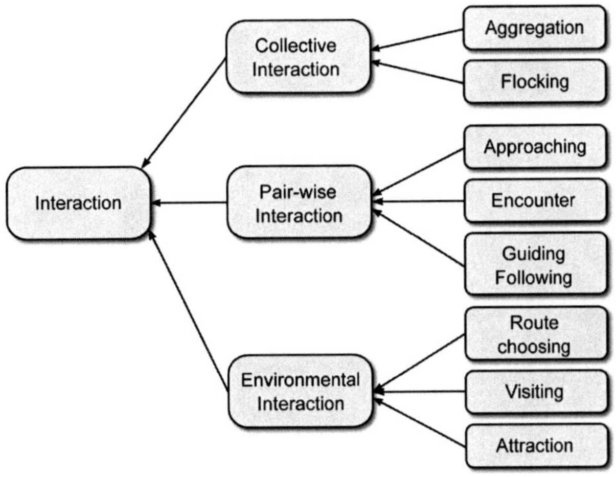

Encounter ≡ is_located_at some (Park_Feature and((has_Attractiveness has Low_Attractiveness) or (has_Attractiveness has Medium_Attractiveness))) And (has_Duration_Class has DC_Long and inter_has_pattern some (SpatioTemporal_Coincidence and(has_Spatial_Threshold has Spatial_Threshold_Small) and (has_Temporal_Threshold has Temporal_Threshold_Short)) or ((has_Duration_Class has DC_Medium) or (has_Duration_Class has DC_Long)and is_located_at some (Park_Feature and has_Constriction_Level has CL_Open)) and (has_Duration_Class has DC_Medium) or (has_Duration_Class has DC_Long)))

Intuitively, this axiom defines the Encounter class as a kind of Interaction that has been demonstrated by a spatial-temporal coincidence pattern if the duration is medium to long and it is not due to environmental features. The conceptualisation of the necessary duration depends on the constriction level of the space. In contrast, the Approaching interaction is characterised by a spatio-temporal coincidence with short or medium temporal threshold and a small spatial threshold. Furthermore it is located within a constrained or semi-constrained space (e.g. a pathway or a square).

Approaching ≡ has_Duration_Class has Short and inter_has_pattern some (SpatioTemporal_Coincidence and ((or (has_Temporal_Threshold has Temporal_Threshold_Medium)) and (has_Spatial_Threshold has Spatial_Threshold_Small)) and is_located_at some (Geographical_Feature and ((has_Constriction_Level has Constrained) or (has_Constriction_Level has Semiconstrained)))

As we can notice in such definitions, each interaction is inferred from movement patterns, characterising them in relation to the context features. In this sense, patterns capture the spatio-temporal essence of the interaction. We have identified a set of possible spatial and spatio-temporal patterns based on their role on interaction inference. An example of a movement pattern (here used on the encounter and approaching interaction) is the SpatioTemporal Coincidence. This pattern describes a pair of trajectories that have some coincidence of time-space (they are in the same place at the same time). The thresholds that define the spatio-temporal coincidence have to be set by the analyst depending on the other contextual parameters (how close is "the same place"? how many meters? how long is a coincidence? Some seconds or some minutes?) The movement pattern taxonomy is depicted in Figure 7.

Figure 7. Movement patterns taxonomy

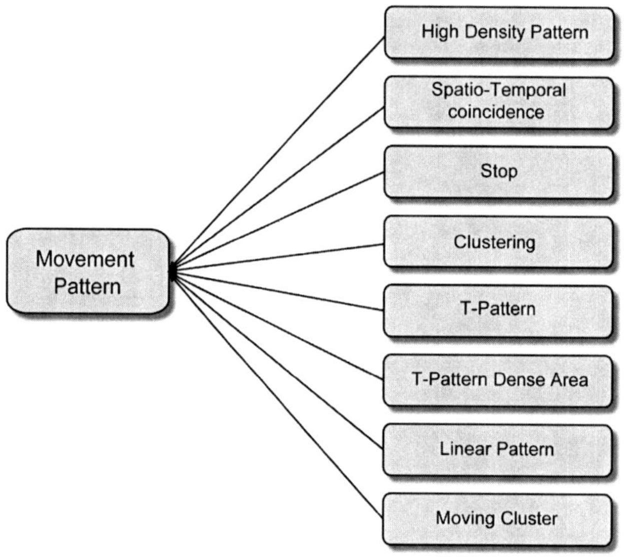

Movement patterns can be computed using several techniques, ranging from spatial analysis tools (Orellana, Wachowicz, Andrienko, N. & Andrienko, G., 2009) and exploratory visual analytics (Thomas & Cook, 2005), to spatial and spatio-temporal data mining (Nanni et al. 2008). An example of pattern detection technique is REMO (RElative MOtion) that compares relative patterns among individuals in a pattern-matching process, focusing on identifying the similarity of one or more movement parameters that are depicted in a movement matrix (Laube et al., 2005, Weaver, 2008). Visual Analytics discovers movement patterns typically found by visual techniques, such as spatio-temporal coincidence (N. Andirenko, G. Andirenko, Wachowicz & Orellana, 2008), stops and moves (N. Andrienko & G. Andrienko, 2008). Data-Mining techniques extract common movements by analysing with a statistical technique a large number of trajectories. Examples of trajectory patterns are T-cluster (Rinzivillo et al., 2008), identified groups of similar trajectories give a similarity measure, and T-patterns (Giannotti et al. 2007) representing the frequently followed sequences of places. Other movement patterns can be discovered by spatial-statistical approaches thus finding linear patterns and dense areas for movement suspension (Orellana et al., 2009).

The other main concept of the interactions ontology is Context. In this work we have defined two different kinds of context: the temporal context, where we defined the time units of interest for the application, and the spatial context, defining the most interesting characteristics that describe the geographical locations where the movement takes place. Therefore, this representation strategy follows the approach known as "Compose-and-Conquer" (Bouquet, Ghidini, Giunchiglia, & Blanzieri, 2003) since it does not use global theories for context, but local theories that represent specific points of view (Figure 8).

It is worth noting that the spatial context deals with geographical features of the environment, such as the constriction level (pedestrian moving in open space, constrained space such as a pathway, or a semi-constrained space like a square), the attraction areas, the accessibility levels, etc. Temporal context relates to the specification of duration and thresholds of interactions. Figure 9 shows the complete taxonomy of interactions

Figure 8. Context class and subclasses

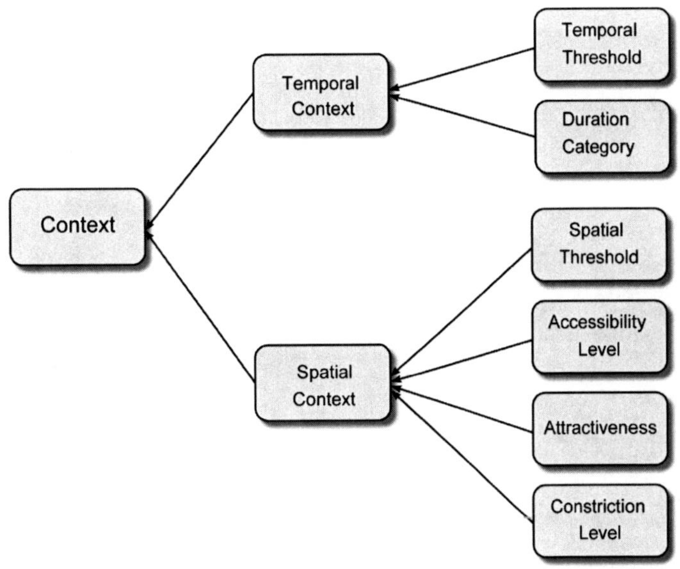

ontology implemented in *Protégé Ontology Editor* v 3.4 (Stanford Centre for Biomedical Informatics Research, 2008).

APPLICATION SCENARIO

The viability of our approach to analysing collective pedestrian movement is experimented through an application scenario about visitor's behaviour in the Dwingelderveld National Park (DNP) in Netherlands. DNP is a natural park of about 3700 ha in the North Eastern part of the Netherlands and receives 2million visitors a year. It is a typical Dutch recreational area with an extensive network of short strolls (60 km of marked trails, each of less than 7 km in length) and long walks for cycling and horse riding. To have a better understanding of how the park is used, park managers assign a specific behaviour to each visitor, depending on their movement behaviour. For example, a visitor who follows only marked trails is named as "follower" whereas a visitor that explicitly does not follow the marked trail is called a "browser" or "explorer" (van Marwijk & Pitt, 2008).

The available dataset comes from three different information sources. A questionnaire that records visitor characteristics; a set of point coordinates captured by GPS receivers given to the visitors; and a GIS data-set containing the path network of DNP and the locations of access point. The questionnaire gives detailed information about the visitor, such as sex and age, and the reason for the visit (to have a walk, to take pictures, etc). The park has five main entrances, from which the visit starts (the beginning of the GPS trajectories). This survey was carried out during 7 days (weekend and weekdays) in spring and summer 2006 for 461 hikers (van Marwijk & Pitt, 2008).

The objective of the current analysis is to characterise visitor behaviour in terms of the specific interactions that can be inferred from their movement patterns. In particular, we exploited the previously defined Interactions ontology, specialised in this particular case study, to represent and infer new knowledge about visitor behaviour.

The overall vision of the proposed approach is depicted in Figure 10. The problem we want to deal with can be stated as: *find visitors of the park*

Figure 9. The taxonomy of interactions ontology in Protégé. Screenshot by the author, reproduced under the terms of "Fair Use".

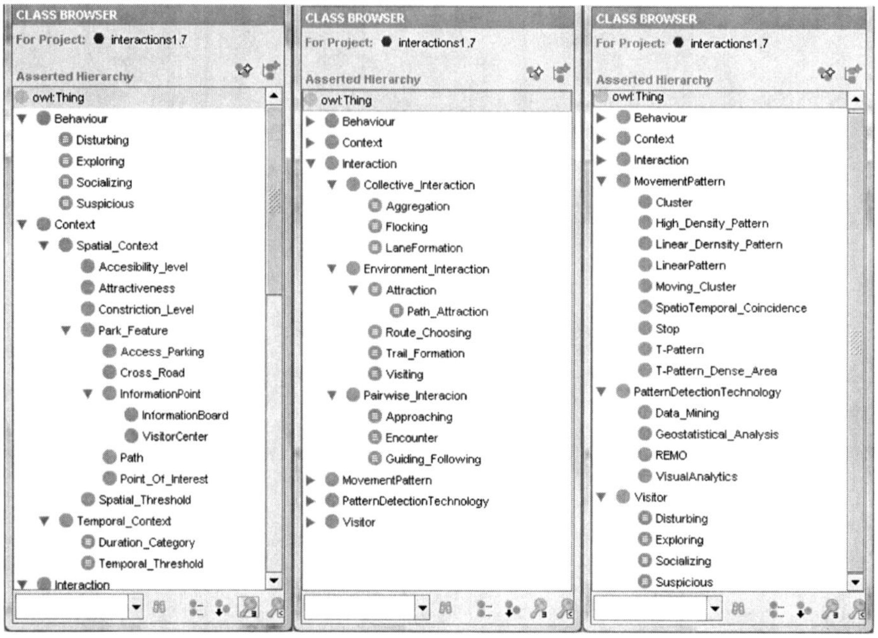

whose behaviour can be classified as "explorer" (or follower, or other). This classification task is actually performed by the ontology reasoning engine, where movement data and patterns are collected and an implicit class defined by OWL axioms specifies each kind of behaviour.

In this experiment, we used Protégé 3.4 (Stanford Center for Biomedical Informatics Research, 2008) with Pellet Reasoner (Sirin, Parsia, Grau, Kalyanpur, & Katz, 2007). As a very first step, the instances were introduced manually in Protégé. However, the feasibility of automatic systems was envisaged, in fact some previous work used ontologies for reasoning with trajectory data exploiting Oracle11g as an efficient reasoning engine over large datasets (Baglioni, Macedo, Renso, & Wachowicz, 2008; Baglioni, Macedo, Renso, Trasarti, & Wachowicz, 2009).

As an example, querying the ontology of the "explorer" behaviour expressed by visitors will give the identifier (ID) of the visitors whose movement has been classified by the ontology reasoning engine (based on the previously defined axioms), as "explorer". In other words, the instances of the ontology class Visitor represent the individual visitor's movement. These visitors are classified by the reasoner into the appropriate classes based on the satisfiability of the axioms.

It is worth noting that pattern finding can be seen as a top-down and bottom-up process. Indeed, extracted patterns can be used in finding interactions, but also interaction definition can drive the analyst to find some new movement patterns. For example, the attraction areas can be some predefined points or regions (the campsite, the café, the radio-telescope, etc.) or can be detected by data mining patterns as the most frequently visited areas, or as the T-Patterns computed dense areas.

To acquire a better understanding of the approach, let us consider the exploring behaviour. As specified above, an explorer is a visitor who tends not to follow the marked trails. Therefore, which kind of interactions can be expected by "explor-

Figure 10. The proposed overall approach

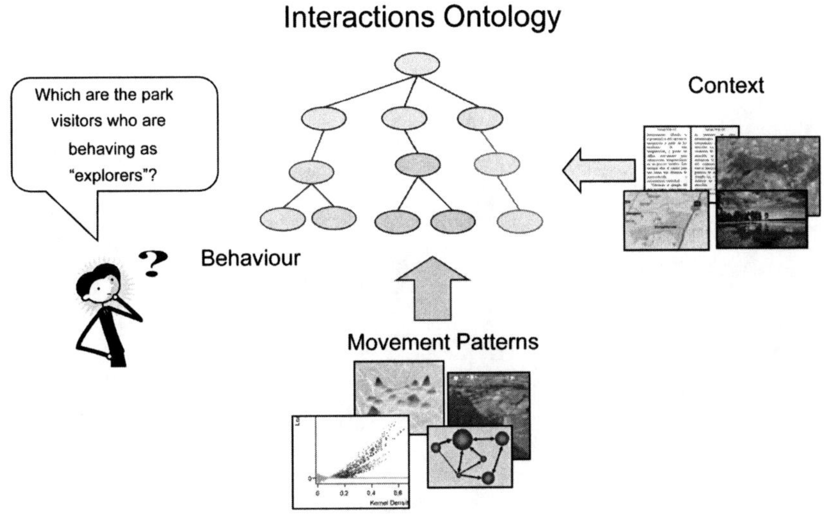

ers"? For example, it is expected that an explorer will tend to stop at the cross-roads to decide which route to take and to consult information points such as the visitor centre or information boards (Figure 11). This definition can be rephrased as "visitor who has interactions of the kind *route-choosing* and *visiting* information points". This means that an *exploring behaviour* is a kind of visitor who can be disclosed by the following axiom:

Exploring ≡ visitor_has_interaction some Route_Choosing visitor_has_interaction some (Visiting and (is_located_at some (Information_Point))

Analogously, a socialising behaviour can be identified by having encounters in places where people usually do not interact, such as the park trails and cross-roads.

Socialising ≡ visitor_has_interaction some (Encounter and (is_located_at some (Cross_Road or Path))) Another example is the axiom below that defines a disturbing behaviour in terms of visiting forbidden park areas.

Disturbing ≡ visitor_has_interaction some (Visiting and (is_located_at some Park_Feature and (has_Accesibility_level has Forbidden)))

Another interesting feature of the ontology is that we can link visitors involved in a pair-wise interaction through a symmetric property called interacts_with. This makes it possible to define a "suspicious" behaviour as a visitor who interacts with someone who is inferred to be a "disturbing" visitor.

Suspicious ≡ Interacts_with some (Disturbing)

These examples can be redefined according to different domain experts, but interactions remain the building blocks used to represent visitor behaviour.

The process outlined in this application scenario can be summarized by the following main steps:

- Visitor information is stored as ontology as instances of the Visitor's class.
- The basic spatio-temporal patterns are detected by means of a number of tools: GIS software such as ArcGIS (ESRI, 2008), or visual analytics tools such as Visual Analytics Toolkit (G. Andrienko, & N.

Figure 11. Part of the spatio-temporal path of a visitor on 06/08/2006. The time is represented by the height of the path; yellow bars represent route choosing interactions and red bars represent visiting interactions; green pins represent points of interest and (i) stands for a information point. The reasoning engine will classify the visitor as "explorer".

Andrienko, 2008; IAIS, 2008), and data mining algorithms (Rinzivillo et al 2008, Giannotti et al., 2007). In particular, here VA Toolkit has been used to find patterns for spatio-temporal coincidence and stops.

- Detected patterns are imported as instances into the ontology. This means that the main features of the patterns are represented as instances of the ontology along with the context information (spatial and temporal threshold, constrained levels in space and so on).
- Once patterns and contextual information are stored in the ontology as instances, the reasoning engine performs an instance-checking task to infer both interactions and movement behaviour.

In the specific case illustrated here, we have focused on *socialising* behaviour. We first detected the spatio-temporal coincidence patterns by means of the VA Toolkit (see, for example, Figure 11). We set the spatial threshold values as 5 meters for small, 10 meters for medium and 20 meters for long. The constrained space is represented by the set of paths, seen easily on the park map, whereas the cross roads and small areas represent the semi-constrained space; the remaining area is considered as open space. We populated the ontology with data referred to a part of the day 06/08/2006 and we ran the reasoning engine. We inferred four instances of visitors to be 'socialising' since they interacted with other visitors (Encounters) when they walked on the paths in the park. One visitor was interpreted as "Explorer" since they stopped and chose a route at several crossroads and stopped in two points that we marked as "Information Points". One visitor was interpreted as "Disturbing" since they were visiting an area that we marked as forbidden. In addition, we inferred one visitor to be "Suspicious" due to the fact that he or she encountered the Disturbing visitor. Figure 12 shows the relevant classes and instances that the reasoner engine used to perform the instance checking for some instances of Interaction and Visitor Classes. The reasoner engine inferred that I3 *is an instance* of Encounter and I6 *is an instance* of Visiting. Furthermore, visitor V9 was inferred as Disturbing since he or she visited a forbidden area and

Figure 12. Two visitors were inferred as Disturbing and Suspicious since they meet the necessary and sufficient conditions stated in the ontology.

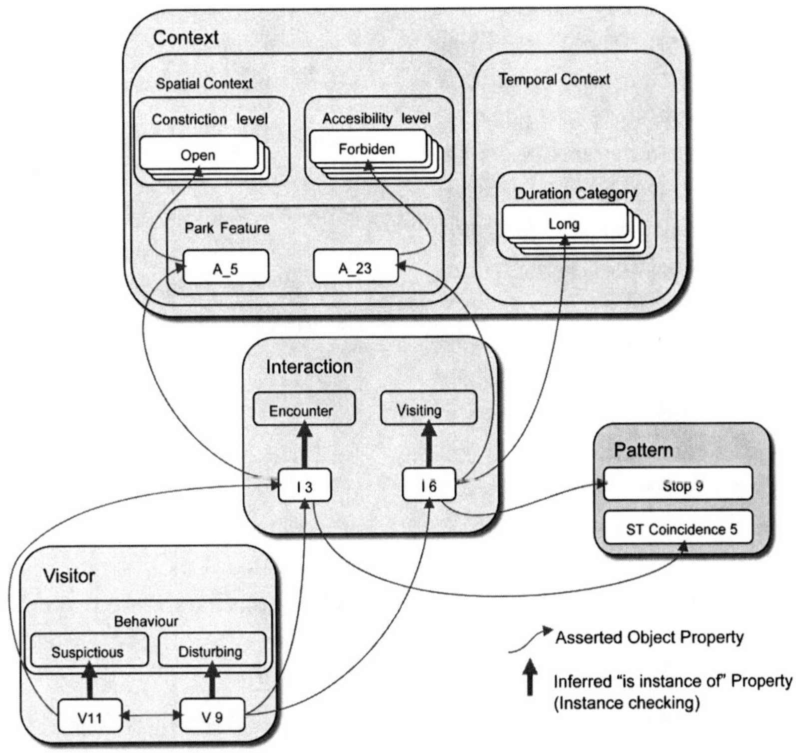

visitor V11 was inferred as Suspicious since he or she interacted with the disturbing V9.

CONCLUSION

In this chapter we proposed a knowledge representation and reasoning approach to characterise pedestrian movement interactions and behaviour. The contribution of this work, in respect to previous literature proposals, is twofold. On the one hand, we have developed a taxonomy of interactions where a number of basic interactions, concepts and their relationships with movement patterns and pedestrian behaviour are defined within a context. On the other hand, we have enriched the introduced taxonomy with reasoning, where, exploiting the ontology formalism OWL, interactions are combined with context to define pedestrian movement behaviour. Furthermore, we have sketched a case study where we experiment the methodology with a real dataset recording visitors' movements in the Dwingelderveld National Park (DNP) in the Netherlands.

A crucial point in the implementation of this experiment is how to populate the ontology with data, such as movement patterns and visitor information. How to link each tool output (namely, an extracted pattern) to a specific ontology class? Common approaches to ontology population map database tables to ontology concepts. This means that at the first stage we have to store both data and extracted movement patterns in a database. The problem then becomes how to map each table to an ontology concept. Several approaches have been proposed in the literature following two main directions: to import each database record into the ontology or dynamically map the ontology to a database. These two approaches offer complementary benefits and drawbacks. Indeed, importing data into the ontology may become impractical when the dataset is very large, since current ontology systems are not scalable to large datasets. On the other hand, dynamic mapping from ontology to database cuts down the degree of expressiveness of the ontology language (Calvanese, Lembo, Lenzerini, Poggi, & Rosati, 2007).

As a future task, we plan to design and implement an architecture where the patterns detected from the various tools are automatically or semi-automatically, inserted into the ontology. For this step we would benefit from approaches like Athena (Baglioni et al., 2009), a system built on top of Oracle 11g that allows for easily import of database tables storing trajectories and contextual information inside ontology concepts.

Although we consider that our approach can potentially achieve a high level of expressivity to represent pedestrian behaviour, we are aware of the shortcomings that it has at this stage. We are planning to further develop the concepts related to pedestrian behaviour applied to new and more interesting scenarios (e.g. including actual collective movement patterns and behaviour).

REFERENCES

W3C Consortium. (2004). *OWL web ontology language.*

Alvares, L. O., Bogorny, V., Kuijpers, B., Macedo, J., Moelans, B., & Vaisman, A. (2007). A model for enriching trajectories with semantic geographical information. In *Proceedings of the 15th Annual ACM International Symposium on Advances in Geographic Information Systems,* (pp. 162-169).

Andersson, M., Gudmundsson, J., Laube, P., & Wolle, T. (2008). Reporting leaders and followers among trajectories of moving point objects. *GeoInformatica, 12*(4), 497–528. doi:10.1007/s10707-007-0037-9

Andirenko, N., Andirenko, G., Wachowicz, M., & Orellana, D. 2008. Uncovering Interactions between Moving Objects. In T.J. Cova, H.J. Miller, K. Beard, A. Frank & M.F. Goodchild, (Eds.), *Geographic Information Science, Extended Abstracts of 5th International Conference, GIScience 2008,* Park City, UT, September 23-26, 2008 (pp. 16-27).

Andrienko, G., & Andrienko, N. (2008). A visual analytics approach to exploration of large amounts of movement data. In M. Sebillo, G. Vitiello & G. Schaefer (Eds.), Visual information systems, 10th international conference, VISUAL 2008 (pp. 1-4). Berlin: Springer-Berlin. doi:10.1007/978-3-540-85891-1_1

Andrienko, G., Andrienko, N., Kopanakis, I., Ligtenberg, A., & Wrobel, S. (2008). Visual analytics methods for movement data. In F. Giannotti, & D. Pedreschi (Eds.), Mobility, data mining and privacy (pp. 375-410). Berlin: Spinger-Verlag.

Andrienko, N., Andrienko, G., Pelekis, N., & Spaccapietra, S. (2008). Basic concepts of movement data. In F. Giannotti, & D. Pedreschi (Eds.), Mobility, data mining and privacy (pp. 15-38). Berlin: Spinger-Verlag.

Antonini, G., Bierlaire, M., & Weber, M. (2006). Discrete choice models of pedestrian walking behavior. *Transportation Research Part B: Methodological, 40*(8), 667–687. doi:. doi:10.1016/j.trb.2005.09.006

Baader, F., Calvanese, D., McGuinness, D., Nardi, D., & Patel-Schneider, P. (2003). The description logic handbook: Theory, implementation and applications. Cambridge, UK: Cambridge University Press.

Baglioni, M., Macedo, J., Renso, C., Trasarti, R., & Wachowicz, M. (2009). Towards a semantic interpretation of movement behaviour. In *Proceedings of 12th AGILE Conference, Lecture Notes in Geoinformation and Cartography*.

Baglioni, M., Macedo, J., Renso, C., & Wachowicz, M. (2008). *An ontology-based approach for the semantic modelling and reasoning on trajectories*. Paper presented at the ER '08: Proceedings of the ER 2008 Workshops (CMLSA, ECDM, FP-UML, M2AS, RIGiM, SeCoGIS, WISM) on Advances in Conceptual Modeling, Barcelona, Spain (pp. 344-353). doi:0.1007/978-3-540-87991-6_41

Batty, M. (2003). Agent-based pedestrian modelling. In M. Batty, & P. A. Longley (Eds.), *Advanced spatial analysis: The CASA book of GIS* (pp. 81-106). Redlands, CA: Esri Press. Retrieved from http://www.casa.ucl.ac.uk/working_papers/paper61.pdf

Batty, M., DeSyllas, J., & Duxbury, E. (2003). The discrete dynamics of small-scale spatial events: Agent-based models of mobility in carnivals and street parades. *International Journal of Geographical Information Science, 17*(7), 673-697. Retrieved from http://www.casa.ucl.ac.uk/working_papers/paper56.pdf

Benkert, M., Gudmundsson, J., Hübner, F., & Wolle, T. (2008). Reporting flock patterns. *Computational Geometry, 41*(3), 111–125. doi:. doi:10.1016/j.comgeo.2007.10.003

Bierlaire, M., Antonini, G., & Weber, M. (2003). *Behavioral dynamics for pedestrians*. Paper presented at the Moving through Nets: The Physical and Social Dimensions of Travel, Lucerne.

Blue, V. J., & Adler, J. L. (2001). Cellular automata microsimulation for modeling bi-directional pedestrian walkways. *Transportation Research Part B: Methodological, 35*(3), 293–312. doi:. doi:10.1016/S0191-2615(99)00052-1

Blythe, P. W., Miller, G. F., & Todd, P. M. (1996). Human simulation of adaptive behavior: Interactive studies of pursuit, evasion, courtship, fighting, and play. Paper presented at the *From Animals to Animats 4: Proceedings of the Fourth International Conference on Simulation of Adaptive Behavior.*

Bouquet, P., Ghidini, C., Giunchiglia, F., & Blanzieri, E. (2003). Theories and uses of context in knowledge representation and reasoning. *Journal of Pragmatics, 35*(3), 455–484. doi:10.1016/S0378-2166(02)00145-5

Burstedde, C., Klauck, K., Schadschneider, A., & Zittartz, J. (2001). Simulation of pedestrian dynamics using a two-dimensional cellular automaton. *Physica A: Statistical Mechanics and its Applications, 295*(3-4), 507-525.

Calvanese, D., Lembo, D., Lenzerini, M., Poggi, A., & Rosati, R. (2007). *Ontology-based database access*. Paper presented at the Proc. of the 15th Italian Conf. on Database Systems SEBD, 324-331.

Daamen, W., & Hoogendoorn, S. P. (2003a). Controlled experiments to derive walking behaviour. *European Journal of Transport and Infrastructure Research, 3*(1), 39–59.

Daamen, W., & Hoogendoorn, S. P. (2003b). Experimental research of pedestrian walking behavior. *Transportation Research Record, 1828*, 20–30. doi:10.3141/1828-03

Davis, R., Shrobe, H., & Szolovits, P. (1993). What is A knowledge representation? *AI Magazine, 14*(1), 17–33.

Dodge, S., Weibel, R., & Lautenschütz, A. K. (2008). Towards a taxonomy of movement patterns. *Information Visualization, 7*, 240–252. doi:. doi:10.1057/palgrave.ivs.9500182

ESRI. (2008). ArcGIS 9.2 [Computer software]

Giannotti, F., Nanni, M., Pinelli, F., & Pedreschi, D. (2007). *Trajectory pattern mining*. Paper presented at the KDD '07: Proceedings of the 13th ACM SIGKDD International Conference on Knowledge Discovery and Data Mining, San Jose, CA. (pp. 330-339). doi:10.1145/1281192.1281230

Giunchiglia, F., & Bouquet, P. (1997). Introduction to contextual reasoning: an artificial intelligence perspective. *Perspectives on Cognitive Science*, *3*, 138–159.

Gruber, T. (2008). Ontology. In L. Liu, & M. T. Özsu (Eds.), Encyclopedia of database systems. Berlin: Springer-Verlag.

Gudmundsson, J., van Kreveld, M., & Speckmann, B. (2004). *Efficient detection of motion patterns in spatio-temporal data sets*. Paper presented at the Proceedings of the 12th Annual ACM International Workshop on Geographic Information Systems, (pp. 250-257).

Güting, R., & Schneider, M. (2005). Moving Objects Databases. San Francisco: Morgan Kaufmann.

Hägerstrand, T. (1970). What about people in regional science? *Papers in Regional Science*, *24*(1), 6–21. doi:10.1007/BF01936872

Helbing, D., Buzna, L., Johansson, A., & Werner, T. (2005). Self-organized pedestrian crowd dynamics: Experiments, simulations, and design solutions. *Transportation Science*, *39*(1), 1–24. doi:. doi:10.1287/trsc.1040.0108

Helbing, D., Molnár, P., Farkas, I. J., & Bolay, K. (2001). Self-organizing pedestrian movement. *Environment and Planning B*, *28*(3), 361–384. doi:10.1068/b2697

Helbing, D., Twining, C. J., Bulsari, A., Oliker, S., Furst, M., & Maimon, O. (1993). A fluid-dynamic model for the movement of pedestrians. *Complex Systems*, *6*, 391–415.

Hillier, B., Penn, A., Hanson, J., Grajewski, T., & Xu, J. (1993). Natural movement: Or, configuration and attraction in urban pedestrian movement. *Environment and Planning. B, Planning & Design*, *20*, 29–66. doi:10.1068/b200029

Hoogendoorn, S. P., Bovy, P. H. L., & Daamen, W. (2002). Microscopic pedestrian wayfinding and dynamics modelling. In M. Schreckenberg, & S. Sharma (Eds.), Pedestrian and evacuation dynamics, (pp. 123-155). Berlin: Springer.

Hornsby, K., & Egenhofer, M. J. (2002). Modeling moving objects over multiple granularities. *Annals of Mathematics and Artificial Intelligence*, *36*, 177–194. doi:10.1023/A:1015812206586

IAIS. (2008). *CommonGIS* [Computer software].

Kalnis, P., Mamoulis, N., & Bakiras, S. (2005). On discovering moving clusters in spatio-temporal data. In C. Bauzer-Medeiros, M. J. Egenhofer & E. Bertino (Eds.), Advances in spatial and temporal databases (pp. 364-381). Berlin: Springer. doi:10.1007/11535331

Laube, P., Imfeld, S., & Weibel, R. (2005). Discovering relative motion patterns in groups of moving point objects. *International Journal of Geographical Information Science*, *19*(6), 639. doi:. doi:10.1080/13658810500105572

Manco, G., Baglioni, M., Giannotti, F., Kuijpers, B., Raffaetà, A., & Renso, C. (2008). Querying and reasoning for spatiotemporal data mining. In F. Giannotti, & D. Pedreschi (Eds.), Mobility, data mining and privacy (pp. 335-374). Berlin: Springer-Verlag. doi.:10.1007/978-3-540-75177-9_13

Miller, H. (2005). A measurement theory for time geography. *Geographical Analysis*, *37*, 17–45. doi:10.1111/j.1538-4632.2005.00575.x

Nanni, M., Kuijpers, B., Korner, C., May, M., & Pedreschi, D. (2008). Spatiotemporal data mining. In F. Giannotti, & D. Pedreschi (Eds.), Mobility, data mining and privacy, (pp. 267-296). Berlin: Springer.

Orellana, D., Wachowicz, M., Andrienko, N., & Andrienko, G. (2009). Uncovering interaction patterns in mobile outdoor gaming. Paper presented at the *Advanced Geographic Information Systems & Web Services, 2009. GEOWS '09*. (pp. 177-182). doi:10.1109/GEOWS.2009.13

Ortale, R., Manco, G., Ritacco, E., Pelekis, N., Trasarti, R., Costa, G., et al. (2008). DAEDALUS: Towards progressively querying and mining movement dataICAR-CNR.

Penn, A., & Turner, A. (2001). Space syntax based agent simulation. In M. Schreckenberg, & S. Sharma (Eds.), Pedestrian and evacuation dynamics (pp. 99-114). Berlin: Springer.

Pfoser, D., & Jensen, C. S. (2001). *Querying the trajectories of on-line mobile objects*. Paper presented at the MobiDe '01: Proceedings of the 2nd ACM International Workshop on Data Engineering for Wireless and Mobile Access, Santa Barbara, CA, (pp. 66-73). doi:10.1145/376868.376902

Renso, C., Puntoni, S., Frentzos, E., Mazzoni, A., Moelans, B., Pelekis, N., et al. (2008). Wireless network data sources: Tracking and synthesizing trajectories. In F. Giannotti, & D. Pedreschi (Eds.), Mobility, data mining and privacy, (pp. 73-100). doi:10.1007/978-3-540-75177-9_4

Rinzivillo, S., Pedreschi, D., Nanni, M., Giannotti, F., Andrienko, N., & Andrienko, G. (2008). Visually–driven analysis of movement data by progressive clustering. *Information Visualization, 7*(3/4), 225–239. doi:. doi:10.1057/palgrave.ivs.9500183

Schlieder, C. & Werner, A. (2003). *Interpretation of intentional behavior in spatial partonomies.*

Shaw, S., & Yu, H. (2009). A GIS-based time-geographic approach of studying individual activities and interactions in a hybrid physical–virtual space. *Journal of Transport Geography*. doi:. doi:10.1016/j.jtrangeo.2008.11.012

Shaw, S., Yu, H., & Bombom, L. (2008). A space-time GIS approach to exploring large individual-based spatiotemporal datasets. *Transactions in GIS, 12*(4), 425–441. doi:10.1111/j.1467-9671.2008.01114.x

Sirin, E., Parsia, B., Grau, B. C., Kalyanpur, A., & Katz, Y. (2007). Pellet: A practical owl-dl reasoner. *Web Semantics: Science, Services and Agents on the World Wide Web, 5*(2), 51-53. Retrieved from http://pellet.owldl.com/papers/sirin05pellet.pdf

Smyth, C. S. (2001). Mining mobile trajectories. In H. J. Miller & J. Han (eds.), Geographic Data Mining and Knowledge Discovery, (pp. 337-361). London: Taylor and Francis.

Spaccapietra, S., Parent, C., Damiani, M. L., Macêdo, J., Porto, F., & Vangenot, C. (2008). A conceptual view on trajectories. *Data & Knowledge Engineering, 65*(1), 126–146. doi:10.1016/j.datak.2007.10.008

Stanford Center for Biomedical Informatics Research. (2008). *Protégé* [Computer software].

Thomas, J. J., & Cook, K. A. (2005). Illuminating the path: The research and development agenda for visual analytics. Washington, DC: IEEE.

Turner, A., Doxa, M., O'Sullivan, D., & Penn, A. (2001). From isovists to visibility graphs: A methodology for the analysis of architectural space. *Environment and Planning. B, Planning & Design, 28*(1), 103–121. doi:10.1068/b2684 van Marwijk, R., & Pitt, D. G. (2008). Where Dutch recreationists walk: Path design, physical features and walker usage. In *The Fourth International Conference on Monitoring and Management of Visitor Flows in Recreational and Protected Areas, Management for Protection and Sustainable Development,* Montecatini Terme, Italy.

Weaver, C. (2008). *Cross-dimensional visual queries for interactive animated analysis of movement.* Paper presented at the Proceedings of Geospatial Visual Analytics Workshop at GIScience 2008, Utah. Retrieved from http://geoanalytics.net/GeoVisualAnalytics08/

Wood, Z., & Galton, A. (2008). *Collectives and how they move: A tale of two classifications.* Paper presented at the Proceedings of the 2nd Workshop on Behaviour Monitoring and Interpretation, BMI'08, Kaiserslautern, (pp. 57-71).

Yu, H., & Shaw, S. (2007). Revisiting Hägerstrand's time-geographic framework for individual activities in the age of instant access. In H. J. Miller (Ed.), Societies and cities in the age of instant access (pp. 103-118). Dordrecht, Germany: Springer. doi:10.1007/1-4020-5427-0.

Chapter 6
GUESS:
On the Prediction of Mobile Users' Movement in Space

Maribel Yasmina Santos
University of Minho, Portugal

Adriano Moreira
University of Minho, Portugal

ABSTRACT

The increasing availability of mobile networks and devices motivate the development of new paradigms of interaction between users, and between a user and the surrounding environment. When the position of a user is known, location-aware applications can adapt their behaviour accordingly to that position. This usual behaviour of location-aware applications can be enhanced in order to turn these applications pro-active, providing services that can be helpful to the user before he/she asks for them. For this kind of behaviour, location-aware applications need to anticipate the user's position, predicting where the user is going to be in the future. This chapter discusses the prediction of movement in space by looking at the topological constraints existing in the geographical space in which the user is moving. The obtained results show that the topological distance can be used as an alternative or as a complement to the geometrical distance normally used in map-matching techniques.

INTRODUCTION

The modelling of the conceptual relationships that exist between spatial objects plays an important role in new application areas based on geographic information. One of these emerging areas is related to context-aware applications. Within this class of applications, the location-aware applications are those that are able to adapt their behaviour accordingly to the location of the user (Dey, 2001). A key aspect of context-aware applications is the easy access to the geographic context of their users. One simple way of describing that context is to describe the place where the mobile user is at in a particular instant of time. The most usual and elementary form of geographic context is therefore the current position of the user, described as a pair of geographic coordinates, or the location, described as an address or other symbolic reference.

DOI: 10.4018/978-1-61520-769-5.ch006

The information about the user location is used by these applications to make available a wide range of services that are, hopefully, useful to the users. The services that are provided depend upon the user location and the geographic context in which the user is mapped by the location-aware application. Actions can only be started, e.g. making a specific service available, after the identification of the geographic context where a user can be located. Another potential contribution to the identification of the location of a user is through the use of predictive models to estimate future movements of a person (Ashbrook & Starner, 2003). Several studies have been carried out in order to predict the next user's location, based on the actual user's location or the previous available locations of the user (Ashbrook & Starner, 2003; Laasonen, 2005; Marmasse & Schmandt, 2002). The most common approaches used to estimate the user's position in a mobile environment follow quantitative approaches, in which the previous positions of the user and the estimation of his/her moving velocity, for example, are used as input. There are also other approaches in which the user's previous behaviour is used to predict the most probable next location. Relaying on user's current position and past behaviour for prediction processes requires the collection of data over a certain period of time, in order to establish his/her movement pattern. Besides the fact the user needs to be tracked, and privacy issues can arise, some learning time is also required. Moreover, systems trained for a certain geographic region may not perform well while the user is visiting a new region for the first time.

Instead of creating a history of the user to predict his/her next position, this work addresses the prediction of such position by resorting to the topological constraints present in the geographical context in which the user is located. Our aim is to focus on the prediction of the near future positions of cars, and on their use in applications that need to reserve resources or request data for future tasks. Examples include the reservation of network bandwidth in the wireless networks that are expected to be deployed along highways and other roads in the near future, and automatic traffic lights control for emergency vehicles (Yang & Wang, 2007).

Two approaches are discussed in this Chapter. The first approach uses the traditional geometric distance in a predictive model that allocates the user to a specific road segment in a road network. The second one uses the concept of topological distance, following a qualitative approach, in the allocation process. In this last approach, the topological spatial relations that exist between a spatially extended point (representing the uncertainty on the position of a mobile user) and a line (representing objects in which movement in space is possible) are identified (Santos & Moreira, 2007; Santos & Moreira, 2009a). Afterwards, and for the prediction task, two conceptual neighbourhood graphs are identified (Santos & Moreira, 2009b): one following the snapshot model, and a second one following the smooth-transition model (Egenhofer & Al-Taha, 1992). A subset of this smooth-transition model is used in this work.

The exploration and discussion of these two approaches, a quantitative and a qualitative, is motivated by the fact that the geometrical approach presents some drawbacks mainly associated to the position of the users (usually integrating some uncertainty). As the topological constraints associated with the geographical context in which the user is located are considered in the proposed prediction model, the formalization of a topological approach seems appropriate as it will be able to catch the relation existing between the user and the road network, and how this relation can progress over time.

The following sections, in outline, include: a) an overview of some existing map-matching techniques and their use in the prediction of next users' positions in a road network; b) the description of the model proposed in this work to predict the next user's positions attending to his/her current position and to the topological constraints present in the geographical context in which the user is located, and; c) a conclusion with some remarks about the work undertaken.

AN OVERVIEW OF MAP-MATCHING TECHNIQUES

The emergence of context-aware mobile environments and in particular the implementation of location-based services (LBS) in those environments lead to the need to identify the current user's position and to assign it a geographical context in which specific services are made available to the user.

When looking at mobile users in a road network, several map-matching techniques have been developed in order to assign a user to a specific road segment. This task uses the current user's position (given by a GPS – Global Positioning Systems – receiver or any similar technique) and a road network. Map-matching algorithms match the user's position (an inaccurate user's position) to a position on a road network by comparing the trajectory of the vehicle with the shapes of the road segments in the road network (Jagadeesh, Srikanthan, & Zhang, 2004).

A simple way to perform map-matching is to assign the user to the nearest road segment. This method, known as point-to-arc method, is easy to implement but it does not consider the historical information associated to the vehicle motion as well as it cannot deal with two road segments that are located at the same distance of the user. The arc-to-arc mapping technique fixes some of the drawbacks of the point-to-arc algorithm by matching the piecewise linear curve formed by a sequence of measured positions to a piecewise linear curve corresponding to a path in the network based on their closeness or similarity (Jagadeesh, et al. 2004). In the case of two linear curves are at the same distance of the user estimated position, the method does not allow a properly assignment of a road segment. Map-matching techniques are even more critical in dense urban road networks where it is difficult to identify the road on which the vehicle is travelling (Jagadeesh et al.). The proposal of Jagadeesh et al. for map-matching includes the identification of the road on which the user is travelling and the calculation of the exact location of the user on the identified road. The identification of the travel road is done using a fuzzy rule based inference system. The determination of the exact location on the road is performed considering the GPS errors characteristics.

Once the user is assigned to a road segment in a road network, several approaches can be used to track the user and to predict his/her future positions. The existing tracking approaches can be classified in point-based tracking, vector-based tracking and segment-based tracking (Civilis, Jensen, & Pakalnis, 2005). In point-based tracking, as defined by these authors, future positions of the user correspond to the recently reported positions. As a result, they correspond to the real reported positions. An update to the user's position is done when the distance to the previously reported position deviates from its current GPS position by a specified threshold. In vector-based tracking the future positions of a user are given by a linear function of time (a start position is used for that) and a velocity vector. Segment-based tracking uses the characteristics of the road network in which the user is moving. In these cases a digital representation of the road network is required as the user is located to a road network through map-matching techniques. The future positions of the user are given by a movement at constant speed along the identified road segment (Civilis et al.). In map-based LBS, if the user provides a destination, a routing algorithm can be used to predict a trajectory and to estimate the time when the user will reach there (Karimi & Liu, 2003). If the user's destination is unknown (what usually happens), conventional algorithms cannot be used and strategies for location prediction must be used.

This is the framework in which the work presented in this chapter is carried out. The destination of the user is not known and the prediction is done knowing only two consecutive positions of the user. No past trajectory of the user or user's preferences/profile is used. This ensures user privacy and motivates a new research area in uncertainty management in location prediction. Although the proposed model avoids the use of any past information about the mobile user, it integrates implicit knowledge about the road network structure: the topological network configuration and the traffic load associated with each possible path in the road network.

The option for avoiding the history of the user is motivated, as already mentioned, by two key reasons. The first one is associated with the need to ensure user privacy. As stated by Pedreschi et al. (2008), the analysis of mobility data creates potential threats to personal privacy, since mobility data is detailed and each record corresponds to a specific individual. For these authors, mobility data can put at risk our right to secrecy in our movements. The disclosure of such information can potentially violate the privacy rights, as specified in national and international laws and regulations (Pedreschi et al.). They also call our attention to the fact that the use of hiding identities is not enough to guarantee privacy protection, since in certain cases it is possible to reconstruct the identities of individuals. The second one is associated with the need to develop tools that are able to predict users' future positions with restrictions in the information that is available about the users. If we think on the services that can be provided by context-aware applications, there will certainly exist situations where the available information is limited, either for the lack of history on those contexts or because the users do not provide, or do not want to provide, information about them.

GUESS: PREDICTING MOVEMENT IN SPACE

For Galton (1995), the phenomenon of movement arises whenever the same object occupies different positions in space at different time moments. The given definition suggests that a theory of movement must include a *theory of time*, a *theory of space*, a *theory of objects* and a *theory of position*. *Time* can be treated as instants or as intervals, and an ordering relation needs to be established, representing temporal sequences. *Space* integrates elements such as points, lines or regions as fundamental entities. *Objects* present variable motion capabilities (or different motion characteristics), or none. *Position* integrates the theory of objects with the theory of space, within each object is occupying a certain part of space at a specific time.

For the proposed model, time is treated as instant timestamps in which the position of the mobile user has been measured or predicted. Space integrates two types of entities, circular spatially extended points and lines. Lines represent objects through which motion is possible and circular spatially extended points represent the position of the moving object as measured by a positioning system and its corresponding uncertainty. The position of each object in space gives us the possibility of prediction of future positions attending to the current position and to the objects through which motion is possible (lines).

In this section we propose a model to predict users' future positions in a road network. By restricting the movement to the road network, the topological characteristics of the space can be exploited to reduce the set of options to the future positions. The representation of that road network, and the information associated with it, is described next.

Representation of a Road Network

A road network can be represented by a graph with vertices and edges, in which an edge represents a road segment (a part of a road) and a vertex represents an intersection or a connection point between road segments. Each road segment (edge) may be characterized by attributes like speed limit or road length. Each road segment has a direction, indicating if it is a one-way road or a two-way road. Intersections usually represent points of uncertainty, since they can lead to different trajectories in the user's motion.

Let us consider the road network shown in Figure 1. This road network can be represented by a graph G, integrating vertices $v_{1,..,j}$ and edges $e_{1,...,k}$. Each edge has a direction associated with it indicating the traffic flow on that road segment. Whenever a road segment allows traffic in both directions that segment is represented by two edges. In this example road segments with their associated directions indicate the paths that can be followed by the users.

In our approach, each road segment has been characterized by a specific traffic load value that is used to support the decision process when more than one road segment can be followed by the user. The values for this traffic load can be obtained from observations of real traffic or estimated. The prediction results presented latter were obtained with an estimate of the traffic load, based on the concept of betweenness. In a graph, the betweenness of an edge corresponds to the number of times that edge is part of the shortest path connecting all pairs of vertices. For large networks, with high number of vertices, the computation effort required to compute the betweenness of all edges is too high. Therefore, our estimate of the traffic load has been computed by considering a large number of randomly generated origin-destination pairs of vertices instead of all the pairs of vertices. In a real scenario, the estimate of the traffic load could be collaboratively constructed by summing the real trajectories uploaded by contributing people (as being exploited for collaboratively map construction).

Prediction Model

The proposed model to predict the next user's positions includes three major steps (Figure 2). The model starts with the calculation of the velocity vector associated with two consecutive GPS readings and with the identification of the most aligned amongst the *k* nearest road segments accordingly to the current position and the velocity vector. The second step assigns the user to a specific road segment, looking at the several geometrical distances calculated between the user and the nearest road segments. The road segment with the shortest geometrical distance is selected, and the user is allocated to the nearest point in

Figure 1. Graph representation of a road network with the traffic load of each road segment

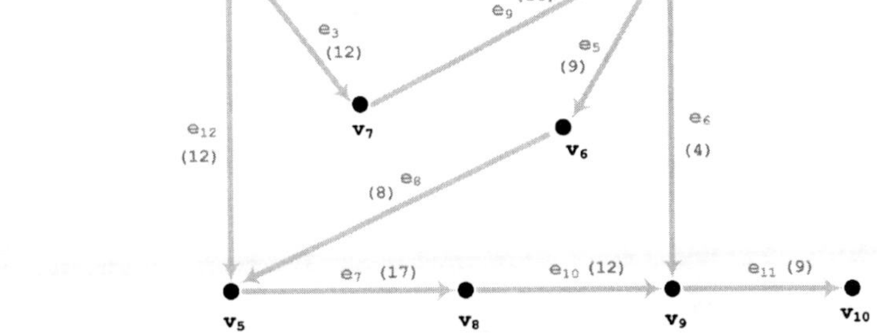

Figure 2. A model for predicting users' future positions in space

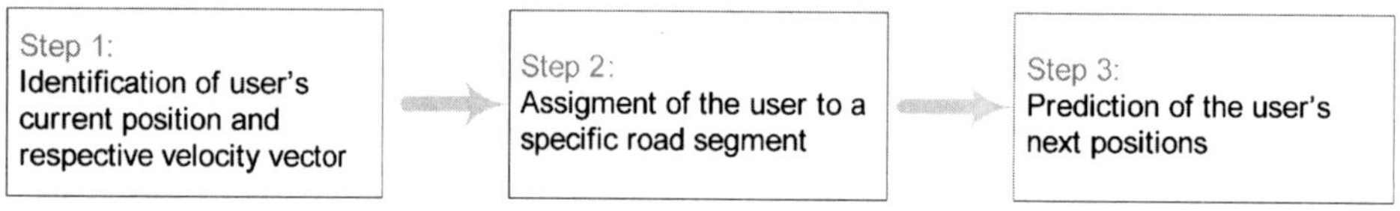

that road segment. The last step is concerned with the prediction of the user's future positions attending to his/her current position, to the topological constraints present in the geographical context in which the user is located and to the traffic metric associated to each road segment of the road network. In general terms, there are two different tasks in the prediction process. If the user is located on a specific road segment, he/she is moved to the end of that road segment at a constant speed. If not, which means that the user is already at the end of the road segment, the topology of the road network and the traffic metric associated to each road segment are used to select the next road segment. In case two or more road segments start in the decision point at which the user is currently located, the road segment with highest traffic metric is selected. This means that the prediction model follows what seems to be the most frequent route.

In step 1, the uncertainty of the current position of the mobile user is taken into account by representing it by a Circular Spatially Extended Point (CSEP), with the radius of the CSEP representing the error of the GPS reading. One must also be aware that the speed value and the direction of movement (bearing) are also estimates of the real situation since they are computed from two consecutive GPS readings with their corresponding uncertainties. By assigning the moving object to a specific road segment, in step 2, the prediction process is trying to minimize the uncertainty resulting from the observation process (the GPS readings). However, in urban dense areas, with a high density of small road segments, the choice of a particular road segment represents a major

Figure 3. One example of a predicted path

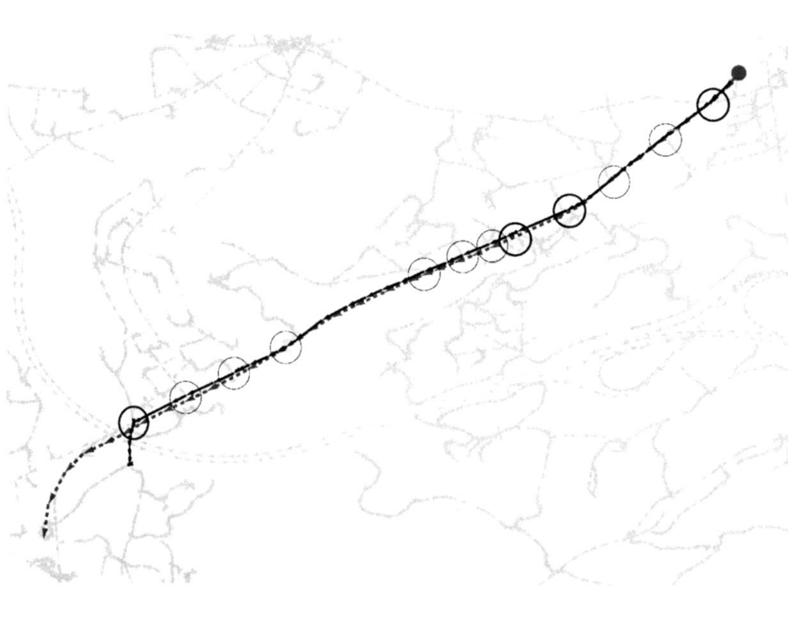

challenge. In this work, we used a commercial road map and observed that the most sinuous roads are frequently represented by a sequence of very small road segments with length equal or smaller than the CSEP radius used to represent the GPS error. Some of the results described in the next section illustrate these difficulties.

Prediction Results

Following the model proposed in the previous section, we computed the future positions for a mobile user over a road network using position measurements obtained from a GPS receiver while the user was on his/her daily life. For each GPS position reading, a future path was computed until it diverges from the real (future) trajectory by a certain maximum distance. Figure 3 shows one example of the real trajectory of a user where the dashed line is the trajectory measured by the GPS receiver and the solid line represents the predicted path. In this example, the model has been able to correctly predict the future path of the user for up to 695 meters. The figure includes several circles that represent the decision points of the predictive model. Bold circles represent situations where three or more alternative paths need to be considered by the prediction model.

We repeated this computation for a large number of GPS points, and calculated the length of the predicted path. A predicted path was considered correct until the distance between the last predicted point and the real trajectory exceeded the GPS error (CSEP radius). A summary of the obtained results is shown in Figure 4, where the solid line represents the percentage of the correct predicted paths that have a length of *m* meters or more for a dense urban only environment. The dashed line represents the results for a mixed urban/rural environment.

These results uncover two interesting aspects. One is that considerable long paths can be correctly predicted even in dense urban environments: 50% of the predicted paths are at least 225 meters long, and 10% of the predictions have found correct paths with 1525 meters or more. In a mixed urban/rural environment, 10% of the predictions have paths lengths with at least 3500 meters. Also, the

Figure 4. Probability of a correct path of length m

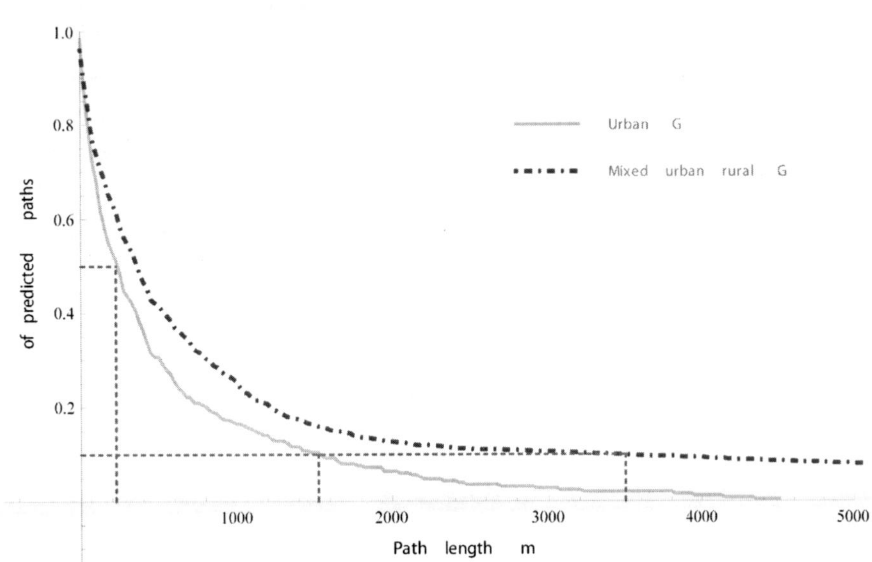

length of the predicted paths is longer in mixed urban/rural areas than in urban only areas, since the spatial density of possible alternatives in urban areas is much higher than in rural areas.

The second aspect emerging from these results is that too many predicted paths are too short. This is a strange result, since we expected to be able to predict the first few tens of meters of the future path in all cases. We therefore analyzed these cases in more detail, and uncover that, in many cases, the initial allocation of the user to a road segment fails. This is mostly due to the low accuracy of the GPS measurements, as depicted in Figure 5. In this figure, the sequence of dashed arrows represents the user trajectory as measured by a GPS receiver. The solid dot is where the user was reported by the GPS at the time he/she was allocated to a road segment, and the large circle is an estimate of the GPS error. The crossed circle represents the result of the allocation process. The dashed arrow departing from this point represents the estimated velocity vector for the new estimated position.

Figure 5. Example of wrong initial allocation

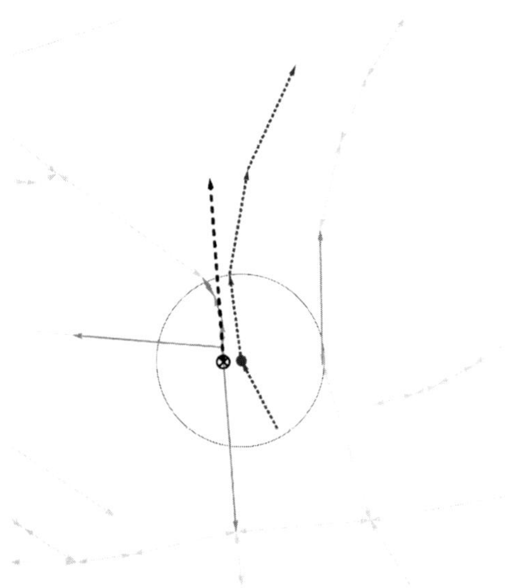

Note that, in this case, many road segments have a length shorter than the estimated error of the GPS reading. The undesirable consequence of this wrong initial allocation is that the prediction of the future path of the user is compromised right at its beginning. In the next subsection, we describe an alternative approach to the initial allocation process that tries to overcome some of the limitations of the approach described so far.

Topological Distance in the Prediction Process

The prediction of users' next positions looking at the geometrical distance that exist between the current position of the user and the nearest road segments presents a failure rate, in the assignment of a user to a specific road segment, that can be improved by looking at the topological distance between the user and the nearest road segments. In this subsection we describe this approach.

The adoption of a qualitative strategy based on the topological distance requires that the objects present in the geographical context (the user and the road segments) be represented in a way that allow us to deal with the uncertainty associated to the position of mobile users. In the following, a CSEP is used to represent the user's position and a certain degree of uncertainty associated to such

Figure 6. A circular spatially extended point and a line

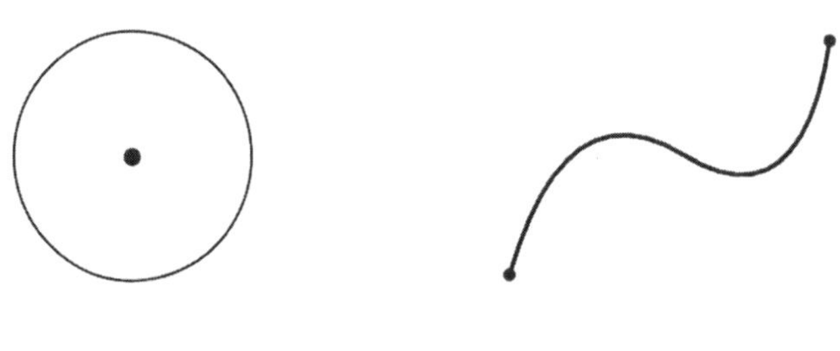

position, and a line is used to represent a road segment in a road network (Figure 6). In commercial digital road maps, the roads are represented by road segments described as pairs of points.

The topological spatial relations that exist between a CSEP and a line were identified (Santos & Moreira, 2007; Santos & Moreira, 2009a) as well as the corresponding conceptual neighbourhood graphs. For the graphs, two approaches were followed: the snapshot model and the smooth-transition model (Santos & Moreira, 2009b). The first one compares two different topological relations with no knowledge of the potential transformations that may have caused the change (Egenhofer & Mark, 1995). The smooth-transition model states that two relations are conceptual neighbours if there is a smooth-transition from one relation to another (Egenhofer & Mark).

From the 38 topological spatial relations that exist between a CSEP and a line (Santos & Moreira, 2007; Santos & Moreira, 2009a), a sub-graph of the smooth-transition model was identified, including only the topological spatial relations that can be verified between a CSEP and a straight line representing a road segment. This sub-graph is shown in Figure 7. The use of the graph obtained by the smooth-transition model is motivated by the fact that a smooth-transition can be defined as an

Figure 7. Conceptual neighbourhood sub-graph for the smooth-transition model (derived from Santos & Moreira, 2009b)

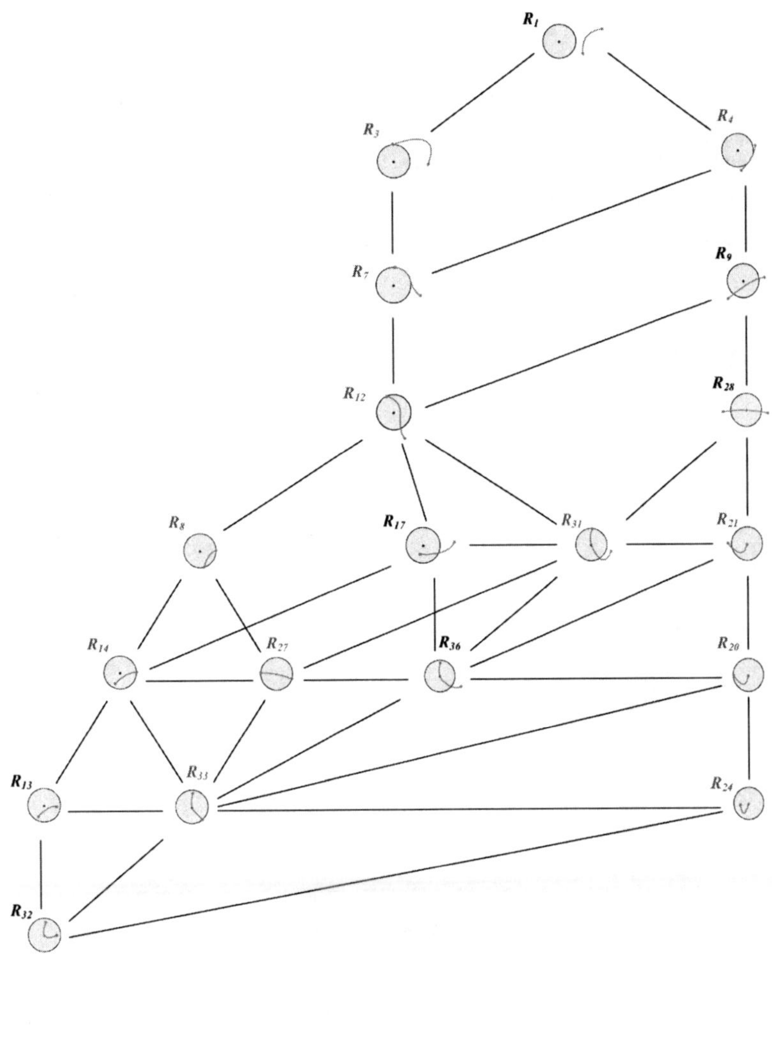

infinitesimally small deformation that changes the topological relation (Egenhofer & Mark, 1995). Those deformations allow us to model the progression of a mobile user along a road segment. This sub-graph includes the topological spatial relations that can be verified between a CSEP and a straight line, and the possible transitions among the relations. Amongst the subset of topological spatial relations depicted in Figure 7, some are considered to be more frequent to occur in a real scenario. The others are less probable since they imply intersections between a CSEP and a line that are difficult to observe in a real context. This analysis allowed the identification of R_3, R_4, R_7, R_8, R_{12}, R_{14}, R_{20}, R_{21}, R_{24}, R_{27}, R_{31} and R_{33} as the less probable spatial relations, and R_1, R_9, R_{13}, R_{17}, R_{28}, R_{32} and R_{36} as the most probable ones.

Even within this last subset, one could think of a further simplification since all road segments are straight line segments and R_{28} and R_{32}, or R_9 and R_{13}, are different only in the line segment length. However, in a real geographic context, as considered in this chapter, the length of the line segments in the digital map vary from values much shorter than the estimated GPS error to values much larger than that (Figure 5). The histogram in Figure 8 shows the frequency of each road segment length for the commercial road map used in this work (supplied by NAVTEQ) and compares it with the estimated GPS error. This context suggests that no further simplification of the conceptual neighbourhood sub-graph could be done.

Back to the sub-graph, each topological relation has associated a 4x3 matrix (Equation 1) with the indication of the intersections verified between the several parts of the objects in analysis. Namely, the pivot ($P^{\bullet}$), interior (P°), boundary (∂P) and exterior (P^{-}) of a CSEP (P), and the interior (L°), boundary (∂L) and exterior (L^{-}) of a line (L) (Santos & Moreira, 2007; Santos & Moreira, 2009a).

$$R(P,L) = \begin{bmatrix} P^{\bullet} \cap L^{\circ} & P^{\bullet} \cap \partial L & P^{\bullet} \cap L^{-} \\ P^{\circ} \cap L^{\circ} & P^{\circ} \cap \partial L & P^{\circ} \cap L^{-} \\ \partial P \cap L^{\circ} & \partial P \cap \partial L & \partial P \cap L^{-} \\ P^{-} \cap L^{\circ} & P^{-} \cap \partial L & P^{-} \cap L^{-} \end{bmatrix} \quad (1)$$

The difference between the values of the matrices allow us to calculate the topological distance (τ) between pairs of topological relations. The definition of topological distance between two spatial relations (R_A and R_B) given by Egenhofer and Al-Taha (1992) is the sum of the absolute values of the differences between corresponding entries of the intersections verified in the corresponding matrices M_A and M_B. The adaptation of

Figure 8. Road segments length

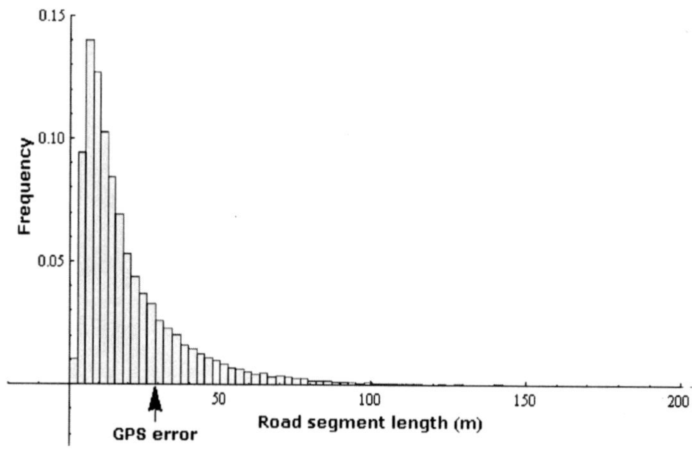

this definition to the context of this work, 12-intersection matrices, leads to the topological distance calculation as described by Equation 2.

$$\tau_{R_A,R_B} = \sum_{i=1}^{4}\sum_{j=1}^{3}\left| M_A\left[i,j\right] - M_B\left[i,j\right] \right| \qquad (2)$$

As an example, consider the topological spatial relations illustrated in Figure 9. Selecting relation 1 (R_1) and relation 9 (R_9), and their corresponding matrices M_1 and M_9, the topological distance between these two topological spatial relations takes the value 2.

Table 1 presents the topological distance for the spatial relations shown in the sub-graph of Figure 7 (the less probable spatial relations are presented in gray).

In the following approach the topological distance is used as an alternative or as a complement to the geometrical distance used in map-matching techniques and also on the predictive model already proposed in this chapter. Here, the two GPS readings previously used to identify the velocity vector associated to the user are also used to identify the topological spatial relations existing between the two corresponding CSEP and the k nearest road segments of the current point (the current GPS user's position). This means that between the previous and the current position of the user, we verify the evolution in terms of topological spatial relations as the user moves from one position to another. This process, which allows the calculation of the total topological distance associated to the user's movement, includes the following steps:

- Identify the topological spatial relations existing between the current point (cP) and the most aligned amongst the k nearest road segments of cP;
- Identify the topological spatial relations existing between the previous point (pP) and the most aligned amongst the k nearest road segments of cP;
- Interpolate three points between cP and pP: iP_1, iP_2 and iP_3;
- Identify the topological spatial relations existing between the three interpolated points (iP_1, iP_2 and iP_3) and the most aligned amongst the k nearest road segments of cP;
- Calculate the topological distance associated to the evolution from pP to cP following the interpolated points. The final topological distance corresponds to the sum of the partial distances: from pP to iP_1, from iP_1 to iP_2, from iP_2 to iP_3, and from iP_3 to cP.

Figure 9. Table of topological distance: An example

$$M_1 - M_9 = \begin{bmatrix} 0 & 0 & 0 \\ -1 & 0 & 0 \\ -1 & 0 & 0 \\ 0 & 0 & 0 \end{bmatrix}$$

$$M_1 = \begin{bmatrix} 0 & 0 & 1 \\ 0 & 0 & 1 \\ 0 & 0 & 1 \\ 1 & 1 & 1 \end{bmatrix} \qquad M_9 = \begin{bmatrix} 0 & 0 & 1 \\ 1 & 0 & 1 \\ 1 & 0 & 1 \\ 1 & 1 & 1 \end{bmatrix} \qquad \tau_{R_1,R_9} = 2$$

Table 1. Topological distances for the conceptual neighbourhood sub-graph

	R_1	R_3	R_4	R_7	R_8	R_9	R_{12}	R_{13}	R_{14}	R_{17}	R_{20}	R_{21}	R_{24}	R_{27}	R_{28}	R_{31}	R_{32}	R_{33}	R_{36}
R_1	0	1	1	2	4	2	3	4	5	3	6	4	6	6	4	5	6	7	5
R_3	1	0	2	1	3	3	2	5	4	4	5	5	7	5	5	4	7	6	6
R_4	1	2	0	1	5	1	2	5	6	2	7	3	7	7	3	4	7	8	4
R_7	2	1	1	0	4	2	1	6	5	3	6	4	8	6	4	3	8	7	5
R_8	4	3	5	4	0	4	3	2	1	5	2	6	4	2	6	5	4	3	7
R_9	2	3	1	2	4	0	1	4	5	1	6	2	6	6	2	3	6	7	3
R_{12}	3	2	2	1	3	1	0	5	4	2	5	3	7	5	3	2	7	6	4
R_{13}	4	5	5	6	2	4	5	0	1	3	4	6	2	4	6	7	2	3	5
R_{14}	5	4	6	5	1	5	4	1	0	4	3	7	3	3	7	6	3	2	6
R_{17}	3	4	2	3	5	1	2	3	4	0	7	3	5	7	3	4	5	6	2
R_{20}	6	5	7	6	2	6	5	4	3	7	0	4	2	2	6	5	4	3	7
R_{21}	4	5	3	4	6	2	3	6	7	3	4	0	4	6	2	3	6	7	3
R_{24}	6	7	7	8	4	6	7	2	3	5	2	4	0	4	6	7	2	3	5
R_{27}	6	5	7	6	2	6	5	4	3	7	2	6	4	0	4	3	2	1	5
R_{28}	4	5	3	4	6	2	3	6	7	3	6	2	6	4	0	1	4	5	1
R_{31}	5	4	4	3	5	3	2	7	6	4	5	3	7	3	1	0	5	4	2
R_{32}	6	7	7	8	4	6	7	2	3	5	4	6	2	2	4	5	0	1	3
R_{33}	7	6	8	7	3	7	6	3	2	6	3	7	3	1	5	4	1	0	4
R_{36}	5	6	4	5	7	3	4	5	6	2	7	3	5	5	1	2	3	4	0

Figure 10. An example of the interpolation process.

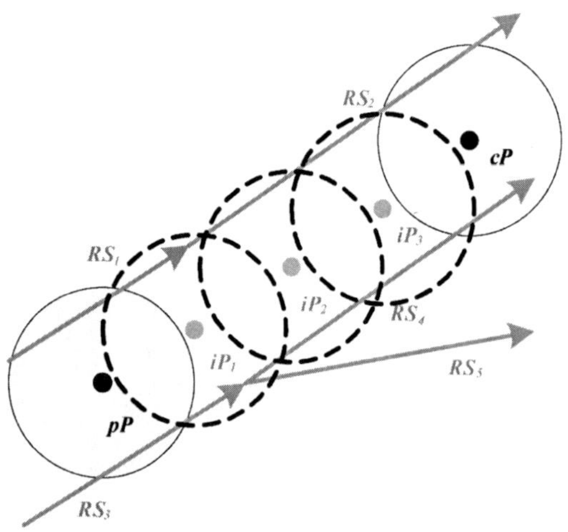

- Assign the user to the road segment (among the *k* nearest ones) with the shortest final topological distance, representing the smoother transition between *pP* and *cP*.

In the evolution process that moves the position of the user from *pP* to *cP*, there are two situations that we need to differentiate: one in a positive way and the other in a negative way. When the movement starts on or leads to a topological spatial relation in which an intersection with the pivot of the CSEP is verified, a topological distance of *-5* is added to the topological distance gathered from Table 1. This means that in the moving process there was a relevant overlapping between the CSEP and the line. As we are dealing with a qualitative approach, this value helps us to tell that the user gets closer to the road segment. To the other side, when the topological spatial relation R_1 is identified in the starting point and also in the destination point, this means that the user is far away from the road segment in analysis, being added a value of

Table 2. Identification of the topological spatial relations and the corresponding topological distances

	RS_1, τD	RS_2, τD	RS_3, τD	RS_4, τD	RS_5, τD
pP	R_9	R_1	R_9	R_1	R_1
iP₁	R_{17}, **1**	R_{17}, **3**	R_{17}, **1**	R_{17}, **3**	R_{17}, **3**
iP₂	R_1, **3**	R_9, **1**	R_1, **3**	R_9, **1**	R_1, **3**
iP₃	R_1, **5**	R_9, **0**	R_1, **5**	R_9, **0**	R_1, **5**
cP	R_1, **5**	R_9, **0**	R_1, **5**	R_{17}, **1**	R_1, **5**
	14	**4**	**14**	**5**	**16**

+5 to the topological distance gathered from Table 1. This value tells us that the user continues far away from the road segment. Figure 10 presents an example of the process described so far. The previous and current points are represented with a CSEP filled in gray and the points obtained through the interpolation process are represented by a white CSEP. The identification of the several topological spatial relations and the corresponding topological distances associated with each movement are shown in Table 2. In this example, and following the criteria of the minimum topological distance (τD) instead of the minimum geometrical distance, the mobile user is assigned to the road segment 2 (RS_2).

The analysis of the previous example shows that the calculation of the topological distance can lead to ties for different road segments. In these cases, the assignment of the user to a specific road segment is random. Specific metrics can be defined to deal with these situations or more interpolation points can be calculated to increase the accuracy of the topological distance. However, this process, as is, showed to be able to overcome some of the limitations of the geometrical distance, namely in the case of huge errors in the GPS readings or in the cases where the user is located in a dense area with many road segments. The next three examples show how the topological approach behaved in those situations. The first one, illustrated in Figure 11, shows the allocation of the user to a road segment following the topological approach. In this case, the prediction model was

Figure 11. One case of correct allocation with the topological distance.

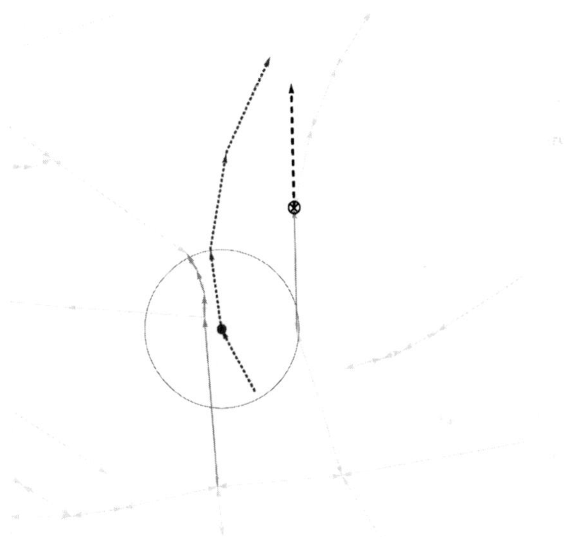

able to correctly assign the user to the road segment in which he/she was actually moving. As the GPS readings have a significant error associated with them, the geometrical approach was unable to properly assign the user to the road segment (as previously showed in Figure 5).

The second example is associated with dense areas in terms of road segments. Mostly in urban environments, there are places in the road network where a huge number of road segments are present. In those situations, the use of the geometrical distance leads in some cases to a wrong allocation as other road segments near the user are not considered. The user is assigned to the nearest one from the set of the *k* nearest road segments, ignoring other possible paths. Using the topological distance, it is possible to consider all the *k* nearest road segments, as the proposed approach verifies the evolution of the user along all these segments. Figure 12 shows the allocation of the user to a road segment in a place with high density of road segments, using the geometrical (left hand-side of the figure) and the topological approaches (right hand-side of the figure). In this case, the topological distance was able to identify the correct road segment.

Another example of a dense urban area where the geometric distance approach failed to correctly assign the user to a road segment is shown in Figure 13. Here, there are a large number of road segments, and their lengths are much shorter than the GPS error. Once again, the geometrical distance approach (in the left) was unable to correctly assign the user, while the topological distance approach (right hand-side) identified the correct road segment.

After the visual analysis of a set of 672 cases, it was possible to verify that the geometrical allocation process fails in around 12% of the cases. Using the topological approach, 45% of these cases are correctly allocated. The majority of them are associated with errors in the GPS readings and with places with high density of road segments, as previously showed in this subsection. On the other hand, the topological approach fails to allocate the user to the correct road segment in 13% of the cases. From these, 49% are correctly allocated by the geometrical process. In conclusion, both approaches have their own strengths and limitations, in such a way that they both perform similarly.

The calculation of the length of the predicted paths using the topological distance approach, as already done using the geometrical approach and which results were shown in Figure 4, leads to the results presented in Figure 14. This figure integrates the results obtained using the geometrical and the topological distance.

As shown, the topological distance is able to behave, in global terms, as good as the geometrical distance. This opens a new possibility of exploration of the capabilities of qualitative approaches in mobile systems/environments, as qualitative

Figure 12. One example of the user allocation in an area with high density of road segments

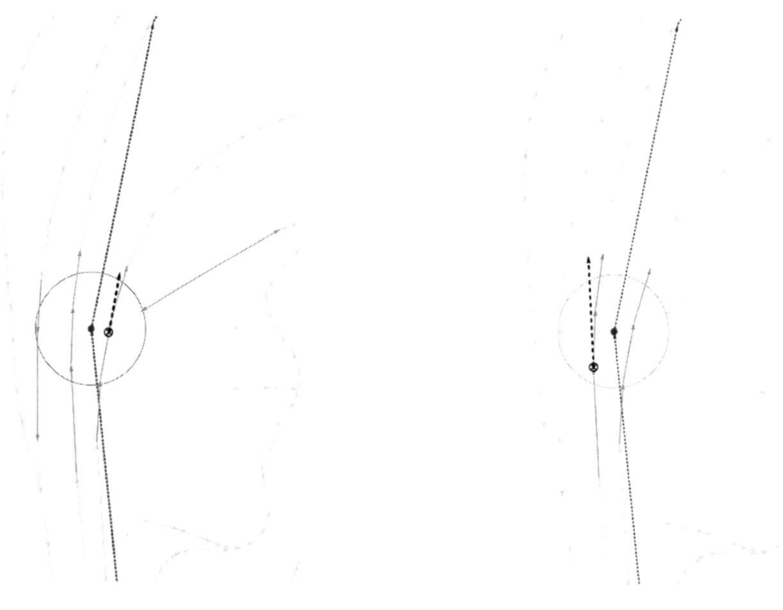

information and qualitative reasoning should be seen as complementary methods of a new synergetic approach (Sharma, Flewelling, & Egenhofer, 1994). It is also important to note that many other domain applications are not linked with a static view of the world and that they need to build theories of how spatial descriptions may change over time, accordingly to some domain-specific constraints (Cohn, 1995). For Cohn (1995), the most obvious constraint is continuity, which can be achieved building a conceptual neighbourhood graph of spatial relations. This was the principle behind the use of the topological distance in the proposed prediction model.

In the work described in this chapter, the topological distance was able to overcome some of the limitations detected with the use of the geometrical distance, yet introducing some other wrong allocations. The analysis of these cases allowed us to verify that these wrong allocations are associated with the huge difference that is verified in the length of the road segments in some places of the road network (as illustrated in Figure 8). In these cases, the final topological distance

Figure 13. Another example of user allocation in areas with high density of road segments

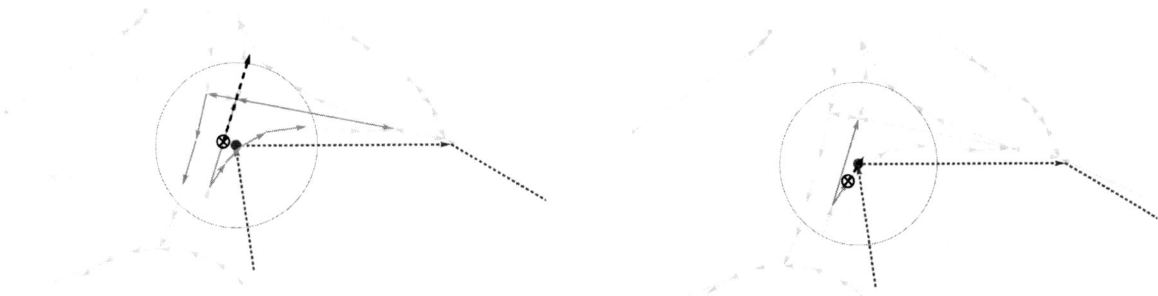

Figure 14. Length of the predicted paths using the geometrical (G) and the topological (T) distances

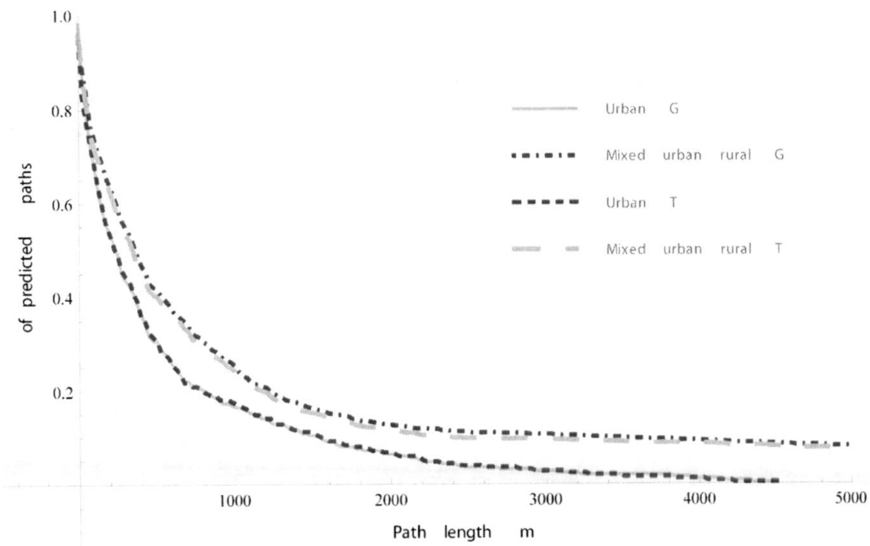

in the moving process, from the previous to the current point, is influenced by the length of the road segments. This scenario is shown in Figure 15, for a case where the length of the *k* nearest road segments is very heterogeneous.

In this example, the topological spatial relations between the two GPS readings and the small road segment do not change in the interpolation process, leading to a topological distance of 0 and to a wrong allocation of the user to the road segment.

CONCLUSION

The development of location-aware applications able to provide services to the users before the users ask for them can substantially benefit from the prediction of their mobile users' future positions. The emergence of this kind of applications opens several issues related to the privacy of the mobile users, as some of the existing approaches need to track the user for some time before they could be able to predict future positions. Moreover, training a system for a particular region may not be sufficient to enable the prediction function in new areas.

This chapter presented a prediction model that attempts to overcome the privacy problems that arise when a person needs to be tracked in order to predict his/her future positions. The proposed model uses two consecutive GPS readings to identify the user's current position and to calculate his/her associated velocity vector. No other information about the user is need in the prediction model. After this step, the model assigns the user to a specific road segment in the road network, allowing the prediction of the future positions by looking at the topological constraints present in the geographical context in which the user is moving.

For the assignment of the user to a specific road segment, two approaches were described. The first one uses the geometrical distance from the current user position to the nearest, most aligned, road segments to identify the road segment with the shortest distance to the user. In the second one, a new approach was proposed, based on the use of the topological distance and the conceptual neighbourhood graph that states the possible transitions that can occur between the topological spatial relations that exist between a circular spatially extended point (representing the position of the user) and a line (representing a road segment).

The results obtained with the two approaches shown to be very similar. The prediction capability of the model is very similar either using the

Figure 15. One example of wrong allocation using the topological distance

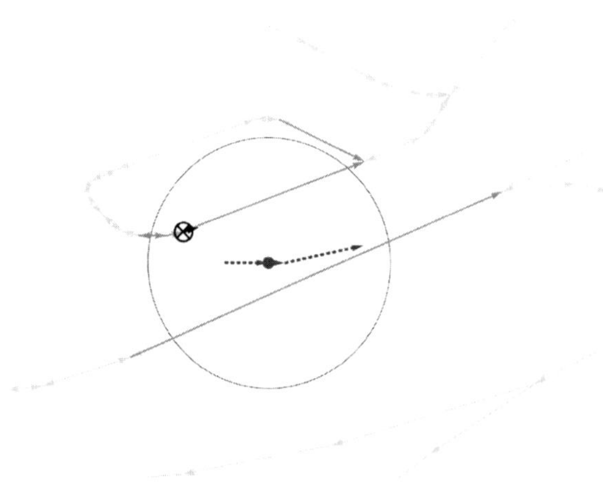

geometrical distance or the topological distance. The topological distance was able to overcome some of the limitations that arise when the GPS readings have associated a significant error and when the user is located in places with high density of short road segments. Although improving the assignment process in such situations, the topological approach introduced some wrong assignments when the lengths of the nearest road segments are very different. The use of these two approaches in a complementary way to improve the overall prediction capabilities of the proposed model is an open issue. Future work includes the study of the synergies between the two approaches and the identification of applications in which they could be used in an isolated way or in a complementary way.

REFERENCES

Ashbrook, D., & Starner, T. (2003). Using GPS to Learn Significant Locations and Predict Movement Across Multiple Users. *Personal and Ubiquitous Computing*, *7*(5), 275–286. doi:10.1007/s00779-003-0240-0

Civilis, A., Jensen, C. S., & Pakalnis, S. (2005). Techniques for Efficient Road-Network-Based Tracking of Moving Objects. *IEEE Transactions on Knowledge and Data Engineering*, *17*(5), 698–712. doi:10.1109/TKDE.2005.80

Cohn, A. G. (1995). The Challenge of Qualitative Spatial Reasoning. *ACM Computing Surveys*, *27*(3), 323–325. doi:10.1145/212094.212112

Dey, A. K. (2001). Understanding and using context. *Personal and Ubiquitous Computing*, *5*(1), 4–7. doi:10.1007/s007790170019

Egenhofer, M. J., & Al-Taha, K. K. (1992). Reasoning about Gradual Changes of Topological Relations. In A. Frank, I. Campari & U. Formentini (Eds.), Theory and Methods of Spatio-Temporal Reasoning in Geographic Space (LNCS 639, pp. 196-219). Berlin: Springer-Verlag.

Egenhofer, M. J., & Mark, D. M. (1995). Modeling Conceptual Neighborhoods of Topological Line-Region Relations. *International Journal of Geographical Information Systems*, *9*(5), 555–565. doi:10.1080/02693799508902056

Galton, A. (1995, September). *Towards a Qualitative Theory of Movement*. Paper presented at the Spatial Information Theory: A Theoretical Basis for GIS - International Conference COSIT'95, Semmering, Austria.

Jagadeesh, G. R., Srikanthan, T., & Zhang, X. D. (2004). A Map Matching Method for GPS Based Real-Time Vehicle Location. *Journal of Navigation*, *57*, 429–440. doi:10.1017/S0373463304002905

Karimi, H. A., & Liu, X. (2003, November). *A Predictive Location Model for Location-Based Services*. Paper presented at the 11th ACM International Symposium on Advances in Geographic Information Systems, New Orleans, USA.

Laasonen, K. (2005, October). *Clustering and Prediction of Mobile User Routes from Cellular Data*. Paper presented at the PKDD'05 - 9th European Conference on Principles and Practice of Knowledge Discovery in Databases, Porto, Portugal.

Marmasse, N., & Schmandt, C. (2002). A User-Centered Location Model. *Personal and Ubiquitous Computing*, *6*, 318–321. doi:10.1007/s007790200035

Pedreschi, D., Bonchi, F., Turini, F., Verykios, V. S., Atzori, M., & Malin, B. (2008). Privacy Protection: Regulations and Technologies, Opportunities and Threats . In Giannotti, F., & Pedreschi, D. (Eds.), *Mobility, Data Mining and Privacy* (pp. 101–119). Berlin: Springer-Verlag. doi:10.1007/978-3-540-75177-9_5

Santos, M. Y., & Moreira, A. (2007, May). *Topological Spatial Relations between a Spatially Extended Point and a Line for Predicting Movement in Space*. Paper presented at the 10th AGILE Conference on Geographic Information Science, Aalborg, Denmark.

Santos, M. Y., & Moreira, A. (2009a, July). *How a Circular Spatially Extended Point is topological related with a Line?* Paper presented at the 2009 International Conference of Computational Intelligence and Intelligent Systems, London, England.

Santos, M. Y., & Moreira, A. (2009b, July). *Conceptual neighborhood graphs for topological spatial relations*. Paper presented at the 2009 International Conference of Computational Intelligence and Intelligent Systems, London, England.

Sharma, J., Flewelling, D. M., & Egenhofer, M. J. (1994, September). *A Qualitative Spatial Reasoner*. Paper presented at the 6th International Symposium on Spatial Data Handling, Edinburgh, Scotland.

Yang, L., & Wang, F.-Y. (2007). Driving into Intelligent Spaces with Pervasive Communications. *IEEE Intelligent Systems*, *22*(1), 12–15. doi:10.1109/MIS.2007.8

Chapter 7
Revealing Taxi Driver's Mobility Intelligence through His Trace

Liang Liu
Massachusetts Institute of Technology, USA

Clio Andris
Massachusetts Institute of Technology, USA

Assaf Biderman
Massachusetts Institute of Technology, USA

Carlo Ratti
Massachusetts Institute of Technology, USA

ABSTRACT

This study develops a methodology for the analysis of taxi drivers' operation behavior in a real urban environment. The research objective is to spatially and temporally quantify, visualize, and examine taxi drivers' operational behavior and skill (as measured by income), which the authors call 'mobility intelligence'. For the first time, taxi drivers' different operation strategies were systematically analyzed through their daily activity traces. Routes and economic behavior data were collected with the use of Global Positioning System (GPS) and a set of spatiotemporal analysis tools were developed. Drivers are categorized by their daily income into top drivers and ordinary drivers. A 3D clustering technique is used to quantitatively analyze the spatiotemporal patterns for top driver and ordinary driver. Also, fractal analysis is employed to quantify tortuosity of movement paths and to explore how top and ordinary drivers operate on different spatial scales at different times, where the primary focus is to reveal top driver mobility intelligence.

INTRODUCTION

In recent years, the deployment of pervasive technologies and their ability to track movement in cities has led to a massive increase in the volume of records of people's spatial traces. These records serve as digital footprints of individual mobility patterns. Also, geo-located websites and services provide new sources of real-world human activity data. For example, Rattenbury et al. (2007) and Girardin et al. (2008) used geo-tagging patterns of

DOI: 10.4018/978-1-61520-769-5.ch007

posted photographs in the popular photo hosting website, *Flickr*, to automatically detect interesting real-world events and draw conclusions about the flow of tourists in a city. In addition, as city-wide urban infrastructure like buses, taxis, subways, public utilities, and roads become digital, these datasets can be used as framework for tracking flows through space and time. Some have capitalized on this technology by using cellular network data to study city dynamics and human mobility (Ratti et al., 2006; Reades et al. 2007; and González et al. 2008). Also McNamara et al. (2008) used data collected from an RFID-enabled subway system to predict co-location patterns amongst mass transit users. These sources of data are ever-expanding and offer large, underexplored opportunities to understand physically-based interactions with the real world.

Recognizing human behavior and understanding a user's mobility from his digital footprints are critical issues in pervasive computing systems. In this chapter, we introduce a novel methodology to measure taxi drivers' ability to earn the most profit by their continuous GPS traces and activity (carrying passenger or not). First, we classify the taxi drivers into top drivers and ordinary drivers by calculating their earnings, earnings rank and 'eigenactivities'; then we compare the spatial-temporal distribution and client frequency of top drivers and ordinary drivers to find why top drivers earn more than ordinary drivers.

It is intuitive that a taxi driver wants to make the most money in the least amount of time. Because of the fare rate system and desire for continuous revenue, a driver hopes to pick up many passengers whose destinations are in places where there are more potential customers for a new pick up, to perpetuate a chaining of constant business. There are several commonsense assumptions about this operation behavior:

- **Assumption 1:** Taxi drivers prefer areas with the most potential customers.
- **Assumption 2:** A taxi, once occupied by a customer at origin zone, will move to the customer's destination zone via the shortest path. Once a customer ride is completed, the taxi becomes vacant and cruises either in the same zone or goes to other zones to seek its next customer. In doing this, each taxi driver is assumed to attempt to minimize his/her expected search time required to meet the next customer.

Assumption 1 reflects the observed behavior of many taxi drivers to aggregate at railway stations, hotels and other places of interest for visitors. Assumption 2 also follows economic intuition, and is bolstered by other scientific urban taxi services models. (See Wong, K. et al. 2001, 2002, 2005; Wong, S. et al. 1998; and Yang et al., 1998, 2001, 2002, 2005a, 2005b).

These assumptions are not definitive, as there are several prevailing arguments about their validity. For assumption 1: An area that is attractive to customers will also be attractive to competition from other taxi drivers and incur increased waiting time and congestion. In some sense, passenger demand levels and time spent in traffic is a tradeoff. As for assumption 2, most of the taxi fare systems are based on trip distance. In order to maximize his revenue, a taxi driver may not always take the shortest path. Bearing these assumptions in mind, our goal is to uncover the difference in taxi driver performance based on when and where they operate in order to reveal what decisions the top driver makes to perform the best.

Next, we introduce the research area and data description. After, we describe how to classify taxi drivers according to income and how to find common features among top drivers. Finally we compare the spatiotemporal operation patterns of top drivers and ordinary drivers to understand the difference and conclude with an outlook on future work.

Figure 1. The layout of Shenzhen and Hong Kong

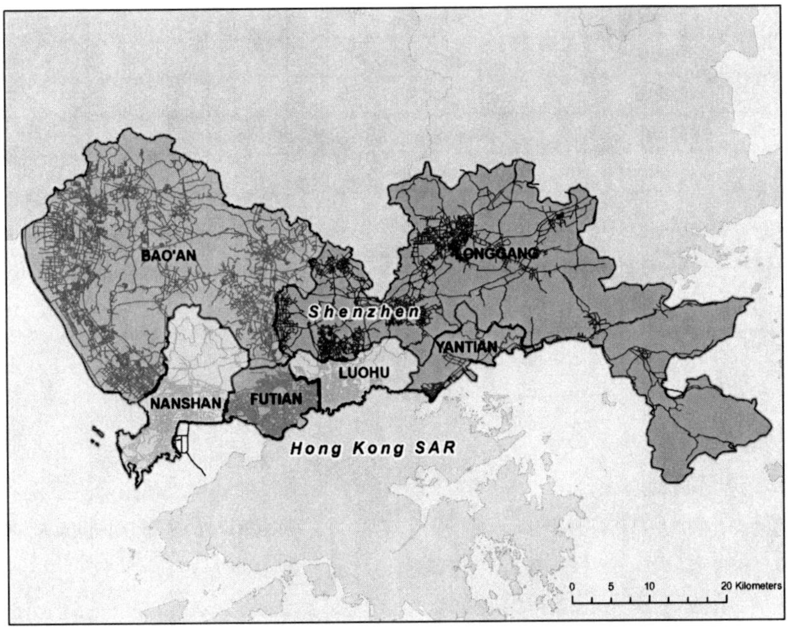

RESEARCH AREA AND DATA

The taxi drivers studied are located in the city of Shenzhen, South China. Shenzhen is the first Special Economic Zone (SEZ) in China to have a close relationship with Hong Kong and has taken advantage of open reform policy and rapid social-economic development for the past 30 years. This community consists of people from different provinces in China with a wide variety of education levels and cultural backgrounds, and is a hub for business and economic activity. We have continuously collected data from 3000 taxi drivers' for 1 year in the form of GPS traces. These traces provide accurate data about taxi location, timestamp and operation status (occupied or empty).

Shenzhen is located in the very south of Guangdong Province. Overlooking Hong Kong to the south and bordering Kowloon, this area is commonly referred to as Hong Kong's "backyard." Virtually all the action is situated around the SEZ, the business and financial center that draws many visitors to Shenzhen. The layout of Shenzhen is showed in Figure 1.

Transit Authority

The Shenzhen Urban Transport Simulation System (SUTSS) is an important component of the Intelligent Transportation System in Shenzhen. Its tasks are to establish the integrated structure of urban transport simulation and common information platforms, improve the planning, construction and management of transportation infrastructure, enhance traffic prediction capabilities, and disseminate real-time traffic information for travelers. These 'smart' new transportation efforts are expected to alleviate road traffic congestion and foster more efficient travel for citizens and visitors. These changes will also be documented in transportation operation reports for system analysis and provide technological support for the government.

Table 1. Description of GPS traces

Field	Instance	Memo
Date	20080415	April 15th, 2008
Time	150236	15:02:36
Vehicle ID	13510354262	Single identification of vehicles
Longitude	114.114117	Degree
Latitude	22.550917	Degree
Speed	66	Spot speed (km/h)
Operation Status	1	0 represents taxi with no passengers 1 represents taxi with passengers
Equipment Status	1	0 represents unavailable 1 represents available

Description of GPS Traces

GPS traces (Data description in Table 1) have been collected from taxis with GPS equipment, for certain intervals, operation status, spot speed, azimuth and equipment status. Data is recorded continuously for 1 year in 2008, for a total of 1200 million GPS records, from which we group and derive 20 million trip records. We string each trace segment together if their tails (segment end) meet a head (segment beginning) within a reasonable time frame. From the GPS traces, we extracted important features for understanding taxi drivers' operation behavior:

- **Data Preprocessing:** GPS data must be preprocessed and two types of outliers must be excluded. The first type is the data which lie out of the range of map. The second type is the data collected when the GPS equipment cannot record whether a passenger is in the car.
- **Coordinate Transformation:** Transform GPS traces from latitude and longitude coordinate to local system.
- **Map Matching:** Matching the GPS trajectories to the road networks.
- **Feature Extraction:** After the GPS trajectories have been mapped to the network, we derive important features from them, including trip distance, trip time and trip zones.

For each trip (either carrying passenger or not), we reconstructed the related features, such as trip time, trip distance, start and end time, and start and end location.

CLASSIFICATION OF TAXI DRIVERS

Calculation of Taxi Fares

Since our trace data does not include fare information, we needed to compute the taxi driver's earnings to progress to the next step in our analysis First, we categorize the taxi drivers and then analyze their operation behavior. The best way to classify a taxi driver is according to his earnings, as it indicates how successful his choices have proven. To calculate income, we extract the operation traces and combine the distance and the standard fare to roughly calculate each taxi's income with the following rate: 12.5 Yuan for the first three kilometers and the 2.4 Yuan for each additional kilometer. According to this standard, we get the formula to calculate the fare for each trip.

Figure 2. Daily income of different taxi drivers (Unit: Yuan)

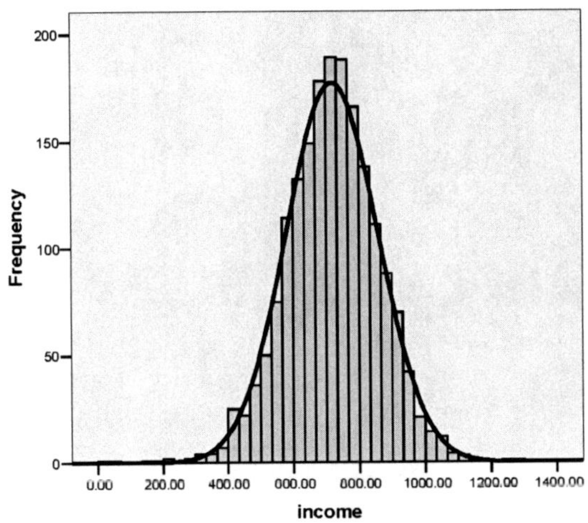

$$\text{income}=\begin{cases}12.5 & \text{when length} \leq 3;\\ 12.5+2.4(\text{length-3}) & \text{when length}>3;\end{cases} \quad (1)$$

According to Equation (1), we calculate the fare for each trip of different taxi drivers and rank them by their total income.(Shown in figure 2)

Taxi drivers' daily earnings follows a normal distribution, as most drivers are centered around the mean value (700-800 YUAN), while several are very high (top driver, 1250 YUAN), and others are very low (200-300 YUAN). The variation for different drivers' income is sizable, as the top driver earns six times as much as the least successful earners.

Operation Activity Eigenvectors

While discrete observations of a taxi driver's idiosyncratic activity can appear almost random, there are typically repeating and easily-identifiable patterns in a driver's routine. These patterns can be evaluated when activity is temporally and spatially contextualized.

However, building models of long-term activity has been hampered due to a lack of contextualized activity data. Additionally, traditional Markov models work well for specific set of activities, but are not ideal for modeling temporal patterns across different timescales (Clarkson, 2002). We present a new methodology for identifying the repeating structures underlying activity. These structures are represented by eigenactivities, an eigenvector of an individual's activity history.

To capture these characteristic activities, we compute the principal components of an individual's activity history. The principal components are a set of vectors that span an 'activity space' and characterize the variation of activity from day to day. An eigenactivity measure is determined by the eigenvectors of the covariance matrix of activity data; the heavily-weighted vectors generally represent a type of repeated activity, such as operating in specific areas or taking breaks at specific times. A linear combination of an individual's eigenactivities is used to represent the driver's activity from each day.

In this study, we characterize different drivers by their temporal location activity. As shown in Figure 3 as B (x,y), a two-dimensional D x 96 array of location information, where D is the total number of days that a driver has been in the study. B contains n labels corresponding to activity, where in our case these labels are {Unknown, Luohu, Futian, Nan shan, Yantian, W Bao'an, N Bao'an, N Longgang}. From Eigenactivities, we cluster different cabdrivers into top drivers, ordinary drivers by his temporal location activity. The horizontal axis represents 96 time slots in a day (5 minutes per slot), and the vertical axis represents 25 sample days. Figure 3 shows great differences between top drivers and ordinary drivers.

Operation Zone Distribution of Top Taxi Drivers

For taxi operation, we are focusing on the top drivers' spatial-temporal distribution because

Figure 3. Eigenactivities of cabdrivers

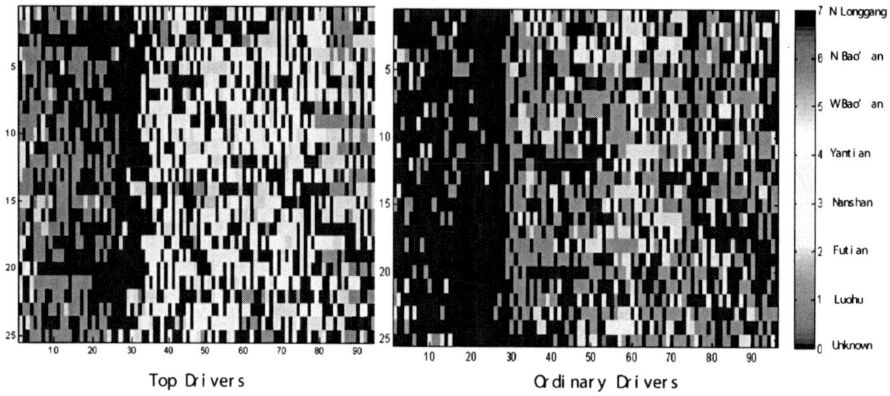

Table 2. Top 5 taxi drivers

Vehicle ID	Average daily income	Overall rank
13510354840	1228	1
13827411446	1081	2
13827411452	1056	3
13510354836	1010	4
13825275014	969	5

they master the operation skills to garner more income. What is the difference between top drivers and ordinary drivers, or what is the top drivers' trick? To answer these questions, we calculated the average rank and income for each taxi driver and rank the average results. Through this process, we can eliminate the random error and select the real top driver to analyze their operation features. We also compute the average daily income and overall rank of the top 5 drivers among the 3000 drivers (Table 2).

The top driver's records show that the good performance is not due to anomalies, but his regularity in mastering the operation skill to keep him as top driver (Table 3). Over 1 year, the top driver ranks No. One 44% of the time, and 76% of the time in the top three. He ranks in the top ten 88% of the time. This trend shows that the taxi driver's operation success is not random, or by luck, but that he has some precision or a replicable strategy. Those who know his secret may have a more fruitful workday.

Spatial Distribution of Top Taxi Drivers

In order to uncover the strategy, we analyze the spatial distribution of the top drivers. Through counting the proportion of different operation zones for the top five taxi drivers, we get the proportion of their operation zones.

Four of the top five drivers spend more time in the Nanshan district than any other. It shows that the Nanshan district is a focal operation area for top drivers, and their choice has proved more lucrative than the choices of those who focus on Luohu-Shangbu and other prominent areas.

Analyzing taxi traffic on the regional level illustrates a picture of a driver's spatial pattern, but these regions still very large for an insightful spatial analysis. If we observe the top three taxi drivers carefully on the more detailed traffic analysis zone (TAZ) scale, we can get a better understanding of their spatial distribution (Figure 4).

We see that top drivers spend at least 1/3 of their time within 10 TAZ's (Table 4). Top 1 and top 2 each operate in zones 103 (Luohuo), 249 (Nanshan), 326 (Nanshan), and 337 (Nanshan), while all three share zones 249 (Nanshan), 326 (Nanshan) and 337 (Nanshan). From this, we conclude that the top drivers have the same criterion when selecting

Figure 4. The focal area of trip generation

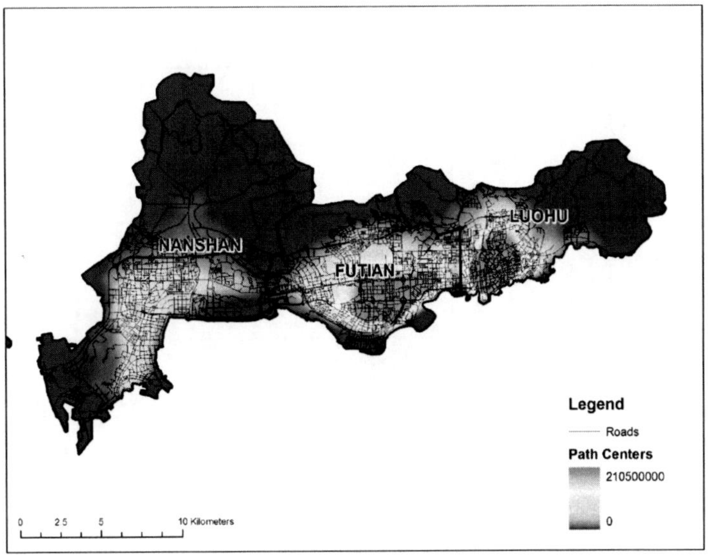

the key operation zones. In this case, zone 249 (Nanshan), 326 (Nanshan) and 337 (Nanshan) are the most profitable areas for the top drivers. The Nanshan area is marked with a good opportunity to pick up passengers in unit time.

Correlation of Top Drivers Space Time Path

Correlation definition. In probability and statistics, correlation (often measured by a correlation coefficient) indicates the strength and direction of a linear relationship between two variables. The correlation coefficient $\rho_{X,Y}$ between two variables X and Y with expected values μ_X and μ_Y and standard deviations σ_X and σ_Y is defined as:

$$\rho_{X,Y} = \frac{\text{cov}(X,Y)}{\sigma_X \sigma_Y} = \frac{E((X-\mu_X)(Y-\mu_Y))}{\sigma_X \sigma_Y} \quad (2)$$

where E is the expected value operator and cov is covariance.

Space time path. Each individual taxi driver user i has the trace sequences $U_i(x_j, y_j, t_j)$, where

Table 3. Top driver's (13510354840) rank frequency

Rank	Frequency	Accumulate frequency
1	0.44	0.44
2	0.2	0.64
3	0.12	0.76
8	0.04	0.8
9	0.04	0.84
10	0.04	0.88
>10	0.12	1
Sum	1	

x_j, y_j, t_j represents the x coordinate, y coordinate and time coordinate of user i in the j-th space time point. The sets of user trace sequence compose the space time path. (Hagerstrand, 1970)

For example, $X(x_j, y_j, t_j)$ represent the sequence of space time points of user X and $Y(x_j, y_j, t_j)$ represent the sequence of space time points of user Y. From these, we can obtain the correlation of top drivers' space time path by using equation 2.

Table 4. Top 10 operation zones for top 3 drivers

13510354840 (1st)			13827411446 (2nd)			13827411452 (3rd)		
TAZ	Percent	District	TAZ	Percent	District	TAZ	Percent	District
103	17.31%	1	106	7.00%	1	106	7.78%	1
326	4.96%	3	326	4.30%	3	326	4.21%	3
249	4.53%	3	249	4.20%	3	337	4.21%	3
337	3.52%	3	337	3.90%	3	249	3.85%	3
301	2.59%	3	105	2.90%	1	202	2.84%	2
328	2.23%	3	102	2.60%	1	328	2.75%	3
315	2.16%	3	224	2.60%	2	303	2.56%	3
111	2.16%	1	103	2.50%	1	306	2.47%	3
306	2.08%	3	211	2.40%	2	307	2.47%	3
210	2.01%	2	202	1.90%	2	301	2.29%	3
Total	43.53%		Total	34.30%		Total	35.44%	

Figure 5. Space-time paths of top 3 drivers

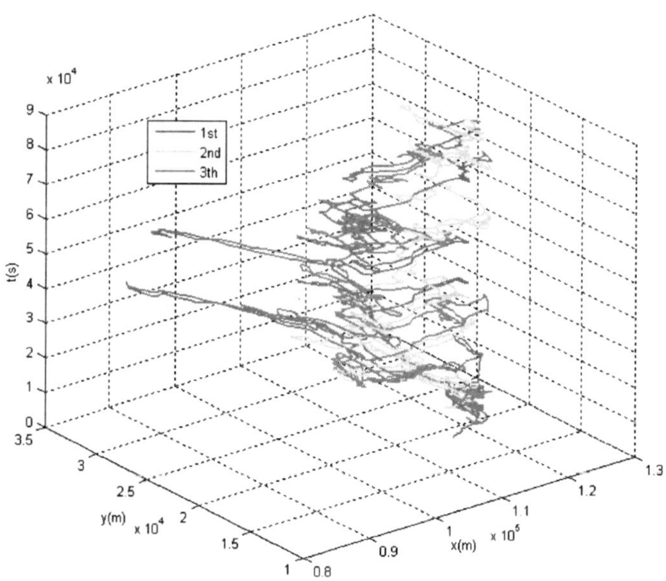

If we plot top drivers' daily space-time paths, we can see that the space-time paths are tightly focused, almost bound to the Nanshan District. The correlation of their space-time paths are above 98 percent (Figure 5).

If we correlate the top 5 taxi drivers with their top 200 TAZs, we find the following law: the greater the correlation coefficient, the closer rank to the No.1 position (Table 5). This means that the closer a driver chooses to mimic the top one driver's strategy of cruising within a tight space, the more revenue he can make. In, theory, the No.1 driver's strategy could set the benchmark that other drivers could follow in terms of maximizing revenue. However, in practice, this may lead to an overcrowding of taxi supply, and may change the feasibility for maximum profit. How to balance individual

Figure 6. Space time path and operation point compare of top drivers (red) and ordinary drivers (green)

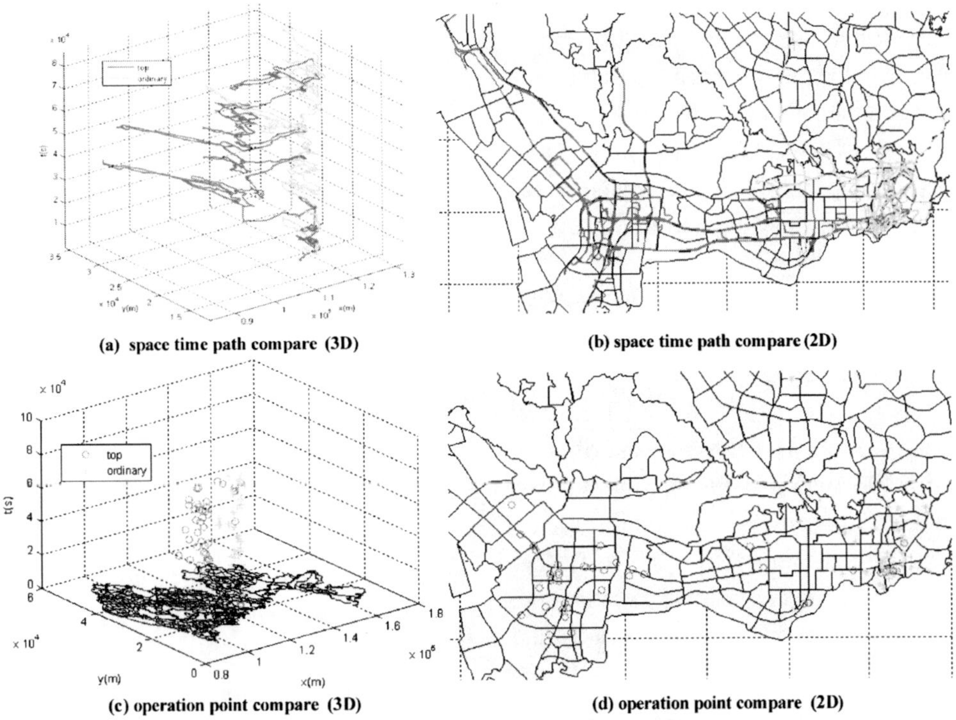

Table 5. The correlation for the top 5 drivers with top one

Rank	Vehicle ID	Correlation to No.1
1	13510354840	1
2	13827411446	0.8065
3	13827411452	0.7541
4	13510354836	0.6530
5	13825275014	0.6209

optimum and global optimum is a topic for future investigation.

COMPARING THE SPATIAL-TEMPORAL OPERATION PATTERN OF TOP AND ORDINARY DRIVERS

Why is the income difference between the top drivers and ordinary drivers so significant? (400 YUAN per day, a 50% percent difference) What is the difference between their spatial and temporal distributions? To find the answers, we select the expert drivers and ordinary drivers according to their eigenactivities to compare their different operation behavior.

Operation Spatial-Temporal Distribution

In order to intuitively understand the difference between top drivers and ordinary drivers in terms of spatiotemporal behavior, we compare the space-time path and operation point (the origin of passenger trip) in 2D and 3D dimensions (Figure 6).

Top driver and ordinary driver's space-time path has obvious differences, where the top driver's operation covers the Luohu, Futian and Nanshan districts, in which Nanshan district is the center, while the ordinary driver focuses on Luohu-Shangbu Area. The fact that the top driver and ordinary driver have different operation behavior indicates that there is a geographic strategy involved in the taxi operation in addition to the fruitfulness of the spatial zone. Had the top drivers all occupied the same zone, we could attribute their profit to their presence in a robust demand area. However, more precise methodology is needed to consider the utility of each action as a function of contextual variables associated with that action.

Operation Center of Different Time Period

By comparing the 3D space-time path distribution and operation generation point distribution, we know that taxi drivers usually operate in different zones at different times. In order to further investigate this underlying law, we calculate the operation centroids over different time periods, and compare the top driver with the ordinary driver. From these, we can get a sense of what determines the different drivers' focal areas.

Cluster Algorithm

In order to obtain the different taxi drivers' operation characteristics, we use a K-means clustering algorithm to calculate operation centroids (cluster center of operation points) in order to represent where drivers most frequently pick up passengers. We list the steps to clustering the centroids:

- Retrieve all the operation points $P_i(x_i, y_i, t_i)$ in a day, where x_i is the horizontal axis, y_i is the vertical axis and t_i is time axis (unit: second)
- Divide the cluster number j into daily segments, where one day is divided into four categories: early morning, morning, afternoon and evening, namely, j=1, 2, 3, 4
- Compute cluster center P_j, for each cluster, ensuring that the sum of $|P_i - P_j|$ squared is minimized, where the distance between 2 points is Euclidian distance.

Cluster Results

According to the K-means clustering algorithm, we obtain the operation centroids for different time

Figure 7. Operation centroids in different time periods for top driver and ordinary driver. Red represents dawn; Green represents Morning; Blue represents Afternoon; Pink represents Evening.

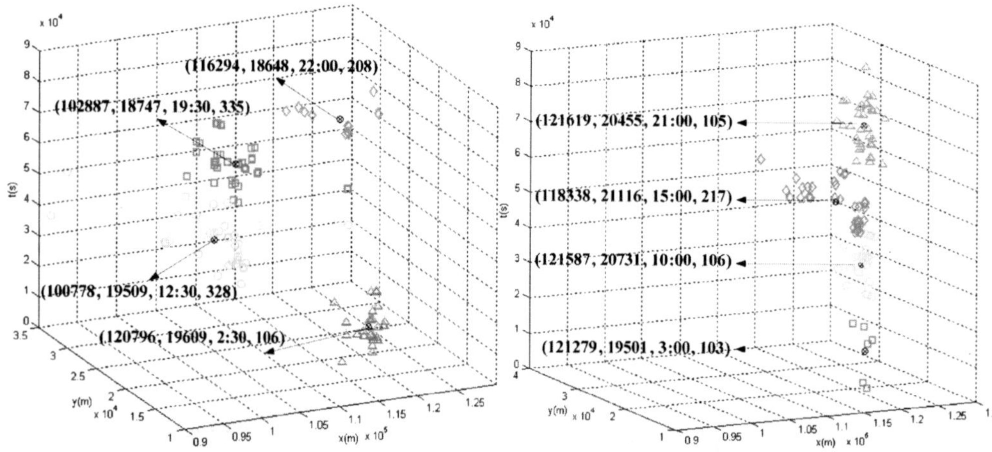

(a) operation centroid in different time for top driver

(b) operation centroid in different time for ordinary driver periods for the top drivers and ordinary drivers (Figure 7 and Tables 6 and 7).

In Figure 7, the first two digits are x,y coordinates, the third is timestamp, the fourth is the zone information. The K Means results show that the four clusters' operation centers are varied in time and space on a typical day. Additionally, we learn that there is an obvious spatial-temporal routine for top driver's operation. Before 6 am, he operates in Guomao and Dongmen (106) district, from 8am to 9pm, he operates on the center of Nanshan district and Science Park (328, 335); from 9pm to 12pm, he operates around Huaqiangbei District (208).

If we consider the traffic congestion conditions (SUTPC 2009) in different districts (Luo Huo, Futian and Nanshan), we learn that the top driver has found an excellent trade-off between travel demand and road conditions in determining their operation zones. During the day, when the traffic is congested, they chooses to operate in the Nanshan district where the traffic is lighter, although the travel demand is less than that of the Luohu-shangbu district, but can satisfy their operation requirement; while during the evening and early morning they operates in the Luohu-Shangbu district where the travel demand is greater, because during this period, the traffic is lighter in this area. Uncovering these econometric conditions give an added dimension to understanding the strategies and motivations of top performers. Similarly, we can now reveal some of the shortcomings of drivers who are slightly less successful.

The ordinary drivers' strategy is less competitive. The geographic cruising range of ordinary taxi drivers is much smaller than that of the top drivers. (Figure 7(b), Table 7) The ordinary drivers, perhaps naively, operate in the Central Business District (Zone 103: Guomao; Zone 105,106: Dongmen) during early morning, morning and evening. In the afternoon, they operate in Huangqiangbei and Shangbu area (217). In this case, ordinary drivers focus more on the traditional city center—Luohu-Shangbu Area, preferring to stay in the area with most travel demand, but ignores the potentially detrimental impacts of traffic conditions. Working backwards, this elastic balance between passenger supply and traffic conditions (Shenzhen Urban Transport Planning Center, 2009) is best explained by the behavior of the top drivers (Figure 8).

Table 6. Operation centroids in different time periods for top drivers

Time Period	X(m)	Y(m)	T(s)	T(h)	Time Point	TAZ
Before 6:00	120796.4	19609.45	9271.40	2.5753	2:30	106
8:00-16:00	100778.9	19509.66	45072.36	12.5201	12:30	328
16:00-21:00	102887.9	18747.68	70018.76	19.4496	19:30	335
21:00-24:00	116294.3	18648.99	79706.23	22.1406	22:00	208

Table 7. Operation centroids of in different time periods for ordinary drivers

Time Period	X(m)	Y(m)	T(s)	T(h)	Time Point	TAZ
Before 9:00	121279.3	19501.9	10720.83	2.9780	3:00	103
9:00-14:30	121587.5	20731.6	34552.37	9.5978	10:00	106
14:30-19:00	118338.5	21116.3	52821.66	14.6726	15:00	217
19:00-24:00	121619.2	20455	73981.89	20.5505	21:00	105

Figure 8. Operation centroid comparison between top and ordinary drivers

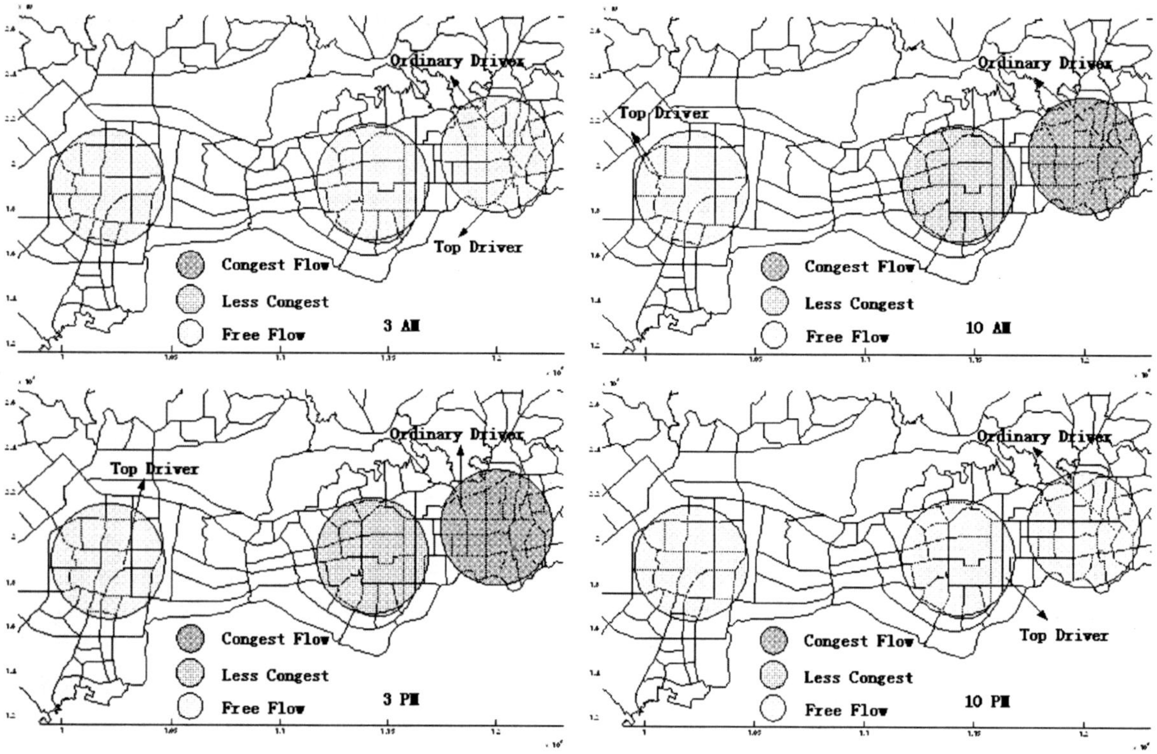

Fractal Analysis of the Traces of Top Driver and Ordinary Driver

Some past research (Wong, K.I. et al., 2001, 2002, 2005 and Yang, 2001, 2002) assumes that the taxi driver chooses the shortest path from which to deliver a passenger to his destination and pick up another passenger, a seemingly sensible idea. However, it does not account for the lucrative decisions taxi drivers could make to capitalize on the condition that taxi fare is usually based on distance, and the opportunities for drivers to make extra profit from a less direct route. This predicament was mentioned briefly as one of our two assumptions.

In order to quantify the whether taxi driver takes the shortest path to deliver passengers, we apply a fractal analysis to quantitatively examine the tortuosity of movement paths (Nams & Bourgeois, 2004; Nams, 2005). Let fractal dimension (D) represent the continuous analog of discrete geometric dimensions (Mandelbrot, 1967). Generally, there is a variety of levels of tortousity, or 'indirectness' a taxi driver can select to maximize his income during his operation.

Fractal Analysis

In this study, we applied the VFractal approach originally proposed by Nams (1996) in order to statistically test for changes in D with various spatial scales. VFractal considers a movement path to be a series of turning angles for each scale. For each V formed by turning angle, various statistics are calculated, and then D is estimated from the mean of these statistics over the whole path. Equation 3 gives an estimate of D for the V formed by one pair of steps:

Figure 9. Fractal value for the top drivers and ordinary drivers' spatial scale

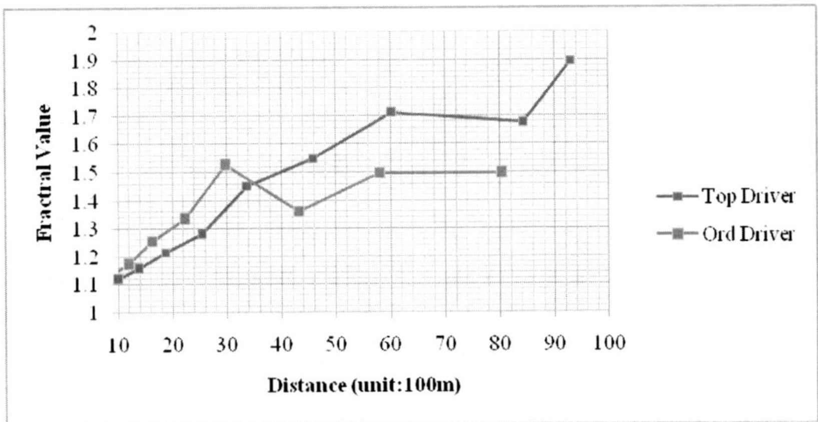

$$D = \frac{\log(2)}{\log(\frac{Net}{s})} = \frac{\theta}{1 + \log_2(\cos\theta + 1)} \qquad (3)$$

where, s is size of one step, Net is the net distance from start of the first step to the end of the second step, and θ is the turning angle between the two steps. To estimate D for the whole path, Nams (1996) proposed four different estimators:

- DMean (V): Estimated D for each V, and then find the mean D for the whole path.
- Net (V): Estimate the mean Net over the whole path, and then estimate overall fractal D by Equation 3, but with using $\overline{Net}$ instead of Net.
- Cos: Estimate the mean $\cos\theta$ over the whole path, and then estimate overall D by the equations described as follows.

$$D = \frac{\theta}{1 + \log_2(\overline{\cos\theta} + 1)} \qquad \theta < 90 \qquad (4)$$

$$D = 2 \qquad \theta \geq 90 \qquad (5)$$

- NetCos: The mean of the estimators of Net (V) and Cos.

Among four estimators, Nams (1996) found that NetCos is the most accurate estimator. Unlike the traditional divider method, confidence intervals for D with these VFractal estimators can be estimated, and thus changes in D can be statistically tested.

Results

Fractal dimensions for each taxi's movement path were calculated with NetCos estimators. According to the calculation method in previous section, we use NetCos method to analyze how the fractal value (tortuous index) varies with different spatial scale (Figure 9).

Within 3.5 km, the ordinary drivers' traces are more tortuous than that of top drivers, while beyond 3.5 km, it is the opposite. Again, the taxi fare rate is 12.5 YUAN within 3 km, while beyond 3km, the fare rate is 2.4 YUAN/km. Thus, the best strategy is: within 3 km, adopt the shortest and most straightforward path to save time; while beyond 3 km, adopt a more tortuous and longer path to increase the travel distance and maximize his income. The top drivers master this strategy, and drive a more tortuous, longer path. On the contrary, the ordinary drivers do not seem to capitalize on this strategy. The results of the fractal analysis show that that taxi driver may not always take the shortest path, which renders the existing model based on this assumption may have some fundamental flaws.

CONCLUSION AND OUTLOOK

As mentioned, a taxi driver's operation strategy is to maximize his income in least amount of time. With this strategy in mind, we consider the taxi driver's behavior within the framework of the unique travel demand patterns in Shenzhen. It would seem that a taxi driver should concentrate on the area with the most travel demand, but after a spatio-temporal analysis of operation behaviors of drivers, we show that he must take other variables into account to make the most profit from skillfully navigating the balance between travel demand and traffic condition. (SUTPC, 2009) The purpose of our research was to follow the digital bread crumbs left by the top drivers in order to understand how they became successful. The results of this analysis showed that they chose a dense place in the Nanshan District, about 10 kilometers east of the traditional Central Business District, which straddles the Futian and Luohu districts. Avoiding the popular business district meant sacrificing the most fruitful potential business, but saved on search and trip efficiency because there was less taxi competition and traffic congestion in the Nanshan District. Drivers who avoided congestion altogether and operated in the suburbs did not see the same kind of 'chaining', or the ability to pick up new customers as often. Conversely, drivers who found passengers easily in the Central Business District were plagued with traffic and competition while cruising with and without passengers. In our findings, we see that the top drivers who operate in the second densest part of Shenzhen had the best luck.

Our findings are helpful for taxi drivers, planning bureaus, city planners and transportation analysts. Also, because of the elasticity component of the drivers' choice to balance variables, the supply/demand dynamics of the competition is also noteworthy for recreational economics. However, one final contribution we hope to make is to provide a methodology for scientists and practitioners to learn about a system of individual dynamics and behavior from large data sets of GPS traces. This research benefits from a robust data set that shows behavior, routes, profit, space and time in order to qualitatively and quantitatively reveals the operation skill of taxi drivers. The methodology that led us to our conclusions may be replicable for others: We classified taxi drivers into top and ordinary drivers according to their income for manageable, nominal classes. We then applied an eigenvector dimensionality reduction to simplify a mass of activities for each group (top and ordinary) into classes of frequent spatial patterns. We then used a clustering algorithm to show that the spatial centers of gravity are different for drivers of varying success levels. Without these two methods, it would be difficult to find patterns in the data set that could show a difference between top drivers' general neighborhood and ordinary drivers' general neighborhood. The eigenvector decomposition and the K-means clustering distinguish significant records from noise and separate masses of records into a few number of groups. The tortuousity analysis also provides a quantitative way to uncover drivers who may find lucrative reason to take the scenic route.

In conclusion, the more closely a driver follows to the top driver's operation sequence, the higher rank he achieves. While area of operation is not necessarily the primary variable for success, as evidenced by the variation of operational zones between the first and second best driver, it still distinguishes top drivers from ordinary drivers. Also, while ordinary drivers operate in fixed spot, the top drivers exhibit more variation over time, probably depending on opportunity at that specific time. Our findings reveal a new pattern that uncovers a significant law dominating their daily operation behavior that has gone generally unnoticed or analyzed. However, the underlying factors that derive this phenomenon are still not completely clear. Future work should explore this research question, to develop a model that describes their operation behavior. At the same time, extra variables like a change of gasoline price and weather have great impact on taxi driver's operation behavior and should also be included. Finally, the mobility intelligence revealed in Shenzhen should be validated with the data from another city for general application.

REFERENCES

Clarkson, B. (2002). *Life Patterns: structure from wearable sensors*. Cambridge, MA: Massachusetts Institute of Technology.

Girardin, F., Calabrese, F., Dal Fiore, F., Ratti, C., & Blat, J. (2008). Digital footprinting: Uncovering tourists with user-generated content. *IEEE Pervasive Computing / IEEE Computer Society [and] IEEE Communications Society, 7*(4), 36–43. doi:10.1109/MPRV.2008.71

González, M. C., Hidalgo, C. A., & Barabási, A. L. (2008). Understanding individual human mobility patterns. *Nature, 453*, 779–782. doi:10.1038/nature06958

Hagerstrand, T. (1970). What about people in regional science? *Papers / Regional Science Association. Regional Science Association. Meeting, 24*, 7–21.

Mandelbrot, B. (1967). How long is the coast of Britain? Statistical self-similarity and fractional dimension. *Science, 156*, 636–638. doi:10.1126/science.156.3775.636

McNamara, L., Mascolo, C., & Capra, L. (2008). Media sharing based on collocation prediction in urban transport. In Proceed. ACM MobiCom '08, San Francisco, CA, Sept. 14 - 19, 58-69.

Nams, V. O. (1996). The VFractal: a new estimator for fractal dimension of animal movement paths. *Landscape Ecology, 11*(5), 289–297. doi:10.1007/BF02059856

Nams, V. O. (2005). Using animal movement paths to measure response to spatial scale. *Oecologia, 143*, 179–188. doi:10.1007/s00442-004-1804-z

Nams, V. O., & Bourgeois, M. (2004). Fractal dimension measures habitat use at different spatial scales: an example with marten. *Canadian Journal of Zoology, 82*, 1738–1747. doi:10.1139/z04-167

Rattenbury, T., Good, N., & Naaman, M. (2007). Towards automatic extraction of event and place semantics from flickr tags. In Proceed. ACM SIGIR '07, Amsterdam, July 23-27, 103-110.

Ratti, C., Pulselli, R. M., Williams, S., & Frenchman, D. (2006). Mobile landscapes: Using location data from cell-phones for urban analysis. *Environment & Planning., 33*(5), 727–748. doi:10.1068/b32047

Reades, J., Calabrese, F., Sevtsuk, A., & Ratti, C. (2007). Cellular census: Explorations in urban data collection. *IEEE Pervasive Computing / IEEE Computer Society [and] IEEE Communications Society, 6*(3), 30–38. doi:10.1109/MPRV.2007.53

Shenzhen Urban Transport Planning Center. (SUTPC). (2009). *2008 Shenzhen Traffic Condition Annual Report* 45-48.

Wong, K., Wong, S., Bell, M., & Yang, H. (2005). Modeling the bilateral micro-searching behavior for urban taxi services using the absorbing Markov chain approach. *Journal of Advanced Transportation, 39*(1), 81–104. doi:10.1002/atr.5670390107

Wong, K., Wong, S., Yang, H., & Tong, C. (2002). A sensitivity-based solution algorithm for the network model of urban taxi services. In M.A.P. Taylor, (Ed.), *Proceedings of the 15th International Symposium on Transportation and Traffic Theory,* Amsterdam: Elsevier Science. 23–42.

Wong, K., Wong, S. C., & Yang, H. (2001). Modeling urban taxi services in congested road networks with elastic demand. *Transportation Research Part B: Methodological, 35*(9), 819–842. doi:10.1016/S0191-2615(00)00021-7

Wong, S., & Yang, H. (1998). Network model of urban taxi services: improved algorithm. *Transportation Research Record, 1623*, 27–30. doi:10.3141/1623-04

Yang, H., Wong, K., & Wong, S. (2001). Modeling urban taxi services in road networks: progress, problem and prospect. *Journal of Advanced Transportation, 35*(3), 237–258. doi:10.1002/atr.5670350305

Yang, H., & Wong, S. (1998). A network equilibrium model of urban taxi services. *Transportation Research Part B: Methodological, 32*(4), 235–246. doi:10.1016/S0191-2615(97)00042-8

Yang, H., Wong, S. C., & Wong, K. I. (2002). Demand-supply equilibrium of taxi services in a network under competition and regulation. *Transportation Research Part B: Methodological, 36*(9), 799–819. doi:10.1016/S0191-2615(01)00031-5

Yang, H., Ye, M., Tang, W. H., & Wong, S. C. (2005a). Regulating taxi services in the presence of congestion externality. *Transportation Research Part A, Policy and Practice, 39*(1), 17–40. doi:10.1016/j.tra.2004.05.004

Yang, H., Ye, M., Tang, W. H. C., & Wong, S. C. (2005b). A multi-period dynamic model of taxi services with endogenous service intensity. *Operations Research, 53*(3), 501–515. doi:10.1287/opre.1040.0181

Chapter 8
Archiving Nature's Heartbeat Using Smartphones

Jinglan Zhang
Queensland University of Technology, Australia

Paul Roe
Queensland University of Technology, Australia

Binh Pham
Queensland University of Technology, Australia

Richard Mason
Queensland University of Technology, Australia

Michael Towsey
Queensland University of Technology, Australia

Jiro Sumitomo
Queensland University of Technology, Australia

ABSTRACT

The impact of urban development and climate change has created the impetus to monitor changes in the environment, particularly, the behaviour, habitat and movement of fauna species. The aim of this chapter is to present the design and development of a sensor network based on Smartphones to automatically collect and analyse acoustic and visual data for environmental monitoring purposes. Due to the communication and sophisticated programming facilities offered by Smartphones, software tools can be developed to allow data to be collected, partially processed and sent to a remote server over the network for storage and further processing. This sensor network which employs a client-server architecture has been deployed in three applications: monitoring a rare bird species near Brisbane Airport, study of koalas behaviour at St Bees Island, and detection of fruit flies. The users of this system include scientists (e.g. ecologists, ornithologists, computer scientists) and community groups participating in

DOI: 10.4018/978-1-61520-769-5.ch008

data collection or reporting on the environment (e.g. students, bird watchers). The chapter focuses on the following aspects of our research: issues involved in using Smartphones as sensors; the overall framework for data acquisition, data quality control, data management and analysis; current and future applications of the Smartphone-based sensor network, and our future research directions.

INTRODUCTION

Increasing urban development and climate change have exerted adverse effects on our environment, especially on the conditions of flora and the behaviour, habitat and movement of fauna species. The key to understanding this problem and its management lies in the ability to monitor, record and detect changes so that evidence-based strategies can be developed. Currently, most ecological sampling and monitoring involve stepping out into the environment to measure physical variables and to manually monitor, count and observe plants and animals. As this process is very time-consuming and labour intensive, it is impractical to carry out large scale studies. Another drawback of this approach is that the presence of the observer affects the behaviour and detection of the subject animals.

Sensor networks provide a viable technology for monitoring the environment. Once deployed, they can monitor their environment automatically with minimal human intervention. A number of sites can be monitored at the same time. Many different environmental characteristics can be monitored, traditionally these include: temperature, humidity, water, pressure, and many other scalar quantities. Other characteristics such as acoustic and visual data provide richer information on flora and fauna monitoring.

Mobile phones have become one of the most important communication tools. They are also pervasive in other sectors and drive market demand. For example, engineers, plumbers, or insurance company staff uses mobile phones to record site images as evidence for approving/rejecting repair or insurance claim. People on the move can use mobile phones to identify their current location and time, or to report on the environment. Mobile phones have also been used as sensors for personal medical sensing or for social networking (Kansel, Goraczko, & Zhao, 2007; Oliver & Flores-Mangas, 2006) and for capturing bird calls (http://owlproject.media.mit.edu/).

This chapter reports an innovative application of mobile phones in the monitoring of environmental changes. We present the design and development of a sensor network which is based on Smartphones for monitoring and archiving the impacts of human and animal movement on the environment as well as the impacts of environmental changes on the behaviour of animals. We choose to use Smartphones because they offer powerful processors and sophisticated programming facilities. Although micro-servers have similar hardware and capabilities, Smartphones are cheaper due to commoditization. The networks have been deployed at a few selected sites in Queensland, Australia. In particular, it is targeted at monitoring a rare bird species, the Lewin's Rail near Brisbane Airport and enhancing koala research at St Bees Island. Frog and fruit-fly population studies are also under experiment at a central city area and a farm for ecological research. To date, the Smartphones communicate with the remote servers only, but we plan to extend the communication capability further, allowing them to communicate with each other at the next stage of development.

This chapter is organised as follows. It presents firstly related work on environmental observatories, the use of different types of sensors and their drawbacks. These drawbacks lead to our approach of using Smartphones as sensors. Issues involved in using Smartphones as sensors are then analysed. This is followed by a detailed description of our system architecture, and the process of data collection and management of data quality. Data analysis and pattern recognition also form an important aspect of our work. However, due to page limitation, we can only briefly describe it here and provide references to other papers which report this aspect of our work in more detail. The chapter then discusses current and future applications of our sensor networks, followed by our future research directions.

BACKGROUND

In this section we discuss related work on environmental observatories and types of sensors deployed.

Automatic and participatory environmental monitoring is an active research area in recent years. Several environmental observing systems have been designed and implemented (Arzberger, 2005). These include the Oceans Observatory Initiative (OOI), the National Ecological Observatory Network (NEON), the Collaborative Large-scale Engineering Analysis Network for Environmental Research (CLEANER), the Hydrologic Observatory Initiative (HOI), the Southern California Earthquake Center (SCEC) (http://www.scec.org/), the Real-time Observatories, Applications, and Data management Network (ROADNet) (http://roadnet.ucsd.edu/), Science Environment for Ecological Knowledge (SEEK) (http://seek.ecoinformatics.org/), and Laboratory for the Ocean Observatory Knowledge INtegration Grid (LOOKING) (http://lookingtosea.ucsd.edu/). The monitoring targets vary including sea (e.g. LOOKING), sky, the earth (e.g. SCEC), the galaxies, plants, seabird nesting environment (Mainwaring, Polastre, Szewczyk, Culler, & Anderson, 2002) and animals (Naumowicz et al., 2008). However, these systems mainly use the mote-based sensors and none of these observatories targets acoustic observation. Mote-based sensors have program memory and data storage capacity and they can communicate with other sensors. However, their memory and data storage capacities are very limited and the communication range is quite short.

There are many webcams around the world pointing to interesting objects ranging from the Eiffel tower to African waterholes (http://www7.nationalgeographic.com/ngm/0509/). Typically, these comprise a few large permanent installations with main or large solar power installations. They mainly just stream data for human viewing. There are also commodity wireless webcams for surveillance use. However, these have low audio and video quality as they are designed for detecting gross human presence and movement.

A number of systems exist for monitoring animal behaviour using acoustic or imagery sensors. For example, Gage's system for monitoring birds' calls (Gage, Ummand, Shortridge, Qi, & Jella, 2004), strapping cameras to crows (Greenemeier, 2007), Deer Net (DeerNet, 2007). These projects use webcams or home-made systems containing webcams and radios as sensors. The drawbacks of these sensors are their short communication distance.

The Owl project at MIT (http://owlproject.media.mit.edu/) used mobile phones to capture bird calls via two different approaches. The first approach allows a phone to be called and sound to be captured through a VOIP connection. The second approach allows sound capture through a custom microphone array connected to Smartphones via Bluetooth. We have found the sound quality of ordinary phone calls (which are filtered and compressed by the telephone hardware and network) to be too poor to use. As mentioned in the Introduction, mobile phones have also been used as sensors for personal medical sensing or for social networking (Kansel et al., 2007; Oliver & Flores-Mangas, 2006). The drawbacks of using mobile phones as sensors are their relatively high communication cost. Mobile phone sensors also need more power than traditional mote sensors.

This research brings together ecologists and information technologists to research, design, build, and deploy an acoustic environmental observatory system using Smartphones as sensors. Unlike other mote-based sensors, Smartphone sensors have relatively large memory, higher processing power, and can communicate with existing mobile networks directly. Relying on existing mobile networks means that we can better use existing networking facilities and avoid the in-network processing in mote-based systems. The sophisticated programming facilities are another advantage arising from the popularity of Smartphones . Micro-servers having all these capabilities are more expensive than Smartphones . By choosing to use Smartphones we benefit from the investment in their hardware capabilities and their relatively cheap price.

MOBILE PHONES AS SENSORS AND PEOPLE AS SITES

Smartphones as Sensors

For acoustic sensing, data loggers provide a simple solution but cannot be remotely controlled nor can data be accessed in real-time. The difficulty of accessing the selected grounds such as an airport or an isolated island makes data loggers undesirable. To control data collection and to make it available in real time requires some kind of networking. Audio and image data is large, so a high bandwidth connection is necessary. Directional WiFi offers a potential solution; however, several powered repeaters would be required. A bigger problem with the airport application is the need to get any high power radio equipment approved by the airport to guarantee that it will not interfere with the airports existing communications infrastructure.

In Australia, Telstra has deployed a state of the art 3G network across both urban and rural areas. This is present at the selected sites (Brisbane airport, Samford Ecological Research Facility, and St Bees Island) and offers an ideal communications medium. The disadvantage of 3G networks over WiFi is the relatively high operating cost. We believe the prices for 3G networking and Smartphones will fall due to the mass production, market competition, and the nature of electronic products.

The sensors themselves need to be powerful enough to record and compress audio, to temporarily store audio data files and to drive a 3G radio. They also need to be programmable and remotely controllable so that scientists can manage and reconfigure them without going to the field. Currently, these requirements favour some kind of micro-server rather than mote type sensors. In this research, rather than using traditional micro-servers, we have used Smartphones because they have similar hardware to micro-servers but are cheaper due to commoditization. Smartphones typically contain a powerful processor, tens of megabytes of memory, SD card storage facility, a lithium ion battery, a microphone and a 3G radio. In addition, they are programmable. Phone software is fixed and cannot be easily changed, so a phone operating system which supports the addition of new services and applications is necessary. We have used phones running the Windows Mobile operating systems, but others could be used. It should be noted that Smartphones can be obtained at a similar price as cheap notebooks. While notebooks require considerably more power than Smartphones, they can be used, at least as a test platform.

We have also found the inbuilt microphones on Smartphones to be very good, with enough sound frequency response to satisfy the needs of ecologists. We also have the option to use the hands-free capability to connect a better quality microphone if required. Unfortunately, the sound is only monaural. We use external microphones for better quality, weatherproof, and flexibility for positioning reasons. The external microphones can also be connected to amplifiers to boost the volume of signals. Due to the size of the acoustic/photographic data obtained from sensors and corresponding power requirements, it is not possible for sensors to maintain continuous recording. Our sensors are designed to take periodic readings according to a schedule dictated by the scientists using the network. The sensor services are implemented using the .NET Compact Framework. This enables the sensors to be simply and easily programmed in a high level language, whilst enabling full and efficient control of the mobile phones.

Fixed Installation Sites

There are two types of fixed site deployment of sensors based on mobile phones. The first one is to fix the mobile phones in specially designed containers and install them in selected sites. The sensors are powered by a solar power configuration comprising a small 15W solar panel, an AGM battery and a charge controller (see Figure 1(a)). This provides enough power storage for the phone to operate over several days with no sun, and to record for several hours per day. This is the main type of installation we deployed so far.

The second approach is to fix the mobile phones in weatherproof containers, attach them to external microphones and put them in selected sites Figure 1 (b). The sensors are powered by four D-size cell batteries. These batteries provide enough power for the phones to operate for about two weeks depending on the phone's recording schedule. In this configuration, the phone records data periodically according to a predefined schedule and stores the data locally on the SD card. The drawback for this deployment type is that scientists need to go back to the field periodically to collect the sensors and download the data. The advantage is that it does not need solar panel to provide power so it is more portable and easier to hide the phones in the field. As data is stored locally, there is no cost for data communication.

People as Sites

We have also developed a prototype service to allow people to report on the environment using mobile phones. We refer this as participatory environmental monitoring, in contrast to the automatic environmental monitoring previously discussed. In this application, participants, such as university or high school students, will record the acoustic and/or visual environmental data using the manual or automatic recording function on the mobile phones. The data recorded are manually or automatically uploaded to a central server via

Figure 1. Sensor deployment (a) A deployed sensor in the field (b) A sensor in a polycarbonate box

WIFI or 3G networking. The technologies used are similar to the ones used in the fully automatic recordings. One difference is that the phones are carried by people on the move, rather than sitting in fixed locations. Another difference is that people can manually record data in addition to the automatic recording. This makes the length of recording vary from time to time, different from automatic recordings where almost all recordings are equal length. This type of deployment is relatively easier to manage as we do not have to worry about weather proof or power supply etc. However, there are significant privacy issues which need to be addressed.

Challenges with Smartphones as Sensors

Although several Smartphone-based sensor networks have been successfully designed and deployed and many challenging issues have been met, some still remain and more are expected to arise when we plan to scale up our system and broaden the applications. The challenges affecting the quality of data recorded are summarized below.

Quality of Sensors

Most sensors work in harsh conditions so the quality of sensors is a concern. Smartphone-based sensors are quite durable as they can easily work continuously for several months. This duration is currently sufficient for the selected applications. The sound and image quality are also very good on modern mobile phones.

One disadvantage of using Smartphones for data collection is their limited interfacing capabilities – they support Bluetooth and Infrared but RS232 or USB hosting for connecting other sensing devices are very limited. This makes it harder if we need to combine several sensed data (e.g. acoustic and weather data) together and conduct a correlation analysis. We anticipate this issue will be solved as soon as mobile phones embed more and more functions.

Sensor Heterogeneity

We have explored a range of Smartphones and pocket PCs as sensors e.g. HTC TyTN phones, DoPod C730, i-mate JASJAM etc. Their prices, processing capabilities and power consumption vary. However, most of them are quite stable. They can all run Windows Mobile 5 operating system. As the sensor software we developed can be used on any Windows Mobile - based phones, other mobile phones can also be used. Heterogeneity is not a major concern at this stage as a client/server system architecture is used.

Internet Connection and Power Consumption

Unlike traditional mote-based sensor networks, our Smartphone-based sensor networks rely on the existing mobile networking facilities and a client-server architecture. When the server is down or the mobile networking is not available, sensors can work autonomously, continue to record data, and save them on a local SD card. However, when the card is full, data collection is stopped.

Energy is needed in sensing and transmitting data. Compared with the AA-battery powered motes, the Smartphone sensors are energy-hungry. To ensure the sensors continue running in low power situations, such as extended cloudy periods, data capture may be paused to allow for recharging. This process has proven effective at preventing complete discharge of the sensor's batteries.

Harsh Environment

With fixed site installations, the sensors have to be protected from flooding or insect infestation. It is easier to make the sensors weatherproof and insect resistant by protecting them with specially designed containers. However, it is quite difficult

Figure 2. Sensor network system structure

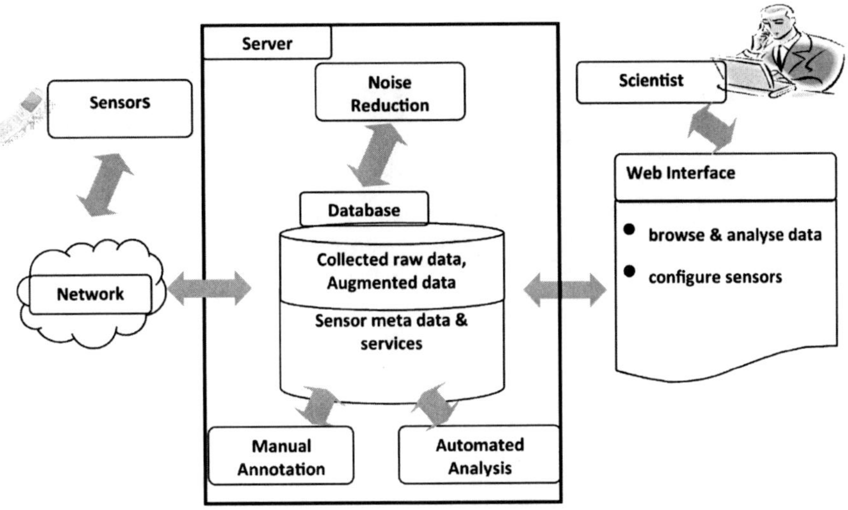

to protect them from animal/human damage or theft, especially when a large solar panel is required for power supply. Data security is also a risk. We have not dealt with the security issue as our target applications are environmental monitoring, where security issue is relatively not as urgent as for some other applications.

In summary, many problems faced in traditional mote-based systems still remain in Smartphone-based sensor networks. However, with Smartphone sensors, the sensing, computing and communication capabilities become better in mobile network reachable areas.

SYSTEM ARCHITECTURE

The overall goal for our research is to build a sensor network system for collecting and analysing the environmental data. The system should be able to combine sensing, computing and wireless communication capabilities. Sensors will monitor the environment, collect data and send them to the server. Sophisticated analysis will detect trends, recognise fauna, weather and human disturbance, and correlate changes over time. This system will allow scientists to access the data and to collaborate through data sharing. We chose Smartphones as sensors as they have relatively large memory and higher processing power. We use existing mobile networks for communication.

The system structure was designed by investigating the data flow in the system from data collection and processing to presentation to the users. A client-server architecture (Figure 2) is employed where a traditional web server is used as the server and mobiles phones as clients. The powerful server is installed in a secure location. Mobile phone sensors collect data, partially process the data and send it to the server over the network for storage and further processing. Sensors communicate with the server using web services.

The software of our framework comprises two parts: a centralized web server running in the lab and an autonomous sensor program running on individual devices in the field.

Server Side Operations

The web server is the core of the framework. On the backend, it is used by the sensors to upload their

Figure 3. Sound playing and selection user interface

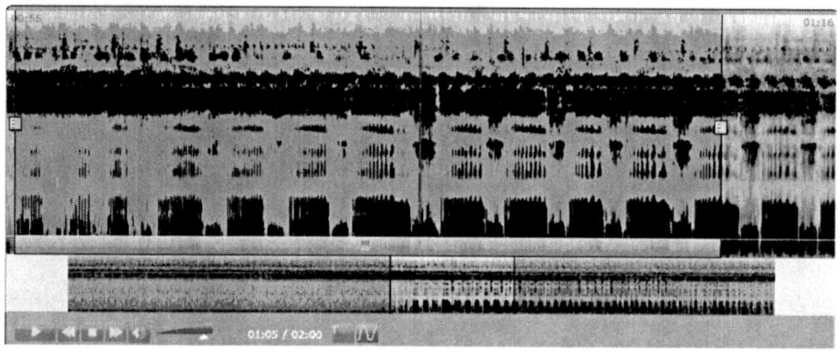

data and to download the schedule for recording. On the frontend (the web interface), it is used by scientists to manage the sensor network and analyse the collected data. Scientists can access data online in real time (see the left hand side of Figure 4 for a list of recorded sound clips displayed as sonograms. The texts show the time and location of the recordings). A web-based interface allows researchers to listen to recordings (Figure 3), to view sonograms/images, to tag or annotate them manually (Figure 4) and to perform data analysis tasks such as signal filtering (see part A in Figure 4), or animal (e.g. bird, koala etc) identification. Users can search certain data according to specified criteria (Figure 5) and download interesting data in different formats such as MP3 and WAV. The web interface also provides administrative tools for authentication, tracking sensor location and monitoring device status such as the time for recording, uploading or sleeping (Figure 6).

The server is implemented in ASP.NET using an AJAX enabled Web Forms application for the frontend and providing SOAP and REST Web Services for communication from the sensors. Most of the facilities provided by the web frontend are also available via web services, allowing for interoperation with third party tools.

Sensor Side Operations

The sensor program is an application written using the .NET Compact Framework. It runs on a variety of Windows Mobile-based devices (Smartphones and PocketPCs). The function of the sensor program is to execute tasks according to a schedule. Typically, these tasks will do

Figure 4. User interface for adding tags for Selected Recordings. (The texts on the right show existing tags.)

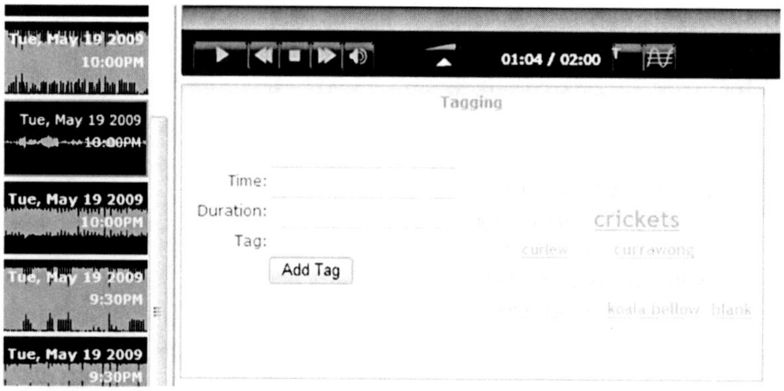

Figure 5. Record searching user interface

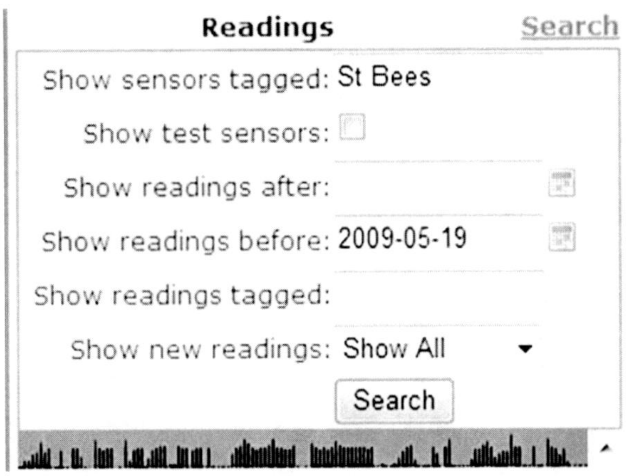

things like recording audio, taking photos, or uploading local data to the server. The first task is to perform an audio recording. For example, scientists can set the audio recordings to start at mid-night and repeat every 30 minutes. The second task is to perform a batch upload of audio/visual recordings at scheduled times. The sensors also perform some general processing as part of their execution. This includes schedule retrieval and a comprehensive logging scheme. Each sensor operates autonomously and follows a recording schedule which is verified each time it contacts the server. Scientists can reschedule sensors by updating this schedule through a web-based interface. If the sensor is unable to retrieve a new schedule from the server (e.g. because of a network failure), then the sensor continues operating with its last known schedule and the recordings are stored in the phone's memory card. Once this becomes full, data capture is paused until older files are uploaded and new space obtained. This allows sensors to be run in a fully disconnected mode by setting up a schedule before deploying the device. It also ensures recording continuity in

Figure 6. Sensors location and status information

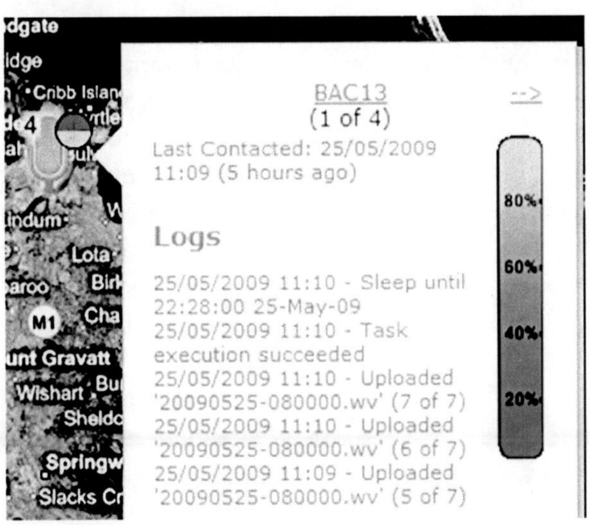

case of a network disruption. The logging scheme logs all the actions the sensor takes, e.g. when it tries to contact the server, what task it will execute next, whether that task executed successfully or not. The logs are routinely uploaded to the server where users can use them to diagnose problems with the sensor (e.g., if a particular task is failing to run and what error message it is giving). In addition to action logs, health data about the sensor (e.g. battery level, memory usage) is periodically reported to the server.

DATA COLLECTION AND QUALITY CONTROL

The sensors are designed to record sound/images where the recording schedule is controlled remotely by scientists. Collected data are uploaded to the server and stored in a relational database. Additional metadata such as sensor identifiers and their locations are also recorded in the database. The quality of the data is controlled by several measures.

Data Quality Control on the Sensor Side

Data is collected periodically according to prespecified schedule. We are interested in animal and environmental sounds up to 11 kHz so audio is sampled at 16-bit 22 kHz mono. To ensure sufficient quality for analysis and to foster future new analysis tools, all recordings are compressed using the lossless WavPack codec (http://www.wavpack.com/) and then communicated and stored on the server.

Incomplete/faulty data are usually caused by the following factors:

- power shortages,
- lack of storage, or
- communication blackouts.

Often these problems were actually related (e.g. exhausting storage during communication black-outs or reduced power due to extended communication periods in low connectivity). These faults could largely be diagnosed from the lab through logging data. Power shortages were highlighted by decreasing battery levels before finally losing contact with the sensor. This is currently managed by measuring the battery level in order to optimize the recording time. However, more accurate measurement is still needed. Lack of storage problems were directly communicated to the server via error messages and health data, and then the sensor pauses recording. Communication black-outs usually cause a sudden loss of contact. In the case of communication blackouts, we simply delay sensor upload or download activities with the reasonable expectation that the communication would recover of its own accord. If the sensor cannot automatically wake up for a period of time after an incident, an SMS message can be sent to wake it up.

All these approaches together roughly ensure that the data can be recorded at a specified time for a specified length on the sensor and the data are safely uploaded to the server.

Data Quality Control on the Server Side

While recorded information is useful, some problems have existed such as recording quality, storage capacity for recordings, retrieval of the recordings, and coordination of timing of recordings. We have managed to control data quality from both hardware and software aspects.

Due to the large size of the data obtained from acoustic/photographic sensors and corresponding power requirements, it is not possible for sensors to maintain continuous recording. Thus sensors will only record periodically. To keep the sensor programs as simple and reliable as possible, we push as much processing and information storage onto the server as possible. The server collects data uploaded from the sensors and stores this in a relational database. Two forms of data are collected and preserved: sensed data such as sounds and images and sensor metadata including health status and deployment information. This enables the performance of sensor hardware over multiple deployments to be tracked and sound data to be re-analyzed using new algorithms or by different scientists at a later date.

Sensor Identification

As we deployed many sensors in different places, correctly identifying the sensor that has collected a reading is essential for the subsequent processing of the data. However, sensors may need to be relocated according to the scientists' needs. Our platform supports this by associating all sensor metadata with a specific sensor deployment. That is, a deployed sensor is identified by a unique deployment identification number. When a reading is stored in the database, it references the deployment it was taken from. When a sensor is relocated, a new deployment is initiated and the appropriate metadata is sent to the database. For example, if data is being analysed to identify where birds are located, the location of the sensor must be known. This sensor metadata is stored in the database and associated with each reading taken by that sensor. Other metadata associated with a specific deployment include sensor hardware identification number, creator, start time, timeout period, description etc.

Noise Reduction

To aid the recognition of interesting signals, we perform some sound filtering on the server to present interesting sounds to scientists. This can reduce the amount of sound which needs to be listened to by up to 90% in some cases.

The quality of sound captured by the mobile sensors is usually quite good and can be directly used for later analysis. However, under rainy or windy conditions or where the monitored site is close to the highway, some background noise exists in the recordings. In order to conduct object recognition, the noise needs to be removed.

Traditionally, data can be cleansed by manual selection and editing process, which is to select a small section of good quality data from the raw data. Consequently, data with low Signal-to-Noise Ratio caused by rain or wind are removed. This approach has been used in the creation of most published, high quality sound databases (Chen & Maher, 2006; Franzen & Gu, 2003; Somervuo, Harma, & Fagerlund, 2006). In our research, the amount of sound data collected is huge so it is impossible to manually pick out the bad data. In some cases, some background noises such as rain or wind are also essential to determine the context of a sound clip. Thus, keeping the noisy data is desirable in some cases.

In the bird song recordings, the Signal-to-Noise Ratio of recorded data varies from 30 dB to -10 dB mainly due to different weather conditions, the distances of birds from the sensors, and surrounding environments. The noise level can change from time to time due to different conditions such as windy weather, highway traffic or flyover airplanes. Therefore, it is essential that the signal enhancement algorithm is able to handle different noises and to track the noise level.

In the past two decades, researchers have developed many noise reduction algorithms including Wiener filter methods (Chen, Benesty, Huang, & Doclo, 2006), signal subspace methods (Lev-Ari & Ephraim, 2003), and statistical methods (Ephraim, 1992). All these algorithms assume that the first few frames of the signal consist of noise only so that the initial noise estimation can be obtained from the first few frames. The noise estimation is usually updated from noise-only frames which are determined by voice (signal) activity detectors. This approach is reasonable for speech enhancement as speakers are cooperative. However, we cannot assume that the first few frames of the recoded data by our sensors are noise only, without bird calls. Sometimes there is almost no gap between bird calls, therefore there are not enough noise-only frames to update noise estimation. Moreover, voice activity detectors do not work well in low Signal-to-Noise Ratio signals. We have developed an effective noise reduction algorithm which is capable of estimating noise from frames with or without signals (i.e. noise only) (Cai, Ee, Pham, & Roe, 2007). Therefore, it does not need a voice activity detector, which is error prone. The basic procedure of this algorithm is shown in Figure 7 below. In this diagram, the input noisy signal is a sound clip recorded by one of our sensors. It includes calls of a powerful owl in the distance together with calls of crickets and other insects. The waveforms of the original sound and the sound after noise reduction shows that the algorithm improves the sound quality significantly even though the Signal-to-Noise Ratio of the original sound is very low (about -10 dB). The filtered data will be used for later analysis task.

Data Augmentation

Although much research has been done in automatic sound and image processing and recognition, the semantic gap between low level features e.g. a sound spectrum and high level semantics e.g. a bird or frog call still exist. This semantic gap can be narrowed with more and better research but it may not be able to be filled completely in the near future.

Collaborative tagging (manually annotate sound clips by many people) is one possible way to get around this challenge. Ultimately, scientists also wish to listen to potentially interesting sound captured from the sensors and to label sounds. For example, the two principal Lewin's Rail calls are known as squeaky door and kek-kek. An example spectrum of kek-kek sound is shown in Figure 8.

To support data labeling, we have developed a web-based annotation interface. This allows a folksonomy style labeling of sound, similar to the popular *flickr* photo sharing site. Scientists can listen to sound clips and if an interesting noise is identified, they can label it using an existing tag or they can create a new tag. Tags can be used for browsing sound and for categorizing sounds for analysis. Rather than presenting scientists with a fixed vocabulary of tags or keywords, we allow scientists to develop their own language to describe the various bird calls they identify.

In order to label sound, a special sound playback control is required which allows the position within a sound stream to be identified. For example, we cannot simply label an hour of sound as containing three bird calls – we need to label where the bird calls are within the sound clip. This is implemented through a customized media player using Microsoft Silverlight (http://

Figure 7. Noise reduction algorithm

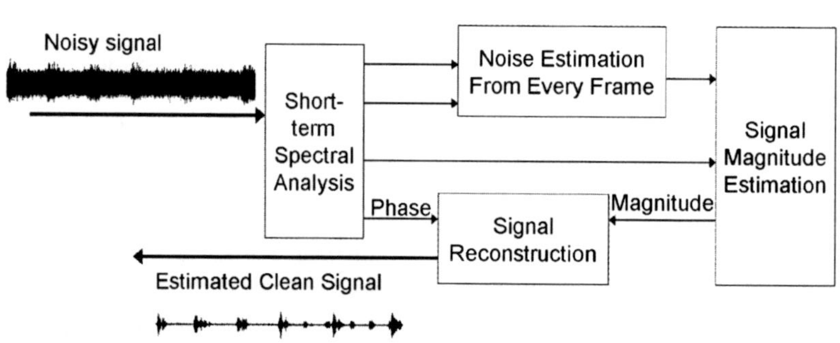

www.microsoft.com/silverlight/). Figure 3 shows the sound playing and sound segment selection interface for labelling a koala bellow sound. Users can select a section of the sound then add tags for this section via interface shown in Figure 4. The tag and the start and end time of the selected sound section are then saved in the database.

Data Analysis and Pattern Recognition

As the amount of data captured is massive, it is not possible for all recordings to be listened to by humans in order to identify and label interesting events such as bird calls. Thus, it is essential to develop automatic analysis tools to filter the signals, recognise interesting events and present them to scientists. We have developed an algorithm for bird call recognition using Mel-Frequency Cepstral Coefficients (MFCC) and Artificial Neural Network (Cai,et al., 2007). The automated analysis of animal calls can be performed in the time-domain (Chenmore & Ohya, 2004), in the frequency domain or in the two dimensional domain of the derived sonograph (Brandes, Naskrecki, & Figueroa, 2006). We present our work on the frequency domain.

As human speech and animal calls are similar in terms of communication and their development (Doupe & Kuhl, 1999), it is natural for researchers to adopt feature extraction methods used in speech analysis for animal call recognition. Recently, almost all speech recognition systems use MFCC as features as they are consistent with human auditory perception, suitable for both aperiodic and periodic signals, able to achieve significant data reduction with reasonable information loss. We also use the MFCC features for bird call recognition and the following four steps are used to derive MFCCs from the signal: 1. use Fast Fourier Transform to obtain power spectrum; 2. use Mel scale to derive Mel Power Spectrum; 3. use Logarithm to obtain log spectrum; 4. perform inverse Discrete Cosine Transform to obtain MFCC. The recognition results to date are very promising, with the accuracy of 86.8% on a data set of 14 bird species.

We have also adapted and developed a number of indices measuring the health of the environment by measuring the amount of acoustic activity in different frequency bands (Mason et al., 2008). This approach is based on the observation that the health of a habitat can be estimated by the prevalence of either natural sounds or manmade noise. In addition, sound distribution among different frequency bands indicates the diversity of species involved. Figure 9 shows the diversity of sound in a wetland area over an 11 kHz band over a 24 hour cycle. We can see that the calls of insects and some birds (6-8 kHz frequency band) get the greatest diversity.

Figure 8. Spectrum of kek-kek sound from Lewin's Rail

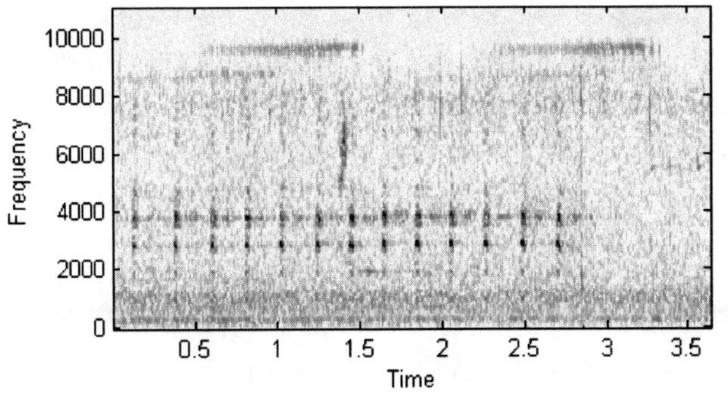

Figure 9. The diversity of acoustic clusters over an 11 kHz band over a 24 hour cycle

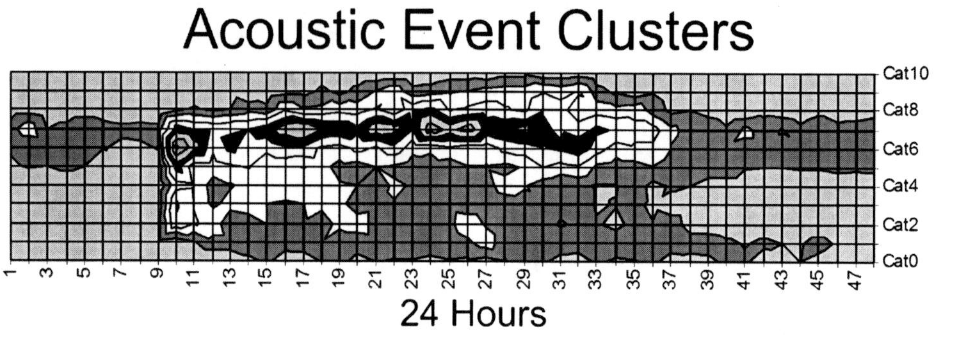

APPLICATIONS

To monitor both specific species and long-term environment change, we have deployed three sensor networks which utilize the techniques discussed in the previous sections. All these sensor networks use an array of Smartphone sensors for environmental monitoring, but with slightly different goals. The advantages of using Smartphones as sensors in these applications are the high quality acoustic and imagery capabilities of these phones as well as their advanced programming facilities.

The first application of this mobile sensor network is to monitor a rare bird species, the Lewin's Rail, which lives in a wetlands area near Brisbane Airport. The main goal is to understand various aspects of the birds' behaviour via recording the environmental acoustics e.g. bird calls (Figure 8), traffic noises etc. It is used as part of a scientific study into the Lewin's Rail - a bird difficult to study due to its shyness, camouflage and ground dwelling nature. Little is known about Lewin's Rail so scientists need to experiment with different strategies for locating birds in a relatively large area of land, in this case, approximately 1000 metres by 30 metres. One approach is to monitor the bird acoustically by humans through identifying and locating its calls in response to played back calls or through the placement of acoustic data loggers (e.g. tape recorders). This had several drawbacks such as forcing scientists to visit the area and perturbing the environment. Using sensor networks to conduct automatic monitoring can avoid these drawbacks and provide a complimentary way to monitor this rare bird. Whilst this study is being undertaken using traditional methods, our network is being used to augment the data collection. This dual-mode study is particularly useful to establishing the validity of acoustic sensors. The data collected have been analysed by the data analysis tools we had developed and a portion of the processed data have been manually validated by scientists.

The same technique is also used for monitoring koalas' activities on the St Bees Island off the coast of Mackay, Queensland, Australia. Scientists are investigating the behaviours of a population of Koalas on this island. The acoustic sensors we provided help scientists to periodically perform remote monitoring of the Koalas' bellowing (Figure 3) day and night. Combined with GPS data from collars on certain animals, koala researchers can gain greater insight into the purpose of Koala vocalisations. Similar to the project monitoring Lewin's Rail, the monitoring of Koalas also uses both the traditional human-based and the modern sensor-network based data gathering in parallel.

In addition to monitoring specific species, the environmental health information can also be obtained by analysing the collected acoustic information as flora and fauna can be important environmental indicators. We use a number of indices to measure the health of the environment (Mason et al., 2008). One application is to monitor the Samford Ecological Reserve (approximately 12 km north east of Brisbane, Australia) whose surrounding areas have been developed due to the expansion of urbanization. This sensor network is used for more general habitat monitoring such as monitoring bird calls, trees and ponds. It is used for ecological studies by QUT researchers.

One similar but slightly different application is fruit fly detection. In Australia, the Queensland fruit fly (B. tryoni), is the most destructive insect pest of horticulture, attacking nearly all fruit and vegetable crops. We have prototyped a system for monitoring fruit flies so that authorities can be alerted when a fly enters a crop in a more timely and informed manner than is currently available. Rather than using acoustic information, the photographic information is used for fruit fly detection. We have designed a new fruit fly trap incorporating a sensor which can capture quality images and developed the automatic fruit fly detection techniques (Liu, Zhang et al., 2009).

All these applications are based on the same general technique - commodity hardware running specialised software. The sensors are Windows Mobile devices – specifically HTC TyTN II and Do-pod C730/HTC S630. These sensors are housed in a polycarbonate box with an external power connector which is attached to a 12V battery/solar panel rig. The external microphones supplied with the devices are adapted for weatherproofing and positioned outside of the box.

The sensors have been actively monitoring the airport site and the Ecological farm continuously since April, 2008 and St Bees Island since September, 2008. By 1:30pm, 18th February, 2009, the total number of automatic readings from the sensors is 51852 and the total duration of audio captured is 632 hours. The most robust sensor is one deployed to the airport site and it has run continuously for 183 days.

FUTURE RESEARCH DIRECTIONS

Our ultimate goal is to build Nature Observatories using Smartphones that can be carried around by people, mounted on poles in the city, or deployed to the wild fields. The entire system will provide an end-to-end application which accommodates data collection, augmentation, analysis and visualisation. There are several areas where our current system needs further research and development to make this nature archive sustainable.

The first area to be investigated is data fusion – currently we can only support simple data correlation based on time and spatial dimensions. Correlation with weather information should not be difficult. However, correlation with other data including qualitative observations made by scientists in the field will be much harder. Secondly, the rich and large audio and image data continuously collected becomes a dominant issue in data storage and retrieval, as well as data understanding. These issues will continuously be researched to develop more efficient or effective indexing and automatic data analysis tools. A third area for investigation involves human-based, collaborative tagging and semantics generation. With greater semantic information, it will be possible to undertake more complex environmental queries. The fourth research direction regards to data publication and access issues such as privacy, security etc, especially when the data are about people or the sensors are carried by people. Finally, the system architecture may need to be improved. Our initial system has a deliberately centralised architecture with a central server and data being uploaded to the server. As the number of sensors is increased, the centralised server will become a bottleneck and a different architecture will be required where data is kept and processed closer to the sensors. Distributed databases and cloud computing techniques may help solve this problem.

Other potential applications of modern sensor network technology to environmental monitoring and agriculture are being discussed with various organizations and professionals, including monitoring birds, frogs, fruit-flies in the field and insects within silos. We will adapt this technology to more application areas as new partnerships emerge.

CONCLUSION

This chapter presents the novel idea of monitoring the environment using Smartphones with the focus on data acquisition and management for environmental changes and animal observations. This innovation makes use of existing mobile technologies and infrastructures to monitor the environment. Smarthphones may be used either manually by people or they may run autonomously as sensor nodes. By using commodity devices, we benefit from cheap hardware which has been highly power optimised and advanced programming facilities offered with such devices. This new dimension of mobile phone applications would add more value to current investments on mobile phone infrastructure and services.

We have described a framework for collecting data via a network of audio-video sensors. This framework has been used for deploying several networks of sensors for environmental monitoring. It works well with the current scale of deployment.

Current experiments show that the quality of the sound collected is very good and suitable for automatic processing. The images collected are relatively low quality but acceptable for human eye reading and simple automatic image processing. These acoustic and visual data have provided scientists with the evidence they are searching for, e.g. the activities of Lewin's Rails in a wetland area, koala's behaviors on St Bees Island, frogs and other animals on an ecology experimentation farm, and fruit flies in a city centre. Through the use of sensors for periodical recordings, site visits can be significantly reduced and the scale of monitoring can be enlarged. Recordings can be immediately uploaded to a central server, reducing the limitations on the number and length of recordings that can be captured. The ongoing monitoring of the activities of animals using these sensor networks will eventually enable scientists to continuously study the change of environment and their impacts on wildlife over a long period of time.

The success of these deployments has enabled scientists to appreciate the potential of the technology, and many more applications are being discussed, including monitoring insect species. We plan to add other devices to enable temperature and humidity to be sensed so that multiple aspects of the environment can be monitored and correlated with each other.

The successful deployment of these Smartphone sensor networks has also raised several challenges in the application of this technology including data quality, storage, interpretation, presentation, security issues, etc. Environmental change detection and prediction by analyzing the collected data demand further research. In addition, managing the terrabyte, even petabyte data is still a challenging task facing the eResearch community.

ACKNOWLEDGMENT

The authors wish to acknowledge support from Microsoft, Queensland State Government, Brisbane Airport Corporation, Telstra and Queensland University of Technology.

REFERENCES

Arzberger, P. (2005). *Sensors for Environmental Observatories. National Science Foundation (NSF)*. Report, Center for Embedded Network Sensing.

Brandes, T. S., Naskrecki, P., & Figueroa, H. K. (2006). Using image processing to detect and classify narrow-band cricket and frog calls. *The Journal of the Acoustical Society of America, 120*, 2950–2957. doi:10.1121/1.2355479

Cai, J., Ee, D., Pham, B., & Roe, P. (2007). Sensor Network for the Monitoring of Ecosystem: Bird Species Recognition. In *ISSNIP Conference*, Melbourne, Australia.

Chen, J., Benesty, J., Huang, Y., & Doclo, S. (2006). New Insights Into the Noise Reduction Wiener Filter. *IEEE Transactions on Audio, Speech, and Language Processing 14*(4).

Chen, Z., & Maher, R. C. (2006). Semi-automatic classification of bird vocalizations using spectral peak tracks Recognition. *The Journal of the Acoustical Society of America, 120*(5), 2974–2984. doi:10.1121/1.2345831

Chenmore, E. D., & Ohya, E. (2004). Automated identification of field-recorded songs of four British grasshoppers using bioacoustic signal recognition. *Bulletin of Entomological Research*, (94): 319–330.

DeerNet. (2007). *DeerNet: Wireless sensor networking for wildlife behavior analysis and interaction modeling*. Retrieved Oct. 29, 2007, from https://winet.ece.ufl.edu/deernet/index.php/Main_Page

Doupe, A. J., & Kuhl, P. K. (1999). Birdsong and Human Speech: Common Themes and Mechanisms. *Annual Review of Neuroscience, 22*, 567–631. doi:10.1146/annurev.neuro.22.1.567

Ephraim, Y. (1992). Statistical model based speech enhancement systems . *Proceedings of the IEEE, 80*(10), 1526–1555. doi:10.1109/5.168664

Franzen, A., & Gu, I. (2003). Classification of Bird Species by Using Key Song Searching: A Comparative Study. In *IEEE International Conference on Systems, Man and Cybernetics.*

Gage, S., Ummand, P., Shortridge, A., Qi, J., & Jella, P. K. (2004). Using GIS to develop a network of acoustic environmental sensors. *ESRI International User Conference*.

Greenemeier, L. (2007). The Secret Lives of Tool-Wielding Crows. *Scientific American*, (October): 4.

Kansal, A., Goraczko, M., & Zhao, F. (2007). Building a sensor network of mobile phones. In *Proceedings of the 6th International Conference on Information Processing in Sensor Networks,* (IPSN '07), 547-548.

Lev-Ari, H., & Ephraim, Y. (2003). Extension of the Signal Subspace Speech Enhancement Approach to Colored Noise. *IEEE Signal Processing Letters, 10*, 103–106. doi:10.1109/LSP.2003.808544

Liu, Y., Zhang, J., et al. (2009). Towards Continuous Surveillance of Fruit Flies Using Sensor Networks and Machine Vision. In *The 5th International Conference on Wireless Communications Networking and Mobile Computing (WiCOM2009),* Beijing, China.

Mainwaring, A., Polastre, J., Szewczyk, R., Culler, D., & Anderson, J. (2002). Wireless Sensor Networks for Habitat Monitoring. In *Proceedings of the 1st ACM international workshop on Wireless sensor networks and applications*. Atlanta, GA, (pp. 88-97).

Mason, R., Roe, P., Towsey, M., Zhang, J., Gibson, J., & Gage, S. (2008). Towards an Acoustic Environmental Observatory. In *Fourth IEEE International Conference on eScience*, Indiana, (pp. 135-142).

Naumowicz, T., Freeman, R., Heil, A., Calsyn, M., Hellmich, E., & Brändle, A. (2008). *Autonomous monitoring of vulnerable habitats using a wireless sensor network. The workshop on real-world wireless sensor networks ACM* (pp. 51–55). New York: ACM.

Oliver, N., & Flores-Mangas, F. (2006). HealthGear: A Real-time Wearable System for Monitoring and Analyzing Physiological Signals. In *Proceedings of the International Workshop on Wearable and Implantable Body Sensor Networks* (BSN'06), (pp. 61-64).

Somervuo, P., Harma, A., & Fagerlund, S. (2006). Parametric Representations of Bird Sounds for Automatic Species Recognition. *IEEE Trans. on Audio, Speech, and Language Processing, 14*(6), 2252–2263. doi:10.1109/TASL.2006.872624

ENDNOTE

1 The authors wish to acknowledge support from Microsoft, Queensland State Government, Brisbane Airport Corporation, Telstra and Queensland University of Technology.

Section 3
Privacy Issues and New Perspectives on Movement-Aware Services

Chapter 9
Law and Technology:
Anonymity and Right to Anonymity in a Connected World

Giusella Finocchiaro
University of Bologna, Italy

Claire Vishik
Intel Corporation, UK

ABSTRACT

In this chapter the authors analyze the concept and definitions of anonymity in the modern connected world. In particular, they explore if modern technology renders complete anonymity impossible and if a new definition of anonymity needs to be adopted. They examine examples of anonymous use of technology that illustrate the complexity of the concept of anonymity and demonstrate that access to anonymity is not uniform for data owners with regard to multiple data controllers and audiences in complex systems and processes. They evaluate legal definitions of "anonymity" and "anonymous data" as well as the right to anonymity provided in the European directives and by some European statutes, observing that anonymity cannot be absolute, that only "relative" anonymity is realistic in the present technological environment, and that different degrees of anonymity exist. They address the issue of measuring these degrees or levels of anonymity in complex systems, in order to provide a new foundation for a nuanced and comprehensive understanding of anonymity. The authors conclude that the concept of relative anonymity can become the basis for a new and more effective approach to personal data protection.

INTRODUCTION

Ubiquitous connectivity today makes it easier for the users of systems and networks to find and use information and increase the efficiency and reach of their activities with digitized processes. But in this environment, it is also possible to track users and devices, frequently in order to ensure the fundamental functionality of protocols that is not connected to tracking. Convergent networks and interoperable integrated systems ensure that a record of transactions is available (although not necessarily stored) in a variety of places and can be more easily aggregated and analyzed. An increasing number of transactions, processes, and assets are digitized, leading to greater efficiency in all areas of life and

DOI: 10.4018/978-1-61520-769-5.ch009

organizational operations. We now use electronic data to keep track of our friends' birthdays and send cards to family and friends, to file taxes, to check on pension accounts, to shop, pay bills, and communicate with friends and members of communities of interest. This environment makes our activities more efficient, but it also makes it harder to preserve the anonymity of a user. In an extreme example, it is possible to make an anonymous cash contribution to a charity without any disclosures if allowed by law. But it is not possible to make a truly anonymous contribution to the same charity online, as several data controllers along the path of a transaction – the ISP, the credit card company-- will know (and have the right to know) the identity of the user

As a result, there is an increased interest in the problem of anonymity addressing both technical and legal implications of the new technological environment. From the legal point of view, the main questions are:

- What does "anonymity" mean today?
- Should "anonymity" be considered relative or absolute?
- Does the right to anonymity exist?

From the technical point of view, we can ask:

- What are the technical means to support the highest possible level of anonymity in the modern computing environment without jeopardizing security?
- Can levels of anonymity be measured in an objective fashion in dynamic computing frameworks?
- Does the existence of different "levels" of anonymity available along the path of a message via different systems and networks represent an issue that needs to be remedied? Or is it a normal outcome of the adoption of modern technologies?

TODAY'S COMPLEX COMPUTING ENVIRONMENT

Modern computing environment is characterized by ubiquitous connectivity with ever increasing interoperability among heterogeneous networks and diverse systems and devices. Increasingly, these devices can connect to networks or other devices using a variety of communications protocols: a PC may be using an Ethernet connection as well as connecting over Wi-Fi or WiMax, with Bluetooth, infrared, and 3G options also available. A mobile handset can use modern cellular phone networks and gateways to connect to TCP/IP networks, and also Bluetooth and infrared interfaces. Some GPS devices permit connectivity via Wi-Fi and can act as mobile phones and MP3 players while also connecting to satellites using appropriate protocols. The list can be continued.

Ubiquitous interoperability and connectivity affect computer systems as well as networks. An email application on a mobile phone may use a different client to access the email server, but the data are harmonized when accessed from different devices, and the user can switch to a PC to continue his email or chat communication. An online banking request submitted from a smart phone flows seamlessly through a variety of applications, networks, and gateways, with each component able to recognize the content of the request and authenticate its user and its origin. These transactions always leave an electronic trail that is frequently not captured unless necessary for troubleshooting, non-repudiation support, or other activities. However, some information is always stored for longer periods of time. Typically, only user facing components are analyzed for privacy support, but back-end data exchanged between organizations can also contain sensitive information important to safeguard the users' privacy. Like user-originated data, back-end data contain a number of heterogeneous identifiers.

The move to open standards and federation of standards and services as well as the growing

Figure 1. Message and ID transfer patterns

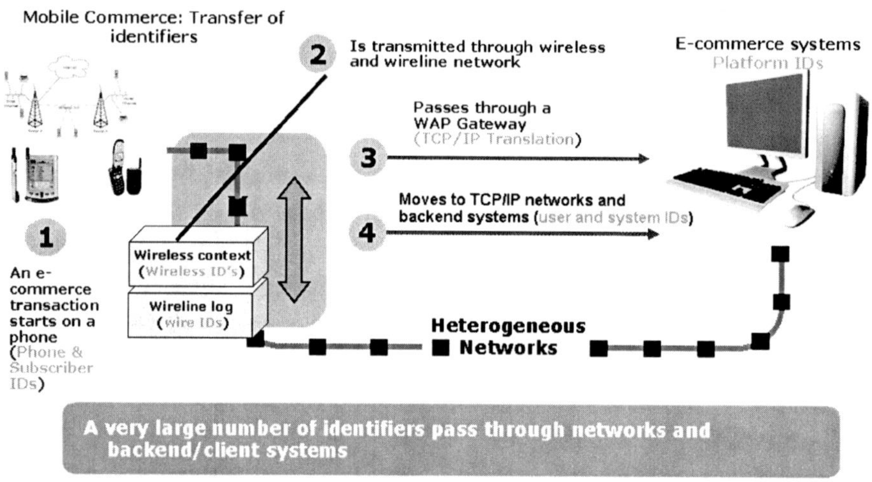

need for integration of systems inside organizations and in multi-organizational environments mean that identifiers and other data have to pass through numerous systems and networks with differing rules before reaching the destination as illustrated in Figure 1.

In the computing world, the trail left by an electronic message as it passes through numerous systems and gateways makes true anonymity difficult to achieve even if the network data are not stored. In order to guarantee some level of anonymity, ubiquitous connectivity and interoperability require a different type of approach to technology and the very concept of anonymity, one that takes the continuum of a transaction into consideration rather than focusing on one component. While we believe that it is necessary to re-think the concept of anonymity, the characteristics of the digital processes and electronic data today that make such a revision necessary also provide benefits for consumers, permitting, for instance, to analyze historical data when needed or seamlessly access multiple services.

The chapter starts with the description of today's computing environment, touches upon definitions of the personal and anonymous data as well as anonymity in European law, discusses relationships between anonymity and security, briefly looks at identifiers, a necessary component of today's systems and network protocols, examines the effect of new technologies on anonymity, and finally defines the concept of "relative" anonymity that can be used with regard to digital data and processes. The chapter evaluates available technologies and anonymity "metrics" as well as changes in the concept and implementation of anonymity brought about by the massive digitization of assets and processes. It is important to note here that anonymity is a very complex topic that has inspired numerous researchers, created a substantial body of knowledge, and led to the development of numerous technologies capable of supporting various degrees of anonymity. It would be impossible to address the whole rich area in one small chapter, and we limit the discussion, for the most part, to the impact of new technologies on anonymity. However, even if some of the important research areas in anonymity are not the direct subject of the chapter, they, without any doubt, influenced the views of the authors.

"Chains of Devices" and Varying Levels of Protection

Simple systems and straightforward connectivity between two endpoints are no longer the foundation of computing, but, rather, represent exceptions from the rule. Today, "chains of devices" become increasingly common, complicating security solutions, privacy approaches, and views on anonymity.

As illustrated by Figure 2 below, modern devices can form complex and dynamic groups that we call "chains". The levels of security protection, security and privacy rules, ability to support anonymous transactions, and technology approaches are different in different components of a chain (Saponas, Lester, Hartung, Sameer, & Tadayoshi, 2007). Although all data from the simplest least protected device doesn't normally need to be transferred to the end of the chain, some identifiers and data have to be transmitted to complete connections and carry out data transfer. Ensuring a uniform level of anonymity in this environment is very difficult because the trail of the message is, in principle, reconstructible. Additionally, although in most environments data from the simple "beginning of the chain" devices are not stored at intermediary points, in some contexts, e.g., healthcare, data received from sensors are aggregated and analyzed, making the determination of its nature (personal or anonymous?) more complex. The choice of anonymization techniques is very limited at the "beginning of the chain": the simplest mono-functional devices don't have the computing power or available infrastructure to execute more complex protocols, and although the ID of a pedometer (in our example on Figure 2) doesn't constitute personal information, the fact that this device is connected to a more attributable and powerful device that is, in its turn, connected to a personal computer with many other potential identifiers, offers opportunities for more complex and more interesting analysis of anonymity in today's environment, which is the purpose of this paper.

Many non-electronic activities where complete anonymity can be achieved (e.g., anonymous contribution to a charity) have electronic equivalents where complete anonymity is impossible as some data controllers have to be aware of the identity of the requestor in order to execute the transaction. If technologies providing varying degrees of anonymity could be automated, thus ensuring that certain data controllers that would

Figure 2. Chains of devices

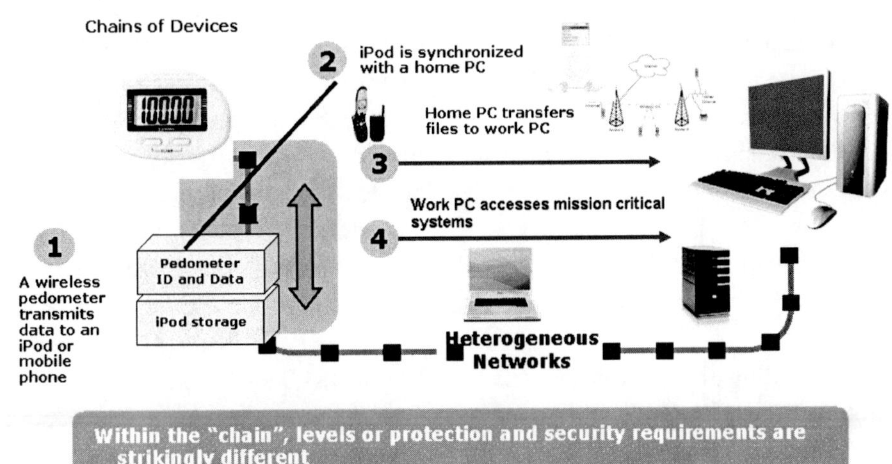

normally know the identity of accessing devices or users could provide the same services anonymously or almost anonymously, the general level of anonymity could increase. Some promising proposals based on Trusted Computing technologies have appeared recently, e.g. Gajek, Sadeghi, Stüble, and Winandy (2007); Jackson, Boneh, and Mitchell (2007), and Xian & Feng (2007). Work on zero knowledge systems continues, offering verification rather than identification options for devices and users (see Camenisch, Groß, & Heydt-Benjamin, 2008; Camenisch & Gross, 2008). The example provided at the beginning of the paragraph is deliberately simple. In reality, individual requirements for anonymity are influenced by many factors and perceptions, and anonymity, in the electronic and real world, is used to support very diverse needs (see, e.g., Jacob Palme's anonymity site[1]). The same user engaged in the same activity or making the same statements may insist on strict anonymity in one environment and require visibility in another. Social networks are good examples of the situational nature of user defined anonymity. It is important, while reading this chapter, to remember that, in some cases, the picture presented here is simplified, in order to better illustrate the influence of new technologies on the true ability of a subject to be anonymous. The previous sections outlined some aspects of the computing environment today and elements of the technologies and approaches available to achieve or measure anonymity as well as clear issues with considering anonymity absolute and uniform in all components of complex electronic processes. We will now move to the evaluation of the legal aspects of anonymity.

Concept of Anonymity and Anonymous Data

The common understanding of anonymity, which comes from the etymology of the term, is "without name". The word denotes an absolute concept: an anonymous person is one, of whom you do not know anything, somebody you cannot recognize or identify. This definition of anonymity as absolute is frequently taken for granted.

However, it is not completely correct in many contexts even in the non-digital world. If you think, for instance, of an anonymous writer, whose right to be anonymous is usually recognized by law, he or she can be anonymous for the readers, but not for the publisher: so, in this case, anonymity is relative. This example helps explain the anonymity framework for electronic transactions today, with different data controllers possessing different degrees of knowledge of the identity of the originator of a transaction or a device at the beginning of the "device chain."

The concept of anonymity has gained particular importance in relation to the application of the European legislation on personal data protection. This regulation is constituted by the Directive 95/46/EC on the protection of individuals with regard to the processing of personal data and on the free movement of such data and by the Directive 2002/58/EC on privacy and electronic communications. The only data which are not "personal data" and therefore are outside the field of the application of the two directives are "anonymous data".

To better define anonymous data according to the Directive, it is necessary to start from the definition of personal data. Personal data are defined in Article 2 of the EU Directive 95/46 as "any information relating to an identified or identifiable natural person ('data subject'); an identifiable person is one who can be identified, directly or indirectly, in particular by reference to an identification number or to one or more factors specific to his physical, physiological, mental, economic, cultural or social identity". It is important to remember that the interpretations of the data subject vary in different Member States of the European Union. Section 46 of the EU Directive 2002/58 specifies that the Directive 95/46/EC covers any form of processing of personal data regardless of the technology used. The key to the definition is constituted by the possibility to identify the data subject, to connect data to a subject. Data can be considered personal if they make a subject identifiable, distinguished from others of the same group, without necessarily referring to his/her name. The subject does not need to be directly identified, but has to be identifiable, i.e. "when, although the person has not been identified yet, it is possible to do it" (p.12). This could mean that the subject can be identified through some characteristics or aggregation of data.

The definition of "personal data" given by the Directive is the starting point to define "anonymous data". The definition of "anonymous data" is not found in the two EU Directives on data protection. However, there is a reference to anonymous data in the "whereas 26" providing that "(...) to determine whether a person is identifiable, account should be taken of all the means likely reasonably to be used either by the controller or by any other person to identify the said person; whereas the principles of protection shall not apply to data rendered anonymous in such a way that the data subject is no longer identifiable; (...)".

Further references to anonymous data and especially to anonymization have been provided in the Directive 58/2002 (see whereas 9, 26, 28, 33 and art. 6, 9). In particular, "whereas 9" states the principle of minimization providing that "the Member States, providers and users concerned, together with the competent Community bodies, should cooperate in introducing and developing relevant technologies where they are necessary to apply the guarantees provided by the Directive and taking particular account of the objectives of minimizing the processing of personal data and of using anonymous or pseudonymous data where possible" and whereas 30 instructs that "Systems for the provision of electronic communications networks and services should be designed to limit the amount of personal data necessary to a strict minimum."

On the definition of anonymous data, the Directive has been interpreted in different ways by different European Member States. The definition of "anonymous data" is present in some national legislations, for instance, in Italy and Germany. The Italian Data Protection Code, adopted by Legislative Decree 30 June 2003, n. 196, in force since January 1, 2004 explicitly defines "anonymous data" as being data, which in origin or after being processed "cannot be associated with an identified or identifiable data subject". Data can be originally anonymous or processed to be rendered anonymous. The German Federal Data Protection Act, as of 15 November 2006, Art. 3 (6) provides that: "'rendering anonymous'" denotes the modification of personal data so that the information concerning personal or material circumstances can no longer or only with a disproportionate amount of time, expense and labor be attribute identifiable individual[2]." Kuner (2007) states that in some local German data protection laws, there is an absolute requirement that it must be impossible for anonymized data to be traced back to an identifiable person.

According to the Opinion 4/2007 on the concept of personal data adopted by Art. 29 Working Party[3], "'anonymous data' can be defined as any information relating to a natural person where the person cannot be identified, whether by the data controller or by any other person, taking account of all the means likely reasonably to be used either by the controller or by any other person to identify that individual" (p.21). The Opinion explicitly refers to a "case-by-case" analysis (p.21) which should be carried out with reference to the means of possible identification as well as: the cost of conducting identification, the intended purpose, the way the processing is structured, the advantage expected by the controller, the interests at stake for the individuals, as well as the risk of organizational dysfunctions (e.g. breaches of confidentiality duties) and technical failures (p.15).

ANONYMITY AND SECURITY ENVIRONMENT

Now, let us determine a more general relationship between requirements for security and anonymity. Similarly to the earlier discussion of personal and anonymous data, we can start the discussion by reiterating several approaches to privacy and anonymity[4] and their relation to security. As was mentioned before, anonymity or pseudonymity of data are considered one of the ways to ensure the protection of the personal data. In the discussion below, we do not attempt to provide a general view on privacy and anonymity, but only detail the characteristics that are relevant to the illustration below, namely, the influence of changes in protocols' design, as exemplified by the privacy and anonymity-friendly features of identifiers, can increase the level of anonymity in resulting systems.

One view, not uncommon in the security community, is that security and anonymity are orthogonal: greater security necessarily demands concessions in the level of anonymity that can be offered (Peissl, Raguse, Meints, & Langfeldt, 2008). This view is in part supported by those working in the area of trust establishment: indeed, in order to establish trust (and make the environment more secure as a result), it is necessary to abandon some level of anonymity by providing more information about one's system (Vishik, Johnson, & Hoffman, 2007). Technologies are now being developed to mitigate the trade-offs in the level of anonymity in mature trust establishment systems. Jackson et al. (2007) or Camenisch and Gross (2008) offer examples of valid approaches in this area.

Another view is that security and anonymity are different: one focuses on protecting digital assets, their movements, and their storage, while the other on safeguarding personal and near-personal data from re-identification (Sweeney, 2007). It is hard to disagree with this opinion, and the references to data minimization in European law demonstrate that the issue is understood by regulators. But one may claim, in relation to data movement within "chains of devices" that characteristics of anonymity-supporting systems are also very important.

Finally, a common view is that the relationship between anonymity and security is context-dependent and has to be evaluated on a case-by-case basis (Gandon & Sadeh, 2004), which, while avoiding strict definitions of anonymity, permits the researchers to take the same approach as 2007/4 Opinion of Art. 29 Working Party referenced above.

We think that the two last approaches as well as remedies developed in conjunction with approach 1 can be combined for a comprehensive analysis of anonymity and its relationship with security. The relationship between security and anonymity in complex computing environments is important because it helps us establish a broader view on anonymity that is beneficial for all the participants in the digital economy, from regulators and technology developers to technology users. Understanding this relationship can enable technologists to support anonymity without sacrificing security, can help consumer advocates explain varying degrees of anonymity obtainable by different available techniques, and can assist regulators in defining approaches to data protection that reflect the advances in today's technologies.

Properties of Identifiers Relevant to Anonymity

Let us look at the characteristics of identifiers that could facilitate a greater degree of anonymity in "chains of devices." We already determined that some devices have limited technological means to ensure anonymity because of the limited processing power and lack of infrastructure to support connection of data confidentiality. Table 1

Table 1. Some features of identifiers

Feature	*Description*	*Ideal Identifier*[5]
Ability to identify	Ability of the ID, or its easily obtainable combinations with other information, to reliably identify an individual or an object with sole ownership by an individual	Has no connection with the individual or easily attributable object in the identifier
Mutability	Frequency of change	Changes very frequently, sometimes used only once
Transferability	Extent to with which the identifier can be transferred to a different object or can be applied to a large group of objects	Can be transferred from object to object with ease
Assignability	Ease of passing the object with identifier to other individuals	Can be easily transferred to other individuals or attributable objects
Accessibility/user control	Ease with which the identifier can be accessed	User exercises control over who/what can access the identifier

describes features of identifiers designed in a way that makes their links to objects and individuals they denote harder to capture.

While incorporating these and similar features into all new protocols and devices is impossible due to technology limitations, some of these characteristics have become common. Dynamic identifiers are ubiquitous, and many applications use random combinations of characters as account IDs or session IDs. Ideally, components of these features can be embedded in protocols to offer, by default, a degree of anonymity and privacy to processes as illustrated in Figure 3. Some of the features (e.g., transferability or mutability) can be supported by new processes and new usage and business models, in addition to technology.

Other methods to add anonymity to identifiers and protocols that they serve have been designed and adopted. Some of them are listed below.

- If the identifier is a cryptographic key, the key hierarchy can be designed in a way to ensure that no mathematical relationship exists between the key identifying the device and the key used in transactions and protocols. This approach is used in Trusted Computing (see. Scarlata et al., 2007) and other disciplines.
- During the process of trust establishment, the designers can ensure that no identifiers are transmitted as part of an attestation or trust establishment processes; instead, if the attestation test has been passed successfully, an affirmative answer can be delivered as part of the protocol.
- Authentication can be designed in a way that only verification is required as opposed to full identification. In this model, the subject is only identified as a member of a group, and his/her membership is sufficient for access to systems (see, e.g. Camenisch et al., 2003). This approach is consistent with the concept of relative anonymity.

Although complete solutions for anonymity have other components in addition to privacy-and anonymity-friendly identifiers, "anonymity by design" is impossible without paying attention to identifiers and their usage, especially when assessing levels of anonymity in complex system. Using IDs as one of the ways to leverage varying degrees of anonymity necessary to support diverse digital processes is an approach that can potentially be applied to most classes of devices and processes.

Figure 3. Characteristics of identifiers

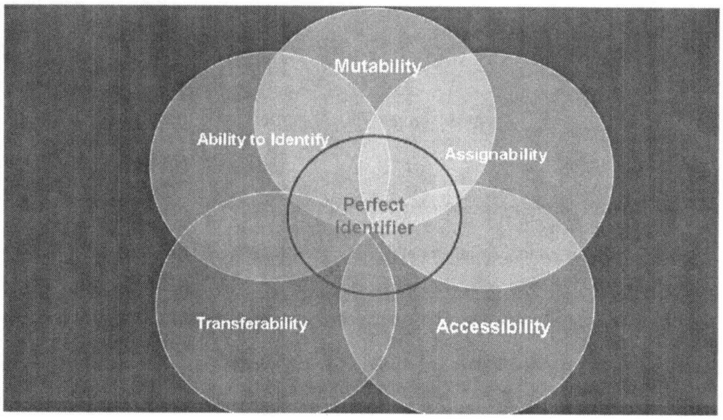

THE DEFINITION OF ANONYMOUS DATA: ABSOLUTE OR RELATIVE?

In view of ubiquitous connectivity, a new analysis of anonymity and anonymous data is necessary. The key concept both in the definition of personal data and in the definition of anonymous data is referability. If information may be referred to a subject, then data are not anonymous but personal: therefore personal data protection legislation applies (i.e. information notice, consent, security measures, etc.).

The crucial question is whether "referability" and therefore anonymity need to be considered absolute or relative. Referability of information to the data subject is measured in relation to the time, cost and technical means necessary to achieve it. The value and sensitivity of the information should also be taken into account. For example, healthcare-related data or tax related data require a higher level of protection. According with the Recommendation of the Council of Europe No. R (97) 5 on medical data protection, information cannot be considered identifiable if identification requires an unreasonable amount of time and manpower. In cases where the individual is not identifiable, data are considered anonymous. Opinion 4/2007 adopted by Art. 29 Working Party states that in order to assess if the subject is identifiable or not, it is necessary to take account of all the means likely and reasonably to be used either by the controller or by any other person to identify that individual.

In both acts mentioned above, the concept used to assess referability is reasonableness. It seems quite clear that anonymity is evaluated as connected to an (un)reasonable amount of time and manpower in the Recommendation of the Council of Europe No. R (97) 5 and as connected to all the means likely reasonably to be used in the Opinion 4/2007 of Art. 29 Working Party. Based on this framework, anonymity appears to be relative and needs to be evaluated case by case, in context. It depends, as mentioned in the Opinion 4/2007, on the means at the disposal of the data controller or any other subject intending to attempt re-identification. Moreover, information can be anonymous for one person and not for another, anonymous for one data controller, but not for another: anonymity is relative in this sense also. Reasonableness is the criterion, on which the referability of information to a person is based. If there is no a possible connection, anonymity is absolute.

Absolute anonymity is particularly difficult to achieve in the digital world, where anonymized data can be re-identified under some circumstances, and re-identification is still possible if the message crosses multiple devices and networks. In this connection, we can state that absolute anonymity is not achievable when the communication is electronic.

To summarize, if there is a possible connection with the originator, in the present or in the future, then anonymity is relative. Anonymity is measured through the criterion of reasonableness, which is not a civil law criterion, but originally a common law principle. Therefore, we can say that there is a graduation of anonymity as well as a graduation of referability. Introducing the "reasonableness" in the definition of anonymity means that we consider anonymity as "relative". We think that anonymity in processing of personal data is not an absolute concept: it is, on the other hand, a relative and functional concept, measured through reasonableness. Data can be anonymous for some, but not for others: relativity is therefore central to the definition. Likewise for functionality, data can be anonymous for certain uses, but not others. Whether anonymity exists or not, is measured through the criterion of reasonableness with a case by case approach. Therefore there are degrees of anonymity as well as degrees of referability. In summary, it is not correct to describe anonymity as a black or white (absolute) concept.

DEGREES OF ANONYMITY, AUDIENCES, AND DATA CONTROLLERS

Let us illustrate varying degrees of anonymity for different data controllers and audiences using the scenario provided in Figure 4. A user adds anonymous comments on an article in an electronic newspaper from a smart phone. His/her level of anonymity varies with respect to various audiences/data controllers that own or use components of the networks or computer systems as the message travels to its destination as illustrated, in a simplified way, in Table 2. The Web site operators know more about the poster, as he/she needed to sign up to be allowed to post comments, so the anonymity is not complete with this data controller. The IP network operator also has some information about the accessing device that can be analyzed in cases of violations as permitted by law, leading to incomplete anonymity for the user in this segment. Finally, the mobile operator can find out the subscriber information for the transaction if required, and therefore the user's level of anonymity is low. Only with the readers of the comments does the user achieve a complete or almost complete level of anonymity.

Technologies to support anonymity have existed for more than 25 years. David Chaum's mix networks were invented in the early 80s (Chaum, 1981) and demonstrated implementable approaches based on chains of proxy servers or re-encryption of messages that continue to be used today. However, the scenario above and its simplified analysis below indicate that, in a more complex environment that is typical today, new technologies (and new user expectations) are required.

Data containing PII (personally identifiable information) are frequently anonymized, as required by law in many countries, and there are techniques and approaches to measure the level of anonymity in static databases as well as the likelihood for data leakage (breach of anonymity). Kelly, Raines, Grimaila, Baldwin, and Mullins (2008) acknowledge that

An ability to confidently measure this information leakage and any changes in anonymity levels plays a crucial role in facilitating the free-flow of cross-organizational network data sharing and promoting wider adoption of anonymization techniques (p.31).

Although several methods of analyzing anonymity exist, such as Anonymity Set Size (ASS), k-anonymity, Individual Anonymity Degree (IAD), or Entropy Anonymity Degree, it is not easy to apply these methodologies to complex "chains of devices" illustrated above where levels of anonymity are frequently dynamic, as opposed to the

Figure 4. Submitting an anonymous Web comment from a smart phone

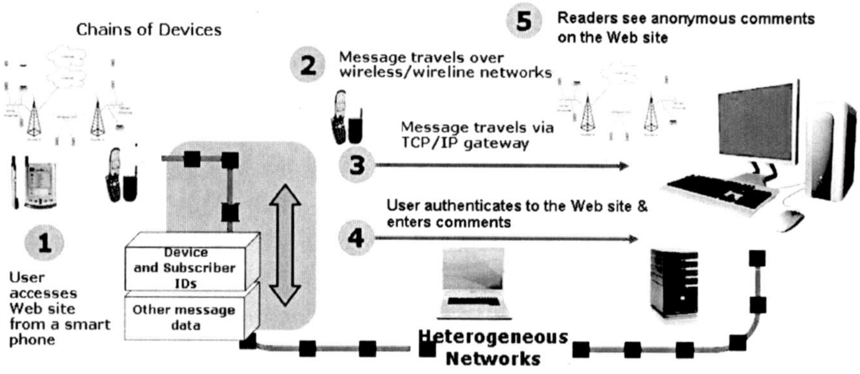

measurements in static databases. In addition to the metrics referenced above, specific techniques have been developed in order to address, e.g., network anonymity metrics, including Combinatorial Anonymity Degree (CAD) or Evidence Theory Anonymity (ETA). These approaches, while very helpful, do not provide universally applicable analysis tools for anonymity metrics. The field of anonymity is active and has supplied answers to many new questions related to anonymity in evolving environments, but more work is required to develop comprehensive techniques to measure anonymity in a way that can inform legal views on anonymity and anonymous data. As we pointed out earlier, if re-identification requires an unreasonable amount of time and manpower, European law tends to consider such data anonymous.

New tests need to be developed to measure "degrees of anonymity" in complex computing environments. While research and standardization activities concentrate on defining policy languages and policy frameworks that can be applied to both anonymity and privacy, much of the work focuses on semantic definitions of policies and analysis of the compliance of the providers with the high quality policies that these providers have developed. This focus led to the establishment of excellent best practices and standards, such as P3P[6] or XACML[7]. However, this approach has not yet led to the development of technologies to support machine-to-machine enforcement of elements of anonymity. When success in automatic policy enforcement is achieved, related areas of research and technology, such as user defined

Table 2. Levels of anonymity in scenario on Figure 4.

Mobile Service Provider (I)	*Telephony Operators (II)*	*TCP/IP Networks (III)*	*Web Site Owner (IV)*	*Electronic Newspaper Readers (V)*
The mobile service provider knows, via subscriber and device IDs, the exact origination of the message and can establish the location of the origination.	The operator has access to wireless IDs and context IDs, but is unable to establish exact origination without other data.	Similar to II: enough information for traffic analysis, but not enough for attribution without additional information	Depending on conditions of registration, similar to I or similar to II and III. E.g., if registration is associated with payment, the level of anonymity will be the same as in I.	For the readers of the comments, the user is likely to possess complete anonymity.
User's level of anonymity: 0%	User's level of anonymity: intermediate	User's level of anonymity: intermediate	From 0% to intermediate.	Close to 100% anonymity

and user controlled privacy and anonymity, will also take off. When examining issues associated with complex processes and "chains of devices," we discover that the analysis is almost always context-dependent. In general, it is easier to distinguish between personal or anonymous data when describing an individual and his/her direct actions. IDs associated with devices used by individuals are harder to classify especially if they constitute a removable and mobile element of a complex environment.

In today's interconnected environments, we can assert that anonymized and minimized data can frequently be traced to the source by using information from associated devices and networks. However, such tracing is complex and time consuming because of different data formats, the fact that some data are not stored for longer periods of time, and that data belong to different data controllers. Sharing data between controllers is not always permitted, further complicating re-identification. Re-identification can be performed either by the controller or by any other person or entity. The identity of the party performing re-identification is an important evaluation criterion, and it is important to distinguish whether all data are under the single or multiple data controllers. In multi-device multi-network processes examined earlier in the paper, elements of data typically fall under different data controllers that are likely to possess varying degrees of insight into the identity of the originator of the message (see Table 2).

The cost of conducting identification is, in an average case, significant, as is the cost of preserving anonymity. The Opinion from 4/2007 (see above) advises to approach the issues of anonymity on a case by case basis. This is important guidance for analyzing the environments described in this paper because "chains of devices" are very diverse, and methods of anonymization and protection are limited in some links. At the beginning of the "chain of devices" as well as in other circumstances, anonymization is not always an option, since identification of a device is necessary for performing the primary function of the technology. In many devices and protocols, eliminating or obfuscating potentially attributable identifiers can result in the inability to use the tool, system, or protocol for its primary intended purpose. A simple medical device (e.g., blood pressure meter) needs to have an identifier to connect the data it transfers to a certain patient or certain medical procedure. A wireless pedometer needs an ID in order to enable the device to connect to a "smarter" tool, such as an iPod or a PDA. A mobile phone requires an IMEI number (device number) in order to connect to the mobile network as defined by applicable standards. A MAC address is necessary to locate the device, to which the connection request has been addressed. These protocol and access requirements lead to different levels of anonymity at different stages of a process as illustrated below.

When device identification is used to perform a primary technological function that benefits a user, with no aggregation and analysis of data involved, with data belonging to multiple controllers, it is likely that the device data could be considered anonymous. But we need to understand that the data are not anonymous to some of the data controllers, that the user's anonymity, in this case, is relative. Opinion 4/2007 of the Article 29 Working Party points out that the tests applied to determine anonymity must be dynamic and consider the state of the art in technology at the time of the processing and the possibilities for development during the period for which the data will be processed. Therefore, if the evolution of technology could lead to future re-identification, this eventuality has to be considered, and additional measures to protect acceptable levels of anonymity need to be adopted. This recommendation points to important research directions for this area located at the intersection of technology innovation and studies of legal frameworks.

FUTURE TRENDS AND RECOMMENDATIONS

Anonymity has become a rare privilege unless we treat it as a relative concept when various degrees of anonymity are possible. The graduation of anonymity can be measured or evaluated, and different classes of anonymous data can be identified.

As identity can be completely defined only in context, as a crossing point of many aspects of information (Mordini), so anonymity is also contextual. From identity to anonymity there is a continuum of information about the subject. The concept of relative anonymity can be used as a guideline to design new technology and develop realistic requirements for data protection. The evolution of technology has made it necessary to move from the original approach of the protection of computers to a new approach of protection of connected environments. It is not effective to impose the same approach used for fixed computers to the mobile and diverse environments of today. New models are needed, and they are already being developed by researchers and technologists.

Can we measure anonymity? A possible quantification of anonymity has been studied by computer scientists, for example with the studies on k-anonymity (Giannotti & Pedreschi, 2008; Sweeney, 2002; Vaidya, Clifton, & Zhu, 2006) or the development of ASS and IAD methodologies referenced above. Additional studies have been conducted on re-identification of data (Sweeney, 2002). The recent study at the University of Washington examined the levels of anonymity afforded by connected devices with limited computing power that cannot support additional security or privacy functionality (Saponas et al., 2007). It is necessary for legal researchers to take into account the results of such studies and work with technologists to develop new methodologies focusing on the measurement of "degrees" of anonymity for modern dynamic computing environments. The evaluation of relative anonymity is very complex, as different types of data cannot be considered equivalent (e.g. healthcare information versus session information). Other criteria for such an evaluation have been already mentioned by the Opinion 4/2007: the cost of conducting identification, its intended purpose, the way the processing is structured, analysis options available for various data controllers, interests at stake for the individuals, as well as the risk of organizational dysfunctions (e.g. breaches of confidentiality duties) and technical failures (p.15). The exchange of various not directly attributable identifiers among devices or between a device and a network, under these guidelines, can guarantee a certain degree of anonymity, varying at different stages of the electronic process.

As pointed out in the Opinion 4/2007 issued by Art. 29 Data Protection Working Party, the test on identifiability is a dynamic concept and should consider the state of the art level in technology at the time of processing. We propose to explicitly introduce the "reasonableness" in the definition, as the directive does implicitly, considering anonymity "relative": the relativity should be, in our opinion, measured through reasonableness. This approach permits us to measure different degrees of anonymity and is beneficial for the technology users, making it possible to set expectations for the true level of anonymity available to them in complex systems. Relative anonymity could be the basis for a more effective control the users exercise for their data. The data subject can be informed about the true level of anonymity of his/her data and become aware of the protection given, while the data controllers can support the level of user anonymity that is expected by the users. We can think of quantifying the degrees of anonymity, associating a numeric value to reasonableness and making reasonableness more objective. The quantification could be as simple as assuming that the value is 0 when identity is known to the data controller (e.g., mobile operator in Table 2) or that the value is 1 when it is reasonable to believe that an audience doesn't know the identity of the user (e.g., readers of the comments in Table 2). This approach, if properly developed, could permit us to measure the level of anonymity in nodes of a complex system or process as well as across that system or process.

The ideal goal is that the users should control their data, both from a technological and a legal point of view, making informed decisions on data processing. In certain cases, anonymity should be balanced with the need for the government control and ability to re-identify the data subject as permitted by law: e.g., for security reasons or for judicial reasons.

A RIGHT TO ANONYMITY?

The last question is whether a right to anonymity exists.

Overall, the European laws do not provide a general and explicit right to anonymity and do not mention a general and explicit right for a subject to be anonymous, not to be identified, not to disclose or reveal themselves either in the digital or in the real world[8]. In general, the right to anonymity is not recognized by European legal systems: it has been stated with references to particular needs. Therefore, the picture is fragmentary. For instance, Italian copyright law provides for the right of an author to publish a work anonymously or with a pseudonym, if included in the contract with the publisher. The right to anonymity can be seen as instrumental to the effective exercise of the right to personal data protection or to the effective exercise of the right to privacy.

Anonymity can have many different characterizations from a legal point of view. Anonymity can be a choice of behavior, irrelevant from the legal point of view, but sometimes it is also a right. In a contract, there are no obstacles to anonymity or pseudonymity, unless the contract is a formal contract (deed) or the identity of the party is essential to the contract. The name in this case is connected to other interests: for instance, ensuring payment. From a theoretical point of view, a contract can be anonymously concluded and performed. In other cases, anonymity can be seen a right of the author: in this case it is instrumental to freedom of expression. In still other cases, anonymity is denied as it is in criminal law, with a particular reference to antiterrorism legislation.

In general, we can affirm that anonymity is not recognized as a general right: it can be a right only in particular cases. It appears to be an extreme way to exercise the right to personal data protection. The right to personal data protection is a right different from the right to privacy. The first one is the right to control personal data. The second one is the right to exclude somebody from knowing personal data.

The right to personal data protection is the right of a subject to exercise control over information regarding themselves: including the right to obscure/hide information: in this sense, anonymity is a way, the strongest way, to exercise the right to personal data protection.

After the European legislation on personal data protection clarified and made more relevant the value of information, the value of anonymity has also increased. In general, in the information society, anonymity is important. It is seen not only as a way of protecting privacy, but as a way of controlling data and information. It can be a way of exercising a positive liberty, i.e. a form of control, as opposed to a form of negative liberty, i.e. the power of excluding others. Anonymity should be seen as a space of liberty in a world where it is almost impossible not to leave digital traces.

In the European Union Charter of Fundamental Rights, the right to personal data protection is defined as a liberty. As the means of controlling individuals increase, anonymity becomes more and more a space of liberty. However, being a form of exercising the right to data protection and the right to privacy, it has to be balanced with other fundamental rights.

CONCLUSION

The idea of anonymity as a relative concept is not easy to accept. The origin of the term and cultural influences induce us to treat anonymity as absolute. However, the relativity of anonymity is mainly a consequence of the deployment of technology. Accepting anonymity as relative helps develop solutions for multi-protocol multi-device environments and processes and for the digital world in general, including new business models that can maintain a higher level of anonymity for those who require it and where allowed by law.

In the past, a person could be anonymous by choice. Today, technology makes it more and more difficult to remain anonymous. Electronic activities always leave a trace: using mobile phones, paying with a credit card, using the PC or a PDA, accessing online services, simply surfing the Web, listening to the i-pod, exercising on electronically enabled equipment is never 100% anonymous.

From a theoretical point of view, it is difficult, if not impossible, to imagine data that are absolutely anonymous for all data controllers and audiences and not re-identifiable under any circumstances. The digital world is characterized by ubiquitous connectivity. Connectivity is frequently not a goal, but simply an ordinary instrument, the normal way to act. Complete anonymity is at variance with the environment where ubiquitous connectivity is normal. But relative anonymity is both possible and acceptable.

In this context, it would be interesting and important to study the level of anonymity that an average user in different contexts considers acceptable as well as the need for business models that permit a user either to regulate, using various technologies, the level of anonymity in transactions, or support a higher level of anonymity. Except for the legal aspects of anonymity and the technical means to ensure or measure anonymity with some (but not all) data controllers and data types, little research exists in this area. More is needed, to adjust our expectations of anonymity to the realities of the digital world. Defining new models and techniques to measure anonymity through reasonableness is the future challenge that is ready to be addressed.

The effectiveness of the right to data protection and of the right to anonymity, where this right is possible, can be guaranteed through technology. Once the law has established fundamental principles and rules, technology has to implement them, enforcing, in this case, support for different levels of anonymity.

REFERENCES

Camenisch, J., & Groß, T. (2008). Efficient attributes for anonymous credentials. In *Proceedings of the 15th ACM Conference on Computer and Communications Security, Alexandria, VA, October 27 - 31, CCS '08*, 345-356. New York: ACM.

Camenisch, J., Groß, T., & Heydt-Benjamin, T. S. (2008). Rethinking accountable privacy supporting services: extended abstract. In *Proceedings of the 4th ACM Workshop on Digital Identity Management,* Alexandria, Virginia, USA, October 31 - 31.

Chaum, D. L. (1981). Untraceable electronic mail, return addresses, and digital pseudonyms. *Communications of the ACM, 24*(2), 84–90. doi:10.1145/358549.358563

Eijck, J v and Orzan, S. (2007). Epistemic Verification of Anonymity. *Electronic Notes in Theoretical Computer Science*, (168), 159-174.

Finocchiaro, G. (Ed.). (2008). *Diritto all'anonimato.* Padova: Cedam.

Gajek, S., Sadeghi, A. R., Stüble, C., & Winandy, M. (2007) Compartmented security for browsers or how to thwart a phisher with trusted computing. In *2nd Intl. Conference on Availability, Reliability and Security (ARES 2007*), 120-127.

Gandon, F. L., & Sadeh, N. M. (2004). Context-awareness, privacy and mobile access: a web semantic and multiagent approach. In *Proceedings of the 1st French-Speaking Conference on Mobility and Ubiquity Computing,* Nice, France, June 01 - 03, UbiMob '04, vol. 64. 123-130. New York: ACM. DOI= http://doi.acm.org/10.1145/1050873.1050901

Garcia, F. D., Hasuo, I., Pieters, W., & van Rossum, P. (2005). Provable anonymity. In *Proceedings of the 2005 ACM Workshop on Formal Methods in Security Engineering* (Fairfax, VA, USA, November 11 - 11, 2005).

Giannotti, F., & Pedreschi, D. (Eds.). (2008). *Mobility, Data Mining and Privacy*. Heidelberg: Springer. doi:10.1007/978-3-540-75177-9

Hu, Y., & Wang, H. (2005). A framework for location privacy in wireless networks . In *ACM SIGCOMM.* Asiba Workshop.

Jackson, C., Boneh, D., & Mitchell, J. (2007). Transaction generators: Root kits for web. In *2nd USENIX Workshop on Hot Topics in Security (HotSec '07).*

Kelly, D. J., Raines, R. A., Grimaila, M. R., Baldwin, R. O., & Mullins, B. E. (2008). A survey of state-of-the-art in anonymity metrics. In *Proceedings of the 1st ACM Workshop on Network Data Anonymization,* Alexandria, VA, October 31 - 31, (NDA '08) (pp. 31-40). New York: ACM.

Kerr, Ian (ed.). (2009) Lessons from the Identity Trail: Anonymity, Privacy, and Identity in a Networked Society, *Oxford University Press.*

Kuner, C. (2007). *European Data Protection Law: Corporate Regulation and Compliance*. Oxford, UK: Oxford University Press.

Nicoll, C.-P. Corien -Van Dellen, M. J. M. (eds.). (2003). Digital Anonymity and the Law, Cambridge University Press, TMC Asser Press,The Hague.

O'Neill, E., Kostakos, V., & Kindberg, T. gen. Schieck, T., Penn, A. et al. (2006). Instrumenting the city: Developing methods for observing and understanding the digital cityscape. In Ubicomp, 2006.

Peissl, W., Raguse, M., Meints, M., & Langfeldt, O. (2008). Criteria for privacy enhancing security technologies. In Deliverable 6.2 of the PRISE project, Vienna, May 2008.

Saponas, S., Lester, J., Hartung, C., Sameer, A., & Tadayoshi, K. (2007). Devices that tell on you: privacy in consumer ubiquitous computing. In *Proceedings of the 16th USENIX Security Symposium.* (Boston, MA, August 06 - 10, 2007). N. Provos, Ed. USENIX Association, Berkeley, CA, 1-16.

Sweeney, L. (2001). Information Explosion . In Zayatz, L., Doyle, P., Theeuwes, J., & Lane, J. (Eds.), *Confidentiality, Disclosure, and Data Access: Theory and Practical Applications for Statistical Agencies*. Washington, DC: Urban Institute.

Sweeney, L. (2007).*Technology Dialectics: Constructing Provably Appropriate Technology*. Retrieved October 7, 2008 from http://privacy.cs.cmu.edu/dataprivacy/projects/dialectics/index.html

Tillwick, H., & Olivier, M. (2005). Towards a framework for connection anonymity. In *Proceedings of the 2005 Annual Research Conference of the South African institute of Computer Scientists and information Technologists on IT Research in Developing Countries* (White River, South Africa, September 20 - 22, 2005). ACM International Conference Proceeding Series, vol. 150. South African Institute for Computer Scientists and Information Technologists, 113-122.

Vaidya, J., Clifton, C., & Zhu, M. (2006). *Privacy Preserving Data Mining*. Berlin: Springer.

Vishik, C., Johnson, S., & Hoffman, D. (2007). Infrastructure for Trusted Environment: In Search of a Solution. In ISSE/SECURE 2007 Securing Electronic Business Processes, 219-227. Wiesbaden, Germany: Vieweg.

Weis, S., Sarma, S., Rivest, R., & Engels, D. (2004). *Security and Privacy Aspects of Low-Cost Radio Frequency Identification Systems*. Security in Pervasive Computing.

Wilhelm, U. G., Staamann, S. M., & Buttyan, L. (2005). A pessimistic approach to trust in mobile agent platforms. *IEEE Internet Computing*, *4*(05), 40–48. doi:10.1109/4236.877485

Xian, H., & Feng, D. (2007). Protecting mobile agents' data using trusted computing technology. *Journal of Communication and Computer*, *4*(3), 44–51.

Yager, R. Entropy and Specificity in a Mathematical Theory of Evidence. (1983). *International Journal of General Systems*, (9), 249-260.

ENDNOTES

1. http://people.dsv.su.se/~jpalme/society/anonymity.html, retrieved January 27 2010
2. http://www.bfdi.bund.de/EN/DataProtectionActs/DataProtectionActs__node.html, retrieved January 27 2010
3. http://ec.europa.eu/justice_home/fsj/privacy/docs/wpdocs/2007/wp136_en.pdf, retrieved January 27 2010
4. In order to simplify the argument, we consider privacy and anonymity as congruent concepts, but they are not identical. Anonymity can help us achieve a maximum level of privacy.
5. An identifier that maximizes the property described in the row of the table.
6. Platform for Privacy Preferences, http://www.w3.org/P3P/. P3P 1.0 is a W3C recommendation, retrieved January 27 2010
7. eXtensible Access Control Mark-up Language, http://www.oasis-open.org/committees/tc_home.php?wg_abbrev=xacml, v.2 at OASIS Open, retrieved January 27 2010
8. See WALDEN, Nicoll, Chris-Prins, Corien -Van Dellen, M. J. M. (eds.), *Digital Anonymity and the Law*, Cambridge University Press, TMC Asser Press,The Hague, 2003

Chapter 10
Show Me My Way:
The Use of Human Spatio-Temporal Behaviour Patterns for Developing Ubiquitous Wayfinding Systems

Alexandra Millonig
Vienna University of Technology & Austrian Institute of Technology, Austria

Georg Gartner
Austrian Institute of Technology, Austria

ABSTRACT

The growing amount of ubiquitously available location based information offers great chances for the development of mobile pedestrian navigation services. Yet, providing spot-on real-time adequate information remains challenging, as relevant information needs to be filtered and group-related behaviour patterns and preferences are still insufficiently known. In this chapter, the authors focus on two aspects. Firstly, they consider the issue of how to comprehensively investigate human spatio-temporal behaviour patterns in order to enable the provision of customised navigation information, and secondly, they discuss potential impacts of ubiquitous wayfinding systems on human perception of space and possible resulting navigation behaviour modifications.

INTRODUCTION

Recent years have seen tremendous progress in the field of developing services, devices and networks for ubiquitous information provision. The idea to utilise rich and interactive content, with devices that are always at hand and over networks that can be accessed anywhere, is becoming increasingly popular. Although this ideal cannot yet be achieved entirely due to several limitations (e.g. limited network mobility or content capabilities), the ability to utilise "any information, anytime, anywhere" comes into reach. It can be expected that e.g. growing amounts of publicly accessible computers that interact with pervasive mobile devices such as mobile phones will enable ubiquitous information access at least in urban environments in the not too far distant future. This is especially interesting regarding the development of mobile wayfinding tools for pedestrians, which will be able to use location based information for richer and more efficient navigation support.

DOI: 10.4018/978-1-61520-769-5.ch010

However, the increasing amount of ubiquitously available environmental and navigational information aggravates effective information extraction for the user and necessitates the customisation of wayfinding instructions and additional environmental information to individual needs. The provision of information tailored to individual preferences and current context situations requires profound knowledge about pedestrian wayfinding behaviour and related determinants. This can be achieved by a comprehensive analysis of movement patterns and underlying preferences and individual attitudes. Recent developments have produced a vast amount of emerging technologies focusing on the collection and analysis of human movement data. One field of particular interest is the use of space by pedestrians, the related influence factors determining movement patterns, and the impact of information and navigation services to the perception of space and wayfinding behaviour. In this respect, our research focuses on the identification of behavioural types of pedestrian mobility styles, which include quantitative-statistical movement data as well as qualitative-interpretative data concerning lifestyle-related influence factors. The identification of core determinants and the classification and description of homogeneous classes of pedestrian motion patterns and type-related interest profiles provide the basis for developing personalised wayfinding services which can offer location-related information tailored to group-related interests and walking preferences.

In particular, two research questions are followed in this contribution. Firstly, we consider the question of how can a comprehensive investigation of human walking patterns and related influence factors (motives, preferences, habits) be used in order to develop customised navigation and information systems. Secondly, we discuss the issue of which impacts on human perception of space and spatio-temporal behaviour can be expected from the use of ubiquitous wayfinding support.

The chapter starts with an overview of previous research in the field of human spatio-temporal behaviour, focusing on influence factors identified in related studies, as well as applicable empirical methods in pedestrian route choice behaviour research. The second part introduces a currently ongoing research project focusing on the development of a model of pedestrian mobility styles describing spatial behaviour patterns and corresponding determinants. The project follows a synergetic approach comprising several complementary quantitative and qualitative empirical methods in order to comprehensively investigate human walking behaviour and related determinants. We present initial results from experiments conducted during the first of two consecutive phases of data collection and analysis in two shopping environments in Vienna, and discuss the suitability of the preliminary outcomes for application in emerging adaptive mobile navigation services for pedestrians. The chapter concludes with a critical discussion of potential impacts of ubiquitous navigation and information systems on human spatio-temporal behaviour.

RELEVANT INFLUENCE FACTORS ON HUMAN SPATIO-TEMPORAL BEHAVIOUR

Studies on environmental preference and route choice behaviour confirm that pedestrians prefer certain routes owing to their environmental qualities, such as relative quietness and greenery (Blivice, 1974). Being the most vulnerable traffic participants, pedestrians are very sensitive to any unfavourable conditions. Hence, in many cases route decisions are not just driven by simply minimising the distance that has to be covered. Usually, the choice of a specific route depends on multiple factors, like the task a user wants to perform, the actual environment and individual preferences associated with personal attitudes and lifestyles. Moreover, an individual user does not always act in the same way; the current choice of a specific route differs according to familiarity with the environment, daytime, weather conditions, actual fitness and many other factors.

This section focuses on the description of relevant influence factors in human spatio-temporal behaviour research: firstly, crucial determinants of route decision processes which have been identified in recent studies are discussed, and secondly we provide an overview about different empirical approaches aiming at the investigation of behaviour patterns and related influence factors using various empirical methods.

Influence Factors Determining Pedestrian Route Choice Behaviour

Many approaches concerning pedestrian behaviour modelling imply that the behaviour of pedestrians is usually determined by utility maximisation, whereas utility is merely defined by distance travelled: A person will therefore try to choose the "optimal" route (which is mainly considered to be the shortest path) among all alternatives to reach a specific destination (Helbing 1991; Joh, Arentze, Harry & Timmermans, 2001). However, this assumption fails to explain certain phenomena in human spatial behaviour, for instance the fact that a large amount of people take different routes to and from a particular place (Conroy Dalton, 2003; Golledge, 1995). Several findings have revealed differences in human navigational and route selection behaviour, suggesting that there seem to be various other factors influencing navigational decisions and walking patterns.

The assumption that individuals normally choose the optimal path is therefore conceded to be merely true for standard situations. Helbing (1991) states that non-optimal behaviour, and therefore differences in spatial behaviour, may also occur because the individual has either not learned the optimal strategy yet, or there may be some emotional or other reasons leading to suboptimal behaviour. Deviations from optimal behaviour (i.e. choosing the shortest path) may also be caused by the fact that an alternative path is more attractive (e.g. less noisy, nicer environment, less waiting time at traffic lights) (Helbing, Molnár, Farkas & Bolay, 2001; Millonig & Schechtner, 2007).

The utility function of a walking subject is obviously not solely defined by minimising the distance. Other features are also taken into account when choosing a walking route. Previous findings indicate that there are two main categories of behaviour influencing factors: internal factors (personal characteristics of individuals) and external factors (characteristics of the environment). Personal characteristics include for example socio-demographic attributes such as gender, age, health, etc. (Daamen & Hoogendoorn, 2003b) as well as culture, lifestyle, level of education, beliefs, and attitudes (Holden, 2000). External factors refer to characteristics of the trip (familiarity, trip length), properties of the infrastructure (type, attractiveness, shelter), and environmental characteristics (weather and other ambient conditions) (Daamen & Hoogendoorn). External factors can also be classified to several dimensions of preferred route qualities: physical (distance, acclivity), emotional (attractiveness, safety), and cognitive qualities (complexity, landmarks) (Millonig & Gartner, 2007).

Common routing services concentrate mainly on physical route qualities, like shortness or accessibility for handicapped people (i.e. avoidance of stairs, usage of elevators). Emotional route qualities, however, also have great importance when choosing a route to a desired destination, as attractiveness and safety possess significant influence on the decision whether to take or avoid a specific route. Cognitive route qualities play a crucial role in the decision process if a person navigates through an unfamiliar environment. The existence of salient landmarks and adequate markers which facilitate the wayfinding task and minimise the risk of getting lost are especially essential for individuals with limited navigation skills and increased fear of disorientation. In this respect, Golledge (1995) found "fewest turns" (simplest path) to be one of the most relevant factors influencing route planning strategies. Other factors include e.g. "least time", "most scenic/ aesthetic", or "different from previous (novelty)". Apart from all the previously mentioned factors influencing human route choice and route planning behaviour, Joh et al. (2001) point out that uncertainty should also be taken into account. Unexpected events may force individuals to reconsider previously scheduled activities, especially under time pressure.

Given the vast amount of factors affecting the spatial behaviour of individuals, it is not surprising that the spatio-temporal behaviour of pedestrians is still poorly understood. Nevertheless, there have already been several approaches to investigate and describe human spatial behaviour applying different empirical methods.

Empirical Techniques for Monitoring Spatio-Temporal Behaviour

Due to the vast amount of critical factors influencing pedestrian walking patterns and route choice behaviour, there only have been few efforts to observe and describe group-specific spatial behaviour so far. The research designs of previous studies predominantly concentrate on a small number of specific aspects, for example the spatio-temporal behaviour of certain age cohorts (Ahrend, 2002), the visible pedestrian movements within a specific area such as a particular urban square (Yan & Forsyth, 2005) or town district (Spek, 2007), the motion patterns of families living in a certain area (Spek, 2008), or the mobile behaviour of people acting within a specific context, like recreation (Götz, Loose, Schmied & Schubert, 2003), tourism (Hartmann, 1984; Shoval & Isaacson, 2007; O'Connor, Zerger & Itami, 2005) or shopping (Koike, Morimoto, Inoue & Kawano, 2003; Kitazawa, Zhao & Shibasaki, 2003).

The complexity of human walking patterns and spatial decision processes have led to the usage of numerous methods in order to monitor, interpret and analyse pedestrian spatial behaviour. Each of the techniques employed possesses its advantages and drawbacks. Many investigations concerning the collection of data on pedestrian route choice use questionnaire survey techniques. Such surveys represent one of the most important data collection techniques in transportation studies and provide detailed information concerning route decisions and individual habits, motives, and intentions (e.g. Blivice, 1974). Questionnaire surveys are relatively cheap, provide comparatively large samples, and allow the collection and analysis of data within a rather short time. However, as human behaviour is never fully determined by verbalised structures (Nisbett & Wilson, 1977), and people tend to adapt their answers – consciously or subconsciously – to what they expect to be socially desired behaviour (Esser, 1985), the accuracy of the results gathered from questionnaires may suffer.

Another frequently used technique in studies concerning human time-space patterns is the time-space budgets technique. This includes several methods: recall diaries in questionnaire or interview form, face-to-face interviews, and self-administered diaries (Shoval & Isaacson, 2007; Thornton, Williams & Shaw, 1997). Recall diaries and interviews can provide rich and detailed information; however, they are strongly dependant on the subject's memory, whereas self-administered diaries (which are written in real-time) require the subjects' cooperation to a very large extent.

While trip diaries and inquiries mainly aim at the investigation of motion behaviour from the agent's perspective, allowing the analysis of motivations and intentions underlying pedestrian activities, other empirical methods focus on the investigation and interpretation of the visible or measurable motion behaviour. A common method used in pedestrian behaviour surveys is direct observation, also known as behavioural mapping or "tracking". Participatory observation involves the observer taking part in the participant's activities, where the researcher can gain insight into the purposes determining the subject's decisions. Nonetheless, this technique faces issues similar to inquiry methods; being aware of the fact that they are under observation, participants may tailor their behaviour to the researcher's expectations.

Non-participatory, unobtrusive observations avoid the risk of such "observer effects". They have first been used for studies concerning the movement behaviour of visitors of museums and exhibitions (Hill, 1984). Tracking involves following the subject at a distance and recording her movements by drawing a line corresponding to the subject's activities on a map of the investigation area. This method yields information concerning the exact routes and activities of pedestrian in urban environments in time and space, while avoiding the risk of influencing the "natural" behaviour of participants (Hill, 1984; Keul & Kühberger, 1996; Keul & Kühberger, 1997). Yet, this technique is very time-consuming and labour intensive, and findings are limited to the observable activities of pedestrians; motives and intentions cannot be revealed.

At present, the majority of pedestrian monitoring studies already applies technological equipment for dataset generation. A common method is the use of video analysis to monitor and interpret pedestrian behaviour. The collection of accurate motion trajectories is essential for several specific purposes. Especially the development of agent-based simulation models requires the collection of real data for calibration and validation in order to confirm the accuracy of simulated human behaviour (Daamen & Hoogendoorn, 2003a; O'Connor et al., 2005). Most studies using video captured data, however, are limited to a very small observation field, although there are approaches observing a larger area by a network of several surveillance cameras (Millonig & Schechtner, 2006). Yet solely still visible behaviour can be investigated, leaving the subjects' intentions and motives as well as most other personal characteristics in the dark, albeit there are attempts to perform modelling approaches not at the level of visible patterns of motion, but at the level of intentions (Dee & Hogg, 2006).

Other technological methods to gather information on the spatial behaviour of pedestrians use digitally based localisation technologies to track individuals. These include for example satellite-based technologies (Global Positioning System, GPS) (Spek, 2008; Spek, 2007), land-based technologies (cell identification) (Ahas, Aasa, Silm & Tiru, 2007, Ray & Schrom-Feiertag, 2007; Svetsuk, 2007; Raento, Oulasvirta & Eagle, 2009), or hybrid solutions (Shoval & Isaacson, 2006). These tracking technologies are especially advantageous when gathering data within a very large study area. Yet, collecting localisation data with the help of tracking technologies can be of a rather invasive nature, if the participants have to be especially equipped with tracking devices, and therefore, again, observer effects may be suspected. On the other hand, the use of data gathered passively from private mobile phones without knowledge of their owners may pose various ethical questions. Apart from that, localisation techniques still are only able to describe observable motion behaviour and have to be complemented with other techniques in order to interpret and understand behaviour patterns.

Being aware of the limitations each method implies, there have been several approaches combining two or more techniques, in order to overcome their drawbacks and maximise their benefits. Examples are the development of activity-based transportation models by collecting data with the help of GPS enhanced self-administered diaries recorded on PDAs (Janssens, Hannes, & Wets, 2007), the combination of unobtrusive tracking methods and inquiries to analyse urban tourism

(Keul & Kühberger, 1996), or the study of tourist behaviour using video and behavioural mapping techniques (Hartmann, 1988).

THE IDENTIFICATION OF PEDESTRIAN MOBILITY STYLES AND THEIR IMPLEMENTATION IN UBIQUITOUS NAVIGATION AND INFORMATION SYSTEMS

In this section we introduce a currently ongoing project using a combination of several empirical methods for monitoring and analysing pedestrian spatio-temporal behaviour. The approach follows the concept of "across-method" triangulation (Jakob, 2001) in order to comprehensively study human spatial activities. Apart from monitoring movement behaviour patterns, we include underlying decision processes and preferences during the analysis. This approach has been chosen based on the idea that mobility is also about positioning within symbolic space, and the way and the places someone is moving is always a sign of social positioning (Götz et al., 2003). This follows the *habitus* concept coined by Pierre Bourdieu (Bourdieu, 1976); a system of durable and transposable *dispositions* (lasting, acquired schemes of perception, thought and action). Differing behaviour of people acting within the same context situation and following the same tasks is supposed to be caused by internal determinants, like individual preferences, habits and attitudes. Such durable schemes determine an individual's behaviour independently from current conditions and form the basis of individual information requirements. We assume that by taking these lifestyle-related factors into account, individual walking patterns and route decision preferences can be classified to create a typology of lifestyle-based pedestrian mobility styles. The typology can then serve as a basis to customise navigational and environmental information to individual needs and to create pedestrian interest profiles in ubiquitous environments.

Methodology and Empirical Set-Up

The study starts with two consecutive stages of empirical data collection; Figure 1 shows the study design including the specific methods we use in each empirical phase of the project. The selection of appropriate methods had to fulfil several conditions: Firstly, we aimed at collecting data of sufficient quality and accuracy in larger environments (indoor and outdoor). Secondly, as it is assumed that people might change their behaviour if they know that they are being observed, an unobtrusive form of monitoring was to be included. Thirdly, visible behaviour patterns were to be combined with interview data in order to allow the identification of relevant underlying intentions, preferences, and lifestyle-related factors. The following three methods have been chosen in order to achieve an optimal combination of empirical data collection techniques in consideration of these preconditions:

- *Unobtrusive Observation (Shadowing):* Observation of the "natural", uninfluenced spatio-temporal behaviour of pedestrians; only visible behaviour, no insight to intentions and motives.
- *Non-disguised Observation (Tracking):* Continuous observation over a long period in combination with standardised interviews (data from both the structural and the agent-centred perspective); observer effects possible.
- *Inquiry (Interviews):* Motivations, self-assessments of individual motion patterns; responses can be incorrect and constructed ex post.

The execution of unobtrusive and non-disguised forms of observation was not feasible in

Figure 1. The overview of the study design

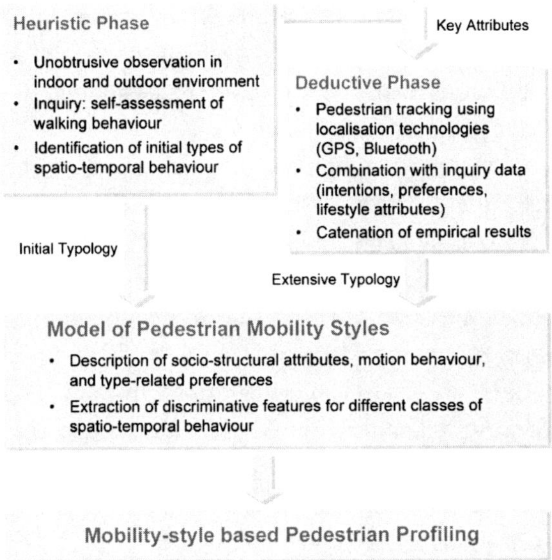

parallel. Hence, a two-step approach was designed, which also offers the chance of using preliminary results for ameliorating the methods and selected features of investigation in the second empirical phase. Additionally, differences in the behaviour of unobtrusively followed pedestrians and pedestrians knowingly participating in the study can be investigated.

Results of both empirical phases are subsequently related to each other, which leads to the development of a typology of pedestrian walking styles and underlying interests and preferences. Discriminative features will be determined and extracted in order to create pedestrian spatio-temporal profiles which can be used for customising spatial information in a mobile wayfinding system.

Following the assumption that the individual behaviour of a person is to a certain extent influenced by the context a person is acting within, all investigations were conducted in shopping environments (a shopping mall and a shopping street in Vienna). This is to minimise the risk of observing differences in behaviour that are largely caused by the influence of various context situations, as individuals in the shopping environments are assumed to be mainly acting within the same context and following similar tasks. The outcomes of the investigations will later be tested with regard to their validity in other context situations.

Data Collection

The unobtrusive observation method ("shadowing") applied in the first phase of the study consisted of random selection of an unaccompanied walking person and following the individual as long as possible while mapping her path on a digital map in order to observe the "natural", uninfluenced spatio-temporal behaviour of shoppers. As the shopping environments were usually quite crowded, subjects generally did not notice the presence of the observer following them. In fact only two observations had to be terminated because the observer had the impression a subject might have noticed the observation. However, this also resulted in the fact that most observations had to be terminated simply because the observer lost sight of the subject after a while. For a detailed description of the observation procedure see Millonig & Gartner (2007). The use of technology in this phase (digital map on a tablet PC, tracking software) offers mainly two major advantages: Firstly, a large investigation area can be covered without having to handle a large paper map, which is more practical and less noticeable, and secondly all points drawn in the map are recorded with time-stamps and map coordinates, which allows the calculation of average speeds and the detection of stops for each trajectory. In total trajectories of 111 individuals with a balanced gender and age ratio have been collected (57 observations on the shopping street, 54 in the shopping mall). Outdoor observations had an average length of 12 minutes (maximum: 62 minutes), indoor observations lasted approximately 16.5 minutes on average (maximum: 57 minutes).

The other method applied during the first empirical phase consisted of brief standardised interviews which have not been taken simultaneously, as a combination of both techniques could not be realised at this stage of the study. The questionnaire consisted of questions concerning socio-demographic attributes (age, gender, education, profession, etc.), goals and time budget for the current activities in the study site, frequency of visits, and questions referring to individual walking habits. Participants were asked to give a self-assessment of their preferences concerning walking behaviour and walking environments (e.g. rather "slow" or "fast", rather "exploring" or "goal-oriented", rather "bustling environments" or "calm environments"). In total, 130 individuals have been interviewed in this phase of the study. 100 interviews have been conducted in the indoor environment, but as pedestrians walking the shopping street were conspicuously less willing to give an interview, only 30 interviews have been given by visitors of the outdoor shopping area. The majority of the participants were female (61.5%). The sample possessed a balanced age distribution with an average age of around 36 years.

In the deductive phase of the project, pedestrians were tracked with the help of localisation technologies: Bluetooth for the indoor environment, GPS in the outdoor area. Participating individuals have been equipped with tracking devices and continuously recording their position, velocity and moving direction. After returning the device to the observer, participants had to complete a semi-standardised face-to-face-interview which explores current intentions, attitudes, and lifestyle and socio-structural attributes.

In the indoor environment, a Bluetooth tracking system has been installed in a major shopping mall, consisting of 49 passive Bluetooth beacons distributed over two floors in one part of the mall. The beacons acted as the located objects and were registered by a tracking software installed on a Bluetooth-enabled smartphone. Single walking visitors of the mall who agreed to participate in the study were equipped with those smartphones. Subsequently, while they followed their planned activities, the tracking software on the phone registered each beacon that had been passed by. In the outdoor environment, tracking datasets have been collected by GPS data loggers providing continuous recordings of positions, velocities and moving directions.

After completing their planned activities, the participants returned the devices (smartphones or GPS data loggers, respectively) to the observer and detailed semi-standardised interviews have been conducted. Customers who did not want or were not able to participate in the tracking study could also give an interview only. In total, 103 tracking and interview datasets have been collected (51 outdoor, 52 indoor); additionally 123 interviews have been conducted without collecting tracking data.

Analysis and Identification of Initial Types of Behaviour

At present datasets collected by shadowing and self-assessments interviews have been analysed and classified. The analysis of datasets collected during the deductive phase of the study has not yet been completed. The following subsections describe the results we obtained from performing multivariate analysis on the collected and normalised datasets from the heuristic phase of the project. Motion data were analysed with respect to velocities (histograms) and stops (frequency, duration, and position of stops), additionally visual attributes like appearance (dress style) and categories of visited facilities have been regarded. Interview data referring to individual walking preferences have been analysed in order to derive classes of self-reported subjective behaviour.

Results Based on Shadowing Datasets

The collected datasets have been analysed according to the velocity computed between each marked point in the observed path, additionally locations and durations of stops within the trajectory have been detected (cf. Figure 2). Subsequently, speed histograms of each trajectory have been compiled, showing the proportional amount of time an individual walked at a velocity within a specific time interval.

In order to gather initial classes of behaviour, the histograms of each investigation area have been classified using clustering algorithms (hierarchical clustering and k-means algorithm). In a second step, socio-demographic attributes and lifestyle-related factors (visual appearance, categories of visited facilities) have been investigated according to the individual cluster membership. We identified three discriminative clusters for the indoor observations, and eight clusters for the outdoor observations. However, half of the clusters derived from the outdoor analysis consist only of a very small number of people; hence these classes have only limited significance. The three homogeneous clusters resulting from the classification process for the indoor datasets contained 10, 14, and 30 individual observations per cluster. The cluster containing the largest number of cases (55.6% of all observed individuals in the shopping centre) shows a high percentage of time spent in front of or inside a shop (77.8%). Individuals falling in this category also walk at a lower speed on average than persons related to the other classes. The behaviour patterns in each of the three clusters can be interpreted as "swift shoppers", "convenient shoppers", and "passionate shoppers":

- *Swift shoppers.* The group contains mainly male participants and represents the youngest group with 50% of all subjects younger than 30 years. They walk at comparatively high speed and stop rarely, for a very short time and mainly at supermarkets.

Figure 2. Velocities and stops detected from a typical trajectory (outdoor area)

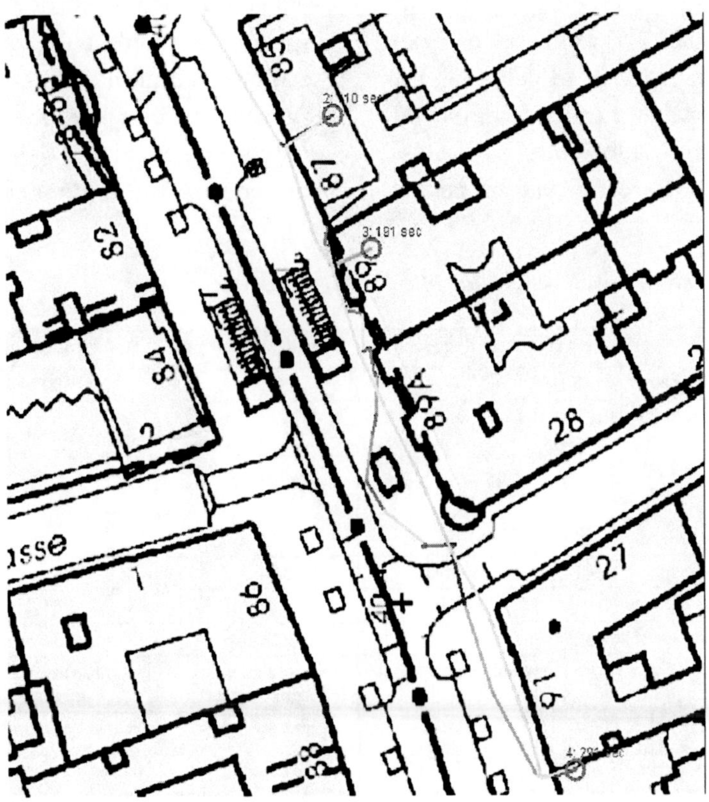

Apparently this group represents people following their daily routine who have little time or dislike extensive shopping.

- *Convenient shoppers.* This group includes almost two thirds of male subjects, the participants are aged around 30 to 35 years with 64% younger than 40. They stop rather shortly but moderately often. They walk slower than the individuals belonging to the first cluster. The behaviour patterns observed within this cluster appear to illustrate mainly male shopping behaviour: Men belonging to this cluster show interest in a greater variety of different shops and facilities than women, and they appear to complete their shopping tasks rather goal-oriented.
- *Passionate shoppers.* This group contains individuals who stop fairly often and comparatively long in front of or inside a shop or other facility. The great number of stops results in a very low average speed. Two thirds of the subjects belonging to this cluster are female. The estimated age lies beneath 35 years for most of them, but the cluster also contains a greater number of elder individuals. Apparently this cluster comprises a large amount of people who are quite keen on shopping or simply have enough time to spend (especially elder and probably retired persons), like to stroll around and to visit a large number of shops, and pause from time to time to have a refreshment. Most interest is attracted by fashion shops (clothing, shoes, and accessories) as well as shops offering culinary specialities.

The analysis of outdoor datasets produced three clusters showing high similarities with the three clusters identified in the indoor environment. Additionally, one further major cluster of specific behaviour has been identified as *Discerning Shoppers.* Like the "passionate shoppers" group, this cluster consists of mainly female subjects and shows a rather balanced age distribution. Individuals walk at comparatively high speed, but still stop comparatively frequently, although the stops are very short on average. Regarding the visited shops and facilities, a slight tendency towards specialised and exclusive shops is observed. Apparently people belonging to this group are goal-oriented and have a particular taste. In Figure 3 the clusters compiled from the indoor analysis are represented; Figure 4 shows the main clusters derived from the outdoor analysis.

Figure 3. Table of clusters based on indoor shadowing datasets

Cluster	*Swift Shoppers*	*Convenient Shoppers*	*Passionate Shoppers*
No. of subjects	10	14	30
Gender: female ■ male □	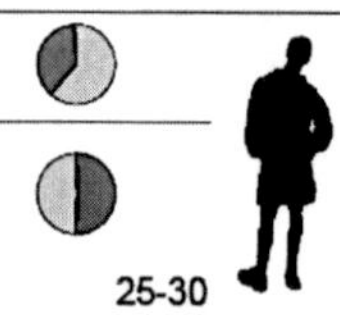	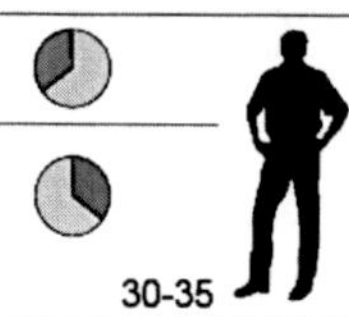	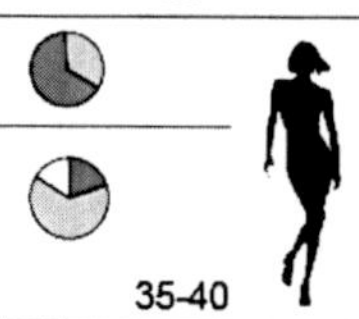
Age: < 30 ■ 30 - 60 □ > 60 □			
Average age	25-30	30-35	35-40
Average speed	1.19 m/s	0.61 m/s	0.24 m/s
Av. no. of stops (max.)	0.3 (3)	1.36 (2)	3.57 (13)
Av. duration of stops	6.97 s	2.58 min	4.66 min
Fashion style	casual or convenient	casual or conservative	casual/trendy or elegant
Visited shops/facilities	food store	no main focus	fashion, specialities, drugstore, bookshop

Results Based on Self-Assessment Interviews

The self-assessment profile interrogated in the interview contained 17 pairs of opposed attributes. To simplify classification processes, common data reduction methods have been applied in order to obtain a smaller number of variables. However, responses turned out to appear quite arbitrary: Although some of the items had very similar meanings, no significant correlation could be observed. This confirms the assumption that pedestrians have only little knowledge about their own behaviour and seem to guess what they prefer. For each of the observation areas the analysis resulted in 4 clusters. In general, the classes derived from the clustering process do not show very distinct patterns when regarding the reported self-assessment of walking preferences. This is presumably caused by the previously discussed limitations of interview data analysis. However, the resulting clusters from the indoor analysis show similarities with the four clusters resulting from the outdoor analysis. Two of the clusters also bear strong resemblances to two of the clusters derived from the classification of observation datasets. Self-assessment cluster 2 resembles to the "swift shoppers", and self-assessment cluster 4 reminds of the "passionate shoppers" cluster. Figure 5 shows the main results for each class compiled from classifying indoor and outdoor questionnaire datasets.

Although the use of questionnaires is very popular in investigating spatio-temporal behaviour and walking preferences, these results do not confirm the opinion that interviews alone belong to the most appropriate methods in pedestrian behaviour research.

Further Analysis and Ongoing Work

The preliminary results of the first empirical phase indicate that a number of homogenous behaviour patterns can be observed, especially in consistent context situations. Though, still further investigations concerning pedestrian behaviour in specific situations is required. As spatial behaviour is largely dependant on the context an individual is acting within, different context situations are believed to result in different behaviour patterns. Shadowing experiments which have been recently conducted in on a university campus in Asia show significantly different motion behaviour patterns than the results from the Viennese experiments.

Figure 4. Table of clusters based on outdoor shadowing datasets

Cluster	*Passionate Sh.*	*Convenient Sh.*	*Discerning Sh.*	*Swift Shoppers*
No. of subjects	7	15	13	14
Gender: female ■ male □				
Age: < 30 ■ 30 - 60 □ > 60 □ Average age	25-30	35-40	30-35	30-35
Average speed	0.25 m/s	0.63 m/s	1.04 m/s	1.35 m/s
Av. no. of stops (max.)	4.14 (11)	2.93 (7)	2.23 (5)	0.71 (4)
Av. duration of stops	4.43 min	1.83 min	32.97 s	11.34 s
Fashion style	convenient or casual	conservative/ elegant (f); convenient/ casual (m)	casual/ convenient or elegant/ proper	casual, trendy
Visited shops/facilities	fashion and accessories	no main focus	tendency to specialised/ exclusive shops	no main focus

Figure 5. Main characteristics of cluster results based on indoor and outdoor interview datasets

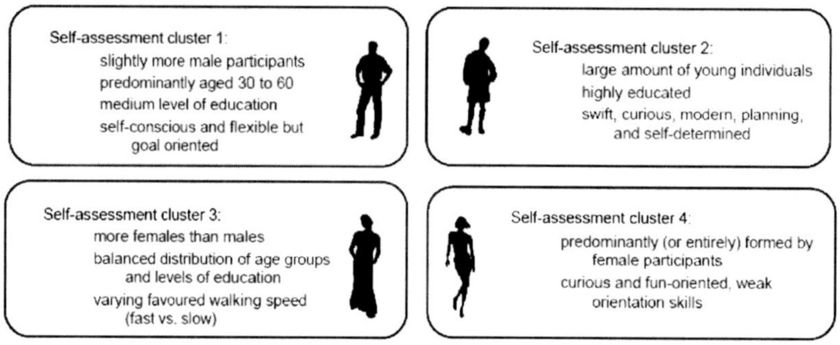

This is assumingly largely determined by the differing context situations.

At present datasets from more than 100 individuals collected during the second phase of the study, where participants have been equipped with tracking devices and interviews have been conducted after the tracking, are analysed. Results will disclose coherences between reported and actual behaviour and the outcomes will be compared with the results of previous analyses. In addition to the investigation of potential classes of behaviour, three main questions are currently considered in particular:

- To what extend do behaviour patterns and preferences reported in interviews correspond with observed behaviour?
- What are the differences in indoor and outdoor behaviour patterns?
- Are there differences in the behaviour of unobtrusively observed pedestrians and participants who know that they are being tracked?

Using Pedestrian Typologies for Customising Mobile Wayfinding Tools

The results derived from analysing data collecting during the heuristic phase of the presented study suggest that distinctive patterns of behaviour based on specific underlying, type-related intentions, attitudes and motivations can already be identified when using individual methods separately. However, the results lack important insight if not combined with complementary techniques. The combination of motion observation and interviews in the ongoing deductive phase of the study will lead to further progress in understanding human spatio-temporal behaviour. Several presuppositions could not be affirmed so far: Individuals seem to be less aware of the factors influencing their route decisions and walking preferences than expected, and context influences can not fully be avoided and barely identified when monitoring human motion behaviour even in assumingly unambiguous context situations (e.g. shopping centres).

Still, the preliminary outcomes indicate that human spatio-temporal behaviour can be sufficiently described by identifying mobility types which can be used as a basis for customising spatial information. Especially for specific situations, the outcomes of the experiments have the potential to form the basis for effective information retrieval in wayfinding services. The key characteristics of each identified type can be easily implemented in applications providing navigational and environmental information. Potential main application fields are shopping and tourism as well as recreation.

The allocation of a user to a specific ideal type by inquiring previously identified key attributes allows the consideration of specific preferences concerning the route choice and navigational information. Analogue to recommending systems used for internet users, navigational systems can learn from the user's behaviour and can be responsive to his individual preferences and requirements during the use of the navigation service. Additionally, collaborative filtering methods and type-related user-generated content can further improve the provision of information tailored to the specific needs of particular types of users.

POTENTIAL IMPACTS OF UBIQUITOUS INFORMATION SERVICES ON MOVEMENT PATTERNS AND THE PERCEPTION OF SPACE

Spatial experience can be obtained in two ways: Firstly through direct physical perception, and secondly through transferred experience communicated via different forms of representation. Communication of space and spatial information can occur in various ways, e.g. via verbal description, achievement of collaborative activities, or written or depictive representations like route descriptions or maps. The preferred form of knowledge transfer strongly affects the acquirement of mental representations of space, and hence influences wayfinding strategies. It can be expected that "anytime, anywhere" available navigational and environmental information will have significant effects on the way pedestrians experience and use the landscape.

Recent years have witnessed rapid advances in the enabling technologies for ubiquitous computing. Ubiquitous computing demands applications that are capable of recognising and adapting to the highly dynamic environments and of placing fewer demands on user's attention (Henricksen, Indulska & Rakotonirainy, 2002). As one type of ubiquitous computing Location Based Services (LBS) should be context-aware and adapt to the users behaviour as well as to the dynamic environment.

Mobile systems such as Location Based Services (LBS) seem to offer the ideal medium for assisting in getting to know an environment. However, several studies show that people who use such devices to interact with an environment perform poorly and acquire hardly any survey knowledge (Krüger, Aslan & Zimmer, 2004; Münzer, Zimmer, D., Schwalm, Baus & Aslan, 2006). In fact, in a recent study, Ishikawa, Fujiwara, Imai and Okabe (2008) showed that subjects using a mobile device that tracked their position and constantly updated this information on a map displayed on the screen of the device performed poorer in following a specific route than people using a paper map or having travelled the route once before. Even though mobile device users seemingly had the best context information (they could always see where they were along the route), they were slower and made more errors than the other subjects. Additionally, they considered the task to be harder than the subjects in the other two groups. There are two main reasons for the failure of current LBS in support of spatial learning, the lack of cognitive processing of spatial information; and the keyhole problem.

Since the LBS devices tell the users where they are and what to do (e.g., in case of navigation systems), the users do not fully process the presented information. This is comparable to the 'backseat driver problem'; people in a car who are driving themselves hardly pay any attention to the followed route. For reasons of cognitive efficiency, humans filter out information that they do not pay attention to. Information not deemed relevant is not integrated in mental representations, i.e., is not stored in memory (e.g. Craik & Lockhart, 1972; Smith & Vela, 2001). What is needed to improve on this problem are interfaces that, while assisting in the task, still force users to mentally process the information, and that adapt the presented information to the current state of knowledge and enhance the information with contents which are relevant to the user according to their interests (cf. also Schmid, 2007; Münzer & Stahl, 2007).

Mobile devices necessarily have small displays. Thus, it is hard to provide any overview (survey) information when using such devices as navigation assistance. This dilemma is termed the 'keyhole problem': it is either possible to display an overview of the surrounding area that is hardly readable as the individual features are too small, or focus on the immediate surrounding of the user. In the latter case, getting an overview (i.e. gain survey knowledge) is only possible by zooming operations (which, in the end, end up in unreadable information again), or by panning operations, which are often hard to control and where users have difficulties in keeping track of the context. The underlying issues of spatial cognition can basically be classified in two categories, the egocentric and the exocentric strategy of spatial cognition. Theories on egocentric spatial cognition deal with aspects of how humans are able to reconstruct a three-dimensional model from the two-dimensional image recorded by the eye. It is stated that three core directions are applied from an egocentric perspective (the vertical body axis for distinguishing above and below, the horizontal body axis for distinguishing behind and in front of and another horizontal body axis in 90° distance for distinguishing left and right (Schweizer & Horn, 2006; Herrmann, Schweizer, Janzen & Katz, 1998). Navigation instructions usually deal with the exocentric spatial cognition, where a point in space is used as origin from where the directions are given (Schweizer, 1997).

Spatial knowledge acquisition is needed to build mental representations which can be referred to for wayfinding. As Platzer (2005) and also Briggs (1973) pointed out, various methods of spatial knowledge acquisition can be used, including the sensual perception of the real world as well as the acquisition of spatial knowledge from models of the real world, such as maps. Another common way of building cognitive representations is the derivation of experiences from comparable situations. In this context Neisser (1976), Downs & Stea (1973), and Platzer (2005) pointed out that the active knowledge acquisition by human senses when moving through real space is resulting in more and better information. Whitaker & Whitaker (1972) and Platzer (2005) mentioned that personal characteristics have a big influence on spatial knowledge acquisition.

Research on the influence on various presentation forms on wayfinding (Radoczky, 2003; Dillemuth, 2007), on the usage of automated navigation systems, and on the influence of the size of displays (Dillemuth, 2007) have demonstrated that the focus is going beyond technical questions when developing navigation systems. There is ongoing work in this context and it can be stated, that mobile devices can be characterised as having small displays, restricting the possible area and size of displayed maps. This has a strong influence on spatial data acquisition and thus on the ability to orientate and navigate in real world (Gartner & Hiller, 2008).

CONCLUSION

Pedestrian spatio-temporal behaviour is determined by a great variety of largely insufficiently known influence factors. To be successful, mobile navigation services for pedestrians must regard the complexity of human wayfinding and route choice behaviour. Ubiquitous computing and smart environments can become the interface between the diversity of pedestrian navigation strategies and route choice behaviour in terms of combining technology (e.g. short-range sensory, peer-to-peer networks, heterogeneous devices) with collaborative filtering methods and user-generated content, as being discussed in this chapter. The overall goal, supporting the navigation process of pedestrians with customised information based on individual mobile behaviour, personal profiles and available ubiquitous infrastructure in the actual surroundings, is hence a matter of linking the relevant findings. Especially findings concerning potential unfavourable impacts such as limited support of spatial learning have to be considered at an early stage of development in order to create efficient and successful systems. The identification of group-specific information requirements is a promising approach for developing wayfinding applications offering higher value for particular types of pedestrians, setting the basis for supporting ubiquitous information for all users according to their individual needs.

REFERENCES

Ahas, R., Aasa, A., Silm, S., & Tiru, M. (2007). Mobile positioning data in tourism studies and monitoring: case study in Tartu, Estonia . In Sigala, M., Mich, L., & Murphy, J. (Eds.), *Information and Communication Technologies in Tourism 2007* (pp. 119–128). Vienna: Springer. doi:10.1007/978-3-211-69566-1_12

Ahrend, C. (2002). *Mobilitätsstrategien zehnjähriger Jungen und Mädchen als Grundlage städtischer Verkehrsplanung. Internationale Hochschulschriften* (*Vol. 381*). Berlin: Waxmann. (in German)

Blivice, S. (1974). *Pedestrian Route Choice: A Study of Walking to Work in Munich.* Unpublished Ph.D. Dissertation, University of Michigan.

Bourdieu, P. (1976). *Entwurf einer Theorie der Praxis auf der ethnologischen Grundlage der kabylischen Gesellschaft.* Frankfurt am Main, Germany: Suhrkamp. (in German)

Briggs, R. (1973). *Urban Cognitive Distance. Image and Environment* (pp. 361–388). Chicago: Aldine Publishing Co.

Conroy Dalton, R. (2003). The secret is to follow your nose: Route path selection and angularity. *Environment and Behavior, 35*(1), 107–131. doi:10.1177/0013916502238867

Craik, F. I. M., & Lockhart, R. S. (1972). Levels of Processing: A Framework for Memory Research. *Journal of Verbal Learning and Verbal Behavior, 11*, 671–684. doi:10.1016/S0022-5371(72)80001-X

Daamen, W., & Hoogendoorn, S. P. (2003a). *Experimental research on pedestrian walking behavior. Transportation Research Board annual meeting.* Washington, DC: National Academy Press.

Daamen, W., & Hoogendoorn, S. P. (2003b). Research on pedestrian traffic flows in the Netherlands. Proceedings Walk 21 IV, (pp. 101-117). Portland, OR: Walk 21 conference.

Dee, H. M. & Hogg, D. C. (2006). *Navigational strategies and surveillance.* Paper presented at the Sixth IEEE International Workshop on Visual Surveillance, Graz, Austria.

Dillemuth, J. (2007). Choosing the scale and extent of maps for navigation with mobile computing systems. *Journal of LBS, 1*(1), 46–62.

Downs, R., & Stea, D. (1973). *Cognitive Maps and Spatial Behavior: Process and Products. Image and Environment* (pp. 8–26). Chicago: Aldine Publishing Co.

Esser, H. (1985). *Befragtenverhalten als „rationales Handeln" – Zur Erklärung von Antwortverzerrungen in Interviews.* ZUMA-Arbeitsbericht Nr. 85/01 (in German).

Gartner, G., & Hiller, W. (2008). Impact of restricted display size on spatial knowledge acquisition in the context of pedestrian navigation . In Gartner, G., & Rehrl, K. (Eds.), *LBS and TeleCartography II.* Heidelberg: Springer.

Golledge, R. G. (1995). *Defining the criteria used in path selection (Technical Report UCTC No. 78)*. University of California Transportation Center.

Götz, K., Loose, W., Schmied, M., & Schubert, S. (2003). *Mobility Styles in Leisure Time*. Paper presented at the 10th International Conference on Travel Behaviour Research, Lucerne, Switzerland.

Hartmann, R. (1984). Ein „Touristentag" in der Stadt – Methodische Betrachtungen zum Städtetourismus, dargestellt am Beispiel der Europareisen junger nordamerikanischer Touristen. [in German]. *Zeitschrift fur Wirtschaftsgeographie*, *28*(3/4), 145–156.

Hartmann, R. (1988). Combining Field Methods in Tourism Research. *Annals of Tourism Research*, *15*, 88–105. doi:10.1016/0160-7383(88)90073-4

Helbing, D. (1991). A mathematical model for the behavior of pedestrians. *Behavioral Science*, *36*, 298–310. doi:10.1002/bs.3830360405

Helbing, D., Molnár, P., Farkas, I. J., & Bolay, K. (2001). Self-organizing pedestrian movement. *Environment and Planning. B, Planning & Design*, *28*, 361–383. doi:10.1068/b2697

Henricksen, K., Indulska, J., & Rakotonirainy, A. (2002). Modeling context information in pervasive computing systems. In F. Mattern & M. Naghshineh (Eds.), *Proceedings of 1st International Conference on Pervasive Computing* (LNCS 2414, pp. 167-180). Berlin: Springer.

Herrmann, T., Schweizer, K., Janzen, G., & Katz, S. (1998). Routen- und Überblickswissen – konzeptuelle Überlegungen. [in German]. *Kognitionswissenschaft*, *7*, 145–159. doi:10.1007/s001970050067

Hill, M. (1984). Stalking the Urban Pedestrian: A Comparison of Questionnaire and Tracking Methodologies for Behavioral Mapping in Large-Scale Environments. *Environment and Behavior*, *16*, 539–550. doi:10.1177/0013916584165001

Holden, A. (2000). *Environment and Tourism*. London: Routledge.

Ishikawa, T., Fujiwara, H., Imai, O., & Okabe, A. (2008). Wayfinding with a GPS-Based Mobile Navigation System: A Comparison with Maps and Direct Experience. *Journal of Environmental Psychology*, *28*, 74–82. doi:10.1016/j.jenvp.2007.09.002

Jakob, A. (2001). Möglichkeiten und Grenzen der Triangulation quantitativer und qualitativer Daten am Beispiel der (Re-) Konstruktion einer Typologie erwerbsbiographischer Sicherheitskonzepte. *Forum: Qualitative Social Research*, *2*(1). Retrieved April 5, 2009, from http://www.qualitative-research.net (in German)

Janssens, D., Hannes, E., & Wets, G. (2007). *Planning interventions in the interactions between individual activity patterns, patterns of functions and infrastructure*. Paper presented at Urbanism On Track – Expert meeting on the application in urban design and planning of GPS-based and other tracking-based research, Delft, The Netherlands.

Joh, C.-H., Arentze, T. A., Harry, J. P., & Timmermans, H. J. P. (2001). Understanding activity scheduling and rescheduling behaviour: Theory and numerical illustration. *GeoJournal*, *53*, 359–371. doi:10.1023/A:1020145214526

Keul, A &, Kühberger, A. (1996). *Die Strasse der Ameisen. Beobachtungen und Interviews zum Salzburger Städtetourismus*. München/Wien, Germany: Profil-Verlag (in German).

Keul, A., & Kühberger, A. (1997). Tracking the Salzburg Tourist. *Annals of Tourism Research, 24*, 1008–1012. doi:10.1016/S0160-7383(97)00038-8

Kitazawa, K., Zhao, H., & Shibasaki, R. (2003). A Study for Agent-based Modeling of Migration Behavior of Shoppers. In *Proceedings of the 8th International Conference on Computers in Urban Planning and Urban Management*, Sendai, Japan.

Koike, H., Morimoto, A., Inoue, T., & Kawano, T. (2003). Studies on the Characteristic Differences of Pedestrian Behaviors among Cities and Between City Districts such as Downtown and Suburban Shopping Centers [Portland, OR.]. *Walk, 21*, IV.

Krüger, A., Aslan, I., & Zimmer, H. (2004). The Effects of Mobile Pedestrian Navigation Systems on the Concurrent Acquisition of Route and Survey Knowledge . In Brewster, S., & Dunlop, M. (Eds.), *Mobile Computer Interaction -- MobileHCI 2004* (pp. 446–450). Berlin: Springer.

Millonig, A., & Gartner, G. (2007). *Monitoring Pedestrian Spatio-Temporal Behaviour* (Technical Report 42). Bremen, Germany: Universität Bremen, TZI Technologie-Zentrum Informatik.

Millonig, A. & Schechtner, K. (2006). Developing Landmark-based Pedestrian Navigation Systems. *IEEE Transactions on Intelligent Transport Systems (ITS), Special Selection on ITSC'05, 8*(1), 43–49.

Millonig, A., & Schechtner, K. (2007). Decision Loads and Route Qualities for Pedestrians – Key Requirements for the Design of Pedestrian Navigation Services . In Waldau, N., Gattermann, P., Knoflacher, H., & Schreckenberg, M. (Eds.), *Pedestrian and Evacuation Dynamics 2005* (pp. 109–118). Berlin: Springer. doi:10.1007/978-3-540-47064-9_10

Münzer, S., & Stahl, C. (2007). Providing Individual Route Instructions for Indoor Wayfinding in Complex, Multi-Level Buildings. In F. Probst & C. Keßler (Eds.), GI-Days 2007 Young Researchers Forum (pp. 241-246). Münster, Germany: ifgi Print series.

Münzer, S., Zimmer, H. D., Schwalm, M., Baus, J., & Aslan, I. (2006). Computer-Assisted Navigation and the Acquisition of Route and Survey Knowledge. *Journal of Environmental Psychology, 26*, 300–308. doi:10.1016/j.jenvp.2006.08.001

Neisser, U. (1976). *Cognition and Reality*. San Francisco: WH Freeman.

Nisbett, R. E., & Wilson, T. D. (1977). Telling more than We can Know: Verbal Reports on Mental Processes. *Psychological Review, 84*, 231–259. doi:10.1037/0033-295X.84.3.231

O'Connor, A., Zerger, A., & Itami, R. (2005). Geo-Temporal Tracking and Analysis of Tourist Movement. *Mathematics and Computers in Simulation, 69*, 135–150. doi:10.1016/j.matcom.2005.02.036

Platzer, E. (2005). *Spatial Cognition research: The human navigation process and its comparability in complex real and virtual environments*. Germany: PhD, Universität der Bundeswehr München.

Radoczky, V. (2003). *Kartographische Unterstützungsmöglichkeiten zur Routenbeschreibung von Fußgängernavigationssystemen im In- und Outdoorbereich. MTh.* TU Vienna. (in German)

Raento, M., Oulasvirta, A., & Eagle, N. (2009). Smartphones: An Emerging Tool for Social Scientists. *Sociological Methods & Research, 37*(3), 426–454. doi:10.1177/0049124108330005

Ray, M., & Schrom-Feiertag, H. (2007). *Cell-based Finding and Classification of Prominent Places of Mobile Phone Users*. Paper presented at The 4th International Symposium on Location Based Services & TeleCartography, Vienna, Austria.

Schmid, F. (2007). *Personalized Maps for Mobile Wayfinding Assistance*. Paper presented at The 4th International Symposium on Location Based Services and Telecartography, Vienna, Austria.

Schweizer, K. (1997). *Räumliche oder zeitliche Wissensorganisation? Zur Repräsentation der Blickpunktsequenzen bei räumlichen Anordnungen*. Lengerich, Germany: Pabst Science Publishers Verlag.

Schweizer, K., & Horn, M. (2006). Raumwahrnehmung und Raumvorstellung. [in German]. *Geographie und Schule*, *164*, 21–26.

Shoval, N., & Isaacson, M. (2007). Tracking Tourists in the Digital Age. *Annals of Tourism Research*, *34*(1), 141–159. doi:10.1016/j.annals.2006.07.007

Smith, S. E., & Vela, E. (2001). Environmental Context-Dependent Memory: A Review and Meta-Analysis. *Psychonomic Bulletin & Review*, *8*, 203–220.

Svetsuk, A. (2007). *Experiments in urban mobility analysis in Rome using mobile phone data*. Paper presented at Urbanism On Track – Expert meeting on the application in urban design and planning of GPS-based and other tracking-based research, Delft, The Netherlands.

Thornton, P., Williams, A., & Shaw, W. G. (1997). Revisiting Time-Space Diaries: An Exploratory Case Study of Tourist Behavior in Cornwall, England. *Environment & Planning A*, *29*, 1847–1867. doi:10.1068/a291847

van der Spek, S. C. (2007). *Legible City – Walkable City – Liveable City: Observation of Walking Patterns in City Centres*. Paper presented at Urbanism On Track – Expert meeting on the application in urban design and planning of GPS-based and other tracking-based research, Delft, The Netherlands.

van der Spek, S. C. (2008). Sensing Human Activity: GPS tracking . In Kooistra, L., & Ligtenberg, A. (Eds.), *Proceedings of International Workshop Sensing a Changing World 2008. CGI report CGI-08-005*. Wageningen, The Netherlands: Wageningen University and Research Centre.

Whittaker, J. O., & Whittaker, S. J. (1972). A cross-cultured study of geocentrism. *Journal of Cross-Cultural Psychology*, *3*, 417–421. doi:10.1177/002202217200300412

Yan, W., & Forsyth, D. A. (2005). *Learning the Behavior of Users in a Public Space through Video Tracking*. Paper presented at the 7th IEEE Workshop on Applications of Computer Vision / IEEE Workshop on Motion and Video Computing (WACV/MOTION 2005), Breckenridge, CO, USA.

Chapter 11
FriendFreight:
Leveraging Real-Time Location Information for a Sustainable Community-Based Goods Delivery Service

Christine Outram
Massachusetts Institute of Technology, USA

Francesco Calabrese
Massachusetts Institute of Technology, USA

ABSTRACT

The relatively new ability to rapidly transfer digital information to people as they move through cities opens up exciting possibilities for services that alleviate some of the issues that dense urban centers face. This chapter examines the potential for one such service – named FriendFreight – to reduce some of the negative effects of goods transportation in cities. FriendFreight operates through exploiting the real-time location information of people and goods and the ability for members of a community to deliver items for others while moving through the city themselves. It aims to lower the number of 'unnecessary trips' that people make to obtain some small goods - groceries, books, documents and dry-cleaning - and reduce what we define as travel demand in the city. However, the success of such a service relies not only on accessing real time location information but also on an understanding of how and why people might deliver goods for each other. Thus, this chapter explores two things: how incentives, trust and reciprocity can be built in services that harness digital information; and how the feasibility of a service like FriendFreight can be established given a particular real world context - Copenhagen and bicycles. Through this, the authors show that access to real-time location and movement information can open up innovative ways of tackling problems in cities from the 'bottom-up' but that essential to this is the nurturing of trust between users of the service. They also demonstrate that it is possible to achieve a significant reduction in travel demand through using FriendFreight for certain types of goods in the context of Copenhagen.

DOI: 10.4018/978-1-61520-769-5.ch011

Figure 1. Common images of dormant freight capacity

INTRODUCTION

There exists an untapped resource in cities: the 'freight capacity' of the vehicles that we use to move around during our daily routes. Our own bodies are the simplest example of a vehicle with 'freight capacity' and we use them for carrying the items that facilitate our daily lives. Similarly, the baskets and racks on bicycles, the trunks and backseats of cars and the storage boxes on motorcycles have an inherent ability to transport items. However, for most of us, most of the time we move around carrying nothing but *pockets of air* in these containers and the *potential* to carry things should the need arise. (Figure 1)

While during our daily routines we are moving large amounts of empty space back and forth, there are teams of people in delivery companies that are dedicated to moving freight across cities. Given the sheer scale of intra-city logistics needs, some small goods such as groceries, books, documents, and dry-cleaning are nearly always likely to be headed in the same direction and to the same general location as we are.

Until recently, there was no way to utilize the available freight capacity of private vehicles to facilitate the movement of small goods in a city. However, the technical constraints for pinpointing people and goods are rapidly disappearing. In terms of people, stringent regulations on location requirements for emergency calls on the 3G network, such as those required by North America's FCC has meant that as of 2003 in 67% of cases mobiles must be able to be located to the nearest 50 meters (using handset-based positioning) and to the nearest 100 meters (for network-based positioning) (Adams, Ashwell, & Baxter, 2003). These types of regulations – although technically difficult at first – have had the additional benefit of creating a stable platform for all sorts of reasonably priced (or free) location based consumer services. Additionally, with real-time inventory software becoming available in stores and the sales of smart phones with applications that allow you to quickly locate the items you need, locating goods that require delivery is becoming easier. Thus, an opportunity now exists: to coordinate the journeys of people and the untapped freight capacity of their personal vehicles with the delivery trajectories of goods.

In this chapter we propose a scenario for a service – what we name FriendFreight – that can take advantage of the dormant freight capacity of one type of vehicle, bicycles, in order to deliver small goods to others in the city of Copenhagen. The aim of the service is to reduce the number of 'unnecessary trips' that people make to obtain some everyday items and in turn to lessen what we define as *travel demand* in the city. This results in a reduction in the amount of energy expended by

Figure 2. A simple system architecture for the friendfreight service

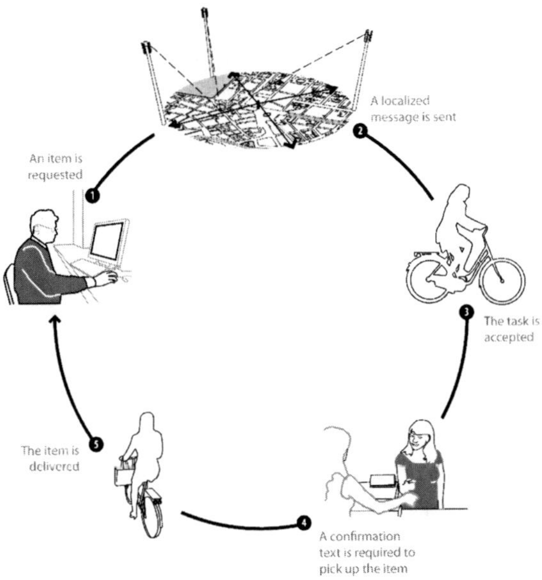

the transportation sector, of congestion and CO_2 emissions. In order to give the reader a sense of the potential of this service and how real-time location and movement information can be innovatively used in this way, we first provide a basic 'system architecture' for how the service may work. As the goal is to reduce some of the negative effects of goods transportation, we next discuss the field of City Logistics: how real time information is used to optimize goods transportation in this field currently and how services such as the one proposed in this chapter provide a new approach to tackling these problems. We then discuss incentives, trust and reciprocity in relation to the FriendFreight service, given that it relies on the real-time dissemination of digital information to members of the community who may or may not know each other. Lastly, we present an evaluation model of FriendFreight that is based in the specific context of Copenhagen to establish the feasibility of using real-time location and movement information to more effectively deliver some small goods in the city and reduce travel demand.

THE FRIENDFREIGHT SCENARIO

Many of the trips we take in order to pick up small every day items are part of our daily movement patterns, such as going to work or school. The trips that a service like FriendFreight targets, however, are those that are stand-alone trips.

The starting point for this service is a person who needs to collect something but cannot do so as part of their usual routine. Perhaps they are going to be in meetings all afternoon and can't leave the office to get the dry cleaning, or maybe they have simply forgotten a vital item for dinner when they were at the supermarket. Whatever the reason, the service offers an alternative delivery method that does not rely on commercial delivery companies, which are expensive, have large time-delivery windows and often restrict themselves to bulk deliveries or office related items. The 'system architecture' for FriendFreight, using bicycles as an example of a mode of transport with excess freight capacity is shown in Figure 2.

The person in need—the requester, who we can assume has already provided some personal information in order to become a member of the service in order to increase their accountability—starts by logging onto the FriendFreight Service website and registering the item they want, the destination they want it delivered to, and the timeframe within which they need it. The system is then able to send a text message - to only those who are enrolled in the service as carriers and are within a certain distance of where this item is located - requesting that somebody delivers the item. Once a carrier accepts the request, they collect the item (after verifying their identity) and deliver it within the specified timeframe. On the assumption that only those who are heading in the direction of the delivery will accept the request, the FriendFreight service generates a reduction in travel demand.

The explanation above represents the 'bare bones' of the service proposed in this chapter. This service relies on real-time location information and the localizing of text messages to specific people. It thus represents a novel approach to goods delivery in cities that would not have been possible even a decade ago. Because the goal of the service is to reduce travel demand, it can be seen as sitting within the field of City Logistics – a relatively new research area that aims to optimize transport and delivery activities within urban areas. However, unlike most work in the City Logistics field – which concentrates on optimizing existing delivery mechanisms through efficient routing or faster modal transitions (a top-down approach) -- the service presented here utilizes the community to deliver goods in urban areas and could thus be considered a bottom-up service.

CITY LOGISTICS: TRADITIONAL APPROACHES TO OPTIMIZING THE MOVEMENT OF GOODS THROUGH THE USE OF REAL-TIME LOCATION INFORMATION

City Logistics is most commonly cited as "the process for totally optimizing the logistics and transport activities by transport companies in urban areas while considering the traffic environment, the traffic congestion and energy consumption within the framework of the market economy" (Taniguchi, Thompson, & Yamada, 2001). The growing importance of this field is a reflection of a number of political, spatial, market and behavioral changes that are taking place in cities. City planners and local governments now recognize the impact that goods transportation in urban areas has on the environment, congestion and energy use. Thus there is an increasing desire to manage these effects in order to achieve more sustainable cities. In Europe, the push for cleaner, more efficient delivery systems for distribution of goods in urban centers, is reflected in the creation of European Commission backed initiatives such as CIVITAS which, through political commitment from cities across the European Union, aims to promote, implement and evaluate approaches to sustainable urban mobility. Elsewhere, Japan, Australia and the USA have been leaders in conducting research and in some cases implementing these optimization systems (Quack, Van Duin, & Visser, 2006).

The use of real-time information in the context of City Logistics research falls predominantly into three categories:

- *Real-Time Routing and Scheduling:* Here, real-time location information is used to optimize distribution schedules particularly when new deliveries are added to the schedule on route. This use of digital information solves what companies and logistics experts call *The Traveling Salesman Problem* and it markedly increases the speed and efficiency at which delivery companies can operate (Kunze, 2004; Thompson, 2004; Yamada, Yoshimura, & Mori,, 2004).
- *Coordinating modal changes at consolidation hubs:* The World Bank estimates that over the last decade, the number of intermodal containers passing through ports worldwide has doubled, with a similar trend occurring in intermodal air, rail and truck traffic. (Nemoto, Browne, Visser, & Castro, 2005) At the same time, there is a disappearance of dedicated spaces for housing goods in the cities - ports, railway terminals and warehouses (Dablanc, 2007) Increasing interest has therefore been shown in using digital information to coordinate modal changeovers at consolidation hubs - also known as urban freight platforms or freight villages. Here digital information can be used to create a real-time inventory of available freight space at any given time of day at these points. This allows a greater coordination between the flow of goods from outside of the city

and the bundling of delivery within a city (Koehler, 2004; Lozano et al., 2005).

- *Enhancing data gathering to improve logistics models:* A rise in devices that can efficiently sense and collect real-time information about the city – congestion and pollution levels, road conditions, parking availabilities for delivery trucks etc – has resulted in increasingly large data sets that can be harnessed to improve the planning, execution and re-planning of City Logistics schemes (Ito et al., 2004; Munuzuri, Larraneta, Onieva, & Cortes, 2004).

Each of these categories uses real-time information to improve an *existing* goods transportation system and could thus be considered a top-down approach to addressing the problem of urban freight logistics. Research proposals that use a bottom-up approach – such as the FriendFreight service proposed in this chapter - are harder to find. A notable exception is Fusco, Tatarelli and Valentini's (2003) research on setting up community drop-points that coincide with daily activities of customers – breakfast bars, newspaper stands, etc. However, while it is clear that the service could be convenient for people, the real impact on alleviating travel demand is difficult to evaluate.

While there has been little work on bottom-up services in City Logistics, we can look towards other examples of web based services, such as eBay, CouchSurfing and LinkedIn among others, that harness the community in a bottom-up and decentralized fashion through the use of distributed digital information. Of importance in these examples is the incentives used to encourage participation and how trust and reciprocity can be built between members of the community so that the life of the service is assured. These issues will now be briefly examined and related back to possible frameworks for the FriendFreight service.

BUILDING INCENTIVE STRUCTURES, TRUST AND RECIPROCITY IN SERVICES THAT RELY ON THE TRANSFER OF DIGITAL INFORMATION

The FriendFreight service can be described as what Pichler (2000) terms a 'distance transaction', whereby text messages are sent to potential carriers of items from a distance and there is initially no face-to-face contact between someone who needs an item (the requester) and the person that delivers it (the carrier).

Without face-to-face interactions, the structure of the service - the type of incentive provided for performing the task, the ability to trust people and the encouragement of reciprocal action - must already be established before it is deployed. Different models for achieving this can be unearthed through looking at other online services that involve distance interactions, such as eBay, Couch Surfing and LinkedIn.

Online services that bring together those at a distance into a transaction usually use one of two incentive structures. The first is a market-based approach where money is used to entice people into doing something for others and the second is a gift-based approach where the building of a common social goal is used to motivate the community, but no money exchange is present. An example of how the former could operate within the FriendFreight scenario can be found by looking at the online auction site *eBay*. Here, in the case of FriendFreight, when the requester wants something delivered, they decide on a maximum amount that they are willing to pay - let's say $11. An initial message can be sent out offering $5 to deliver the item. If no one accepts the task, the price can rise until the incentive is enough to get someone to deliver the package.

The building of a common social goal is an alternative to the market-based incentive approach. Online services such as *Couch Surfing* show how this can operate successfully and at a large scale. Couch Surfing, which is a transaction between strangers that does not involve money, aims to match up those in need of a place to stay with those who have a spare couch, bed or piece of floor. Members of the community create a profile on the website detailing what type of free accommodation they can offer. In return, when these people are traveling, they use the service to find free accommodation with other members. The common social goal that binds this community together is the spirit of traveling: the desire to meet other people and additionally the saving of money. In the case of FriendFreight, we could imagine a common social goal being the carrying items for the elderly or infirmed. Structuring the service in this way would create a community of like-minded people who are willing to deliver items because they already know that a person is in need. Alternatively, if the service was structured around those that already know each other, then the common social goal might simply be 'helping a friend'.

There has been much research surrounding the building of trust in online services. Besides identity verification when a new user enrolls in the service, the most likely method for establishing trust *between* strangers who are engaged in a distance transaction is through a reputation system (initially developed for eBay). Reputation systems employ a self-policing model in which the behavior of a participant is recorded by other users – usually in the form of ratings from 1-5 with the option for additional comments. Once created, it is attached to a persons profile and is available for other members of the service to see. A study by Resnick, Zeckhauser, Swanson, and Lockwood (2006) on the reputation system used by eBay demonstrates the effectiveness of this system by proving that sellers with good reputations are both more trusted and more used than those with low scores (although, interestingly, one or two negative responses were not sufficient to affect the willingness of buyers to use a distributor). This simple peer-to-peer feedback system could be especially useful to the service proposed in this chapter as it could quickly weed out those with a bad performance record for delivering goods.

In order to ensure the ongoing life of services that function without money as a motivator, reciprocity, or the encouragement of 'returning the favor', plays a vital role. In the case of Couch Surfing, reciprocity is encouraged through the language of the website and highlighting the rewards of being a member – the host is not simply hosting but in a certain sense is traveling vicariously through their visitor. At the same time, the visitor is getting an authentic experience with a local and also a free place to stay. Making a conscious statement of this dual role enhances reciprocity and is a way of binding the community together. It also works to exclude those who are unwilling or unable to reciprocate. (Molz, 2007)

Where online interactions occur between people that already have established social links, degrees of trust and expectations of reciprocity are often already established. The professional networking site *LinkedIn* – which makes apparent the network of business connections between people so that members can use this to further their professional goals - additionally shows that just as we have a trusted group of close relationships, we are more likely to trust those who are connected to our trusted inner circle, even though we might not know them personally. This type of trust network, which nurtures accountability, offers an alternative framework that could be used in the FriendFreight service proposed in this chapter. Here, it could be imagined that when signing up to the FriendFreight service, members nominate five people that they trust to deliver things to them and for whom they would also return the favor. If these people were not available when the member needed something, the system could offer alternative people who are nearby the item and are connected to those within the 'inner circle'. If each member had five trusted friends, within three degrees of separation, a network of 155 'trusted' people would be available. If the power of this type of network was also combined with a common social goal, then it is feasible to imagine that the FriendFreight service could operate without the use of money.

A point of difference between the FriendFreight service and the online services that have thus far been examined is the collection, use and storing of location information (vital to the operation and future optimization of the service). From a technical standpoint, privacy when using this information has become a hot topic in recent years (see Beresford & Stajano, 2003; Gedik & Liu, 2005). Scholars have also outlined the potential hazards of continual access to real-time location information as detailed in the paper *GeoSlavery* (Dobson & Fisher, 2003). A simplistic and often standard approach to location-oriented services relies on people opting-in to having their location details used by the system. Applications like *Loopt*, where you share with friends location information and your activity status rely on the fact that there is perceived value in having your location known and that a reciprocal arrangement of trust exists between friends when it comes to knowing location information. The idea that we might trade privacy for convenience, cost savings or perceived social value, however, is not new. In his book, *Civilizing Cyberspace: Policy, Power and the Information Superhighway,* Miller (1995) notes that people routinely trade personal information if they see a benefit to it for themselves. When these benefits are common goals that can be shared across a community – such as the desire to have goods delivered more effectively and cheaply – the sharing of location information may not be seen as a threat. Adding to this, the option of opting-out from location information being collected over time and transparent statements for how the data is used by the system could provide an additional buffer of perceived trust. As Brin (1998, p.152) states: "feelings of distrust and fatalism do not tend to dominate when transparency is reciprocal, nor when people retain a sense of participation and control".

The issues and approaches highlighted above, represent a few of the challenges that would need to be addressed should the FriendFreight service be implemented. Due to the length of this chapter, other issues such as building a legal framework, getting people to provide real information when they sign up to the system so as to promote accountability, the building of the databases to manage the location and movement information, the monitoring of cyclists' movements so that request messages can be targeted towards likely carriers and the use of the system to predict goods demand in cities are not covered, although it is felt that all of these can be incorporated to improve the likelihood that a service like this could succeed.

TESTING THE FEASIBILITY OF THE FRIENDFREIGHT SERVICE

The final section of this chapter involves the building of a computer model to test the feasibility of the FriendFreight service within a real-world context. In order to reduce the number of variables, we ground our work by choosing one type of personal mobility vehicle and one city: bicycles in the city of Copenhagen. Bicycles are the ideal carriers for small goods since, unlike cars, they can be easily parked and are virtually non-polluting, but still have a carrying capacity that matches many desired loads. With 50% of people in the city of Copenhagen using bicycles on a daily basis, and over 175,000 bikes crossing the city limits each day, this is an ideal test-case city to implement the work. Additionally, it should be noted that the future testing of this research is made possible through a 3-year research partnership currently operating between our research group, the Senseable City Lab, MIT and the City of Copenhagen. Through this partnership we have been investigating how small amounts of technology on bicycles (in particular location awareness) can be used as a platform for creating services that not only improve the biking experience but are also tackling issues of sustainability in cities.

The geographical area that we are dealing with is shown in Figure 3: it is approximately 11 x 12 kilometers and includes the municipality of Copenhagen, as well as the neighboring municipality of Frederiksberg that lies within the boundaries of the city of Copenhagen. Although we include both areas in the model, we will refer to it as the Copenhagen area.

The Establishment of Freight Capacity

The first step in gathering data for the model is to establish that there is adequate freight capacity on bicycles in Copenhagen. A survey of bicycles was therefore conducted at three busy intersections in January 2009 for half an hour at a time. During this period, a total of 795 bikes were counted and of these a significant proportion - 68% - had what we can term available freight carrying capacity, which is, there was enough room for additional items to be carried. The freight carrying capacity by type is summarized in Table 1.

Figure 3. Map of copenhagen/frederiksberg area

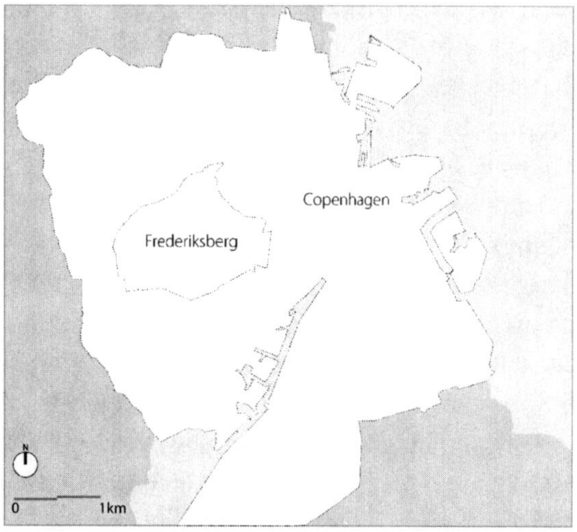

Goods

Types of Goods in High Demand That Can Be Carried by Bikes

In order to reduce the complexity of the model, a limited number of goods are initially chosen. The criteria for choosing these goods is that they can be carried by bike and have a high daily demand—we want to target stand-alone trips for items (those trips that are not combined with other daily activities) and we have assumed that the number of stand-alone trips is proportional to the demand for the item.

The categories of goods are determined through an examination of average household consumption (in kroner) of different types of goods in the Copenhagen area – information that is publicly available through *Statistikbanken,* the Danish government's statistics website (Statistikbanken, 2006/2007). Initially, we eliminated all goods that are considered sensitive (easily breakable, containing poisonous substances etc) in nature or cannot be carried by bike because of excess volume or weight, and then concentrated on establishing which goods households require most often. We did this by dividing the total amount spent per year for each goods category by their average price – as found online – for each likely good type. This gives us an estimate of the number of goods required per type/per year/per total number of households. Although it is recognized that demand for goods has a temporal aspect (that is some goods might be bought more on a Sunday than a Monday or in a certain month), a lack of statistics forced us to simply divide the yearly figures by 365 to establish the daily averages on which we can base our model.

Four categories of goods emerged from this process: Flowers, Books, Dry Cleaning and Medical Items The average numbers of goods per category required each day across all households (277,624) in the Copenhagen area is found in Table 2.

Table 1.

	Basket	Rack	Buggy	Side Basket	Rack and Basket	Nothing	TOTAL
No. of bikes	114	276	18	6	258	123	795
No. with available freight capacity	54	228	9	0	252	0	543

Table 2.

	Flowers	Books	Dry Cleaning	Medical Items
Amount/day/total number of households	2728	5670	3109	6170

Figure 4. Demand for goods over the course of a typical day

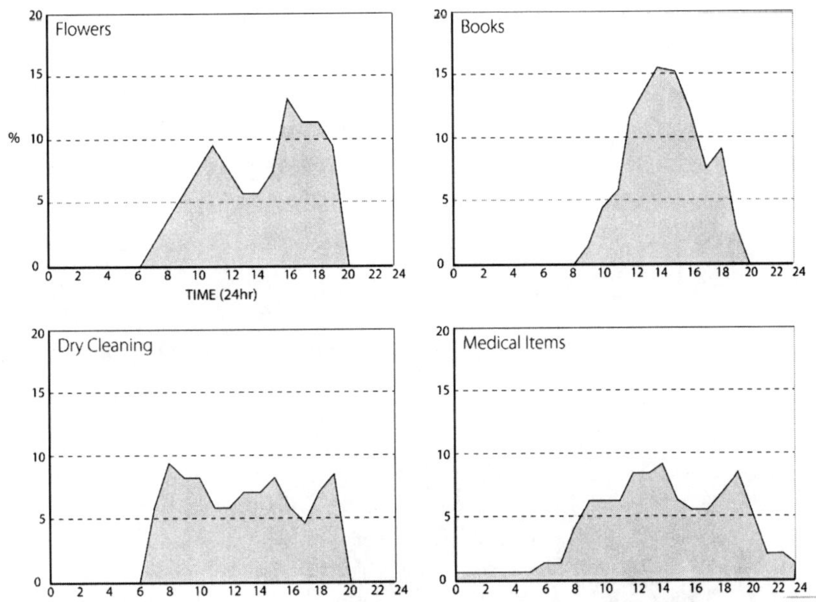

Average Hourly Demand for These Types of Goods

We were also interest in establishing the demand for goods over time, in our case a typical day. The distribution of this demand was established through an analysis of shop opening hours and mobility patterns as found through statistics available from the City of Copenhagen, the results of which are shown in Figure 4.

Locations Where These Goods are Needed

In a real world scenario, goods are delivered everywhere – to residences, businesses, schools and institutions. In this model only residences are considered, as households generate many more stand-alone trips than businesses or schools, which often place bulk orders for items. GIS land use data sets can determine the spatial distribution and density of households across the Copenhagen area. (see Figure 5). Such information will be used as a proxy for demand for goods as distributed spatially in the model.

Locations Where These Goods can be Found

The next step in the data gathering was to determine the locations of stores that sell the four categories of goods in the Copenhagen area. The addresses of these stores are found through Internet directories. A GPS visualizer is then used to convert the addresses into discrete sets of latitude and longitude coordinates.

Average Distances Between Points-of-Sale and Residences

In the model, it is assumed that when someone requests a good it will be ordered from the closest shop that has that type of good for sale. The distance between these two points provides a measure of the number of kilometers that must be traveled to bring the item from the point of sale to its final destination, a place of residence. The distance between every residence and its closest shops can be calculated by combining the coordinates of shops with the centroid coordinates of each polygon that is designated as residential in the GIS database. It is recognized that this is an "as the crow flies" measure of distance but it will still be useful in establishing a benchmark for total kilometers traveled and the reduction of kilometers traveled when using the FriendFreight service. The spatial distribution of the above distances are shown in Figure 6 where the color of each cell relates to how far each cell is from the location of a shop with a given category of goods:

The distances range from a few hundred meters to up to 6 km, with a mean of 1.63km for flower shops, 2.2km for book stores, 2.33km for dry cleaners, and 1.86km for pharmacies. All of these distances are within normal range for bike

Figure 5. Cells designated as residential units in the Copenhagen area. The color indicates the number of houses per 100x100 meters

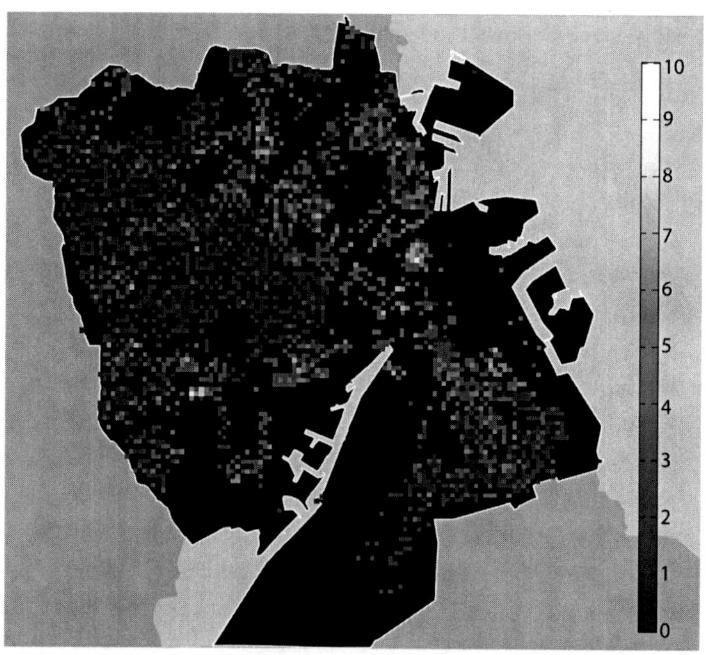

riding, confirming that bikes are a suitable mode of delivery for these services.

Cyclists

To assess whether we could provide adequate levels of service through this system, we needed to understand how many cyclists would be potentially available to carry goods at different times of the day as well as the typical trajectories of their journeys. Since different types of people have different temporal and spatial movement patterns, it seems sensible to involve diverse groups of people, such as university students and workers with varying shift or work patterns.

According to Danish statistics (Statisbanken, 2007), the Copenhagen area has 78,125 students that live and study within the municipality and 450,140 people who work in this area. Thirty-six percent of people commute to work by bike and 50% of the population use a bike on a daily basis. This figure is even higher in the inner city, but as a starting point we assumed that 36% of each category of people would ride their bicycles daily. Aggregating the number of people in each category according to the percentage that move over the course of a day, gave us a total number of potential carriers per hour. This total however, needs to be diminished as only two-thirds of the bike population has available freight capacity as established above. Data on bicycle mobility trends from the City of Copenhagen allowed us to estimate the variability in the number of potential carriers and how this changes depending on the group of people considered (see Figure 7).

To establish which places in the city people would be most available for becoming carriers, we consider, for both categories, that they are best

Figure 6. Spatial distribution of distance of residential cells from shops

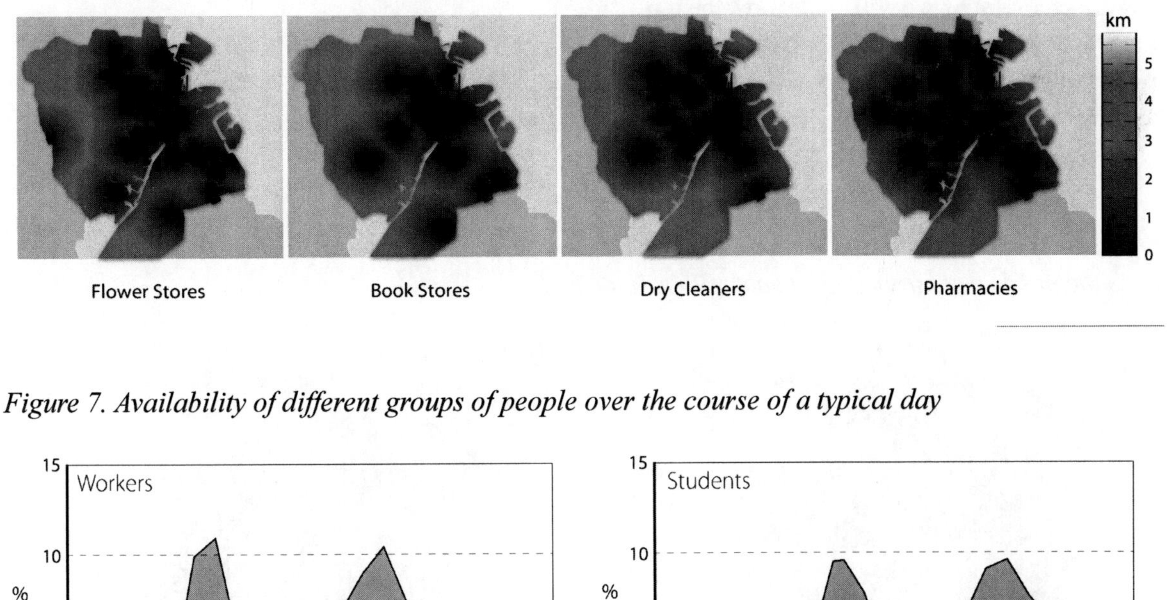

Figure 7. Availability of different groups of people over the course of a typical day

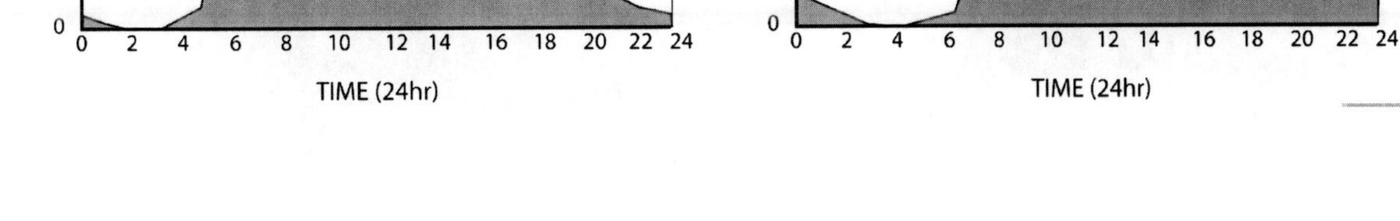

able to deliver goods when they are traveling from their place of residence to their other significant location (work or university), and vice versa. Consequently, the location of universities and businesses are taken to be a good proxy for the spatial availability of possible carriers. Figure 8 shows this spatial distribution. As many students live close-by to the institution they attend, for universities, we consider a circle with a radius of 800m as the spatial distribution of students in the area.

MODELING THE LOGISTICS SYSTEM

In this section we propose a simple model to evaluate the performance of the logistic system used for the delivery of small goods. We will call *Business as Usual* (BAU) the current situation in which the service is not in operation. The BAU condition is going to provide us with a certain travel demand.

As a second step, we will model how the introduction of the delivery service would change the logistics system, and how the travel demand will be affected. Herewith, we will call this *FriendFreight Travel Demand* (FFTD).

Modeling the BAU condition as well as our new FriendFreight condition allows us to:

- evaluate the theoretical effectiveness of this service in terms of reduction of kilometers traveled. From this the reduction in CO_2 emitted and number of vehicles on the road could also be inferred.
- Establish the minimum number of people required to deliver a certain percentage of goods.
- Establish which category of goods can be delivered most effectively given its temporal need and the spatial distribution of shops that provide these goods.
- Establish how many people from each category will best satisfy demand given their temporal availabilities and the spatial mobility differences between each group.
- Establish for which spatial area of Copenhagen this service would be most effective.

To model the FriendFreight service effectively, a number of assumptions must be made. These are summarized as follows:

Figure 8. Spatial distribution of universities and businesses. In the case of businesses, the color represents the number of businesses in each 100 x 100m cell

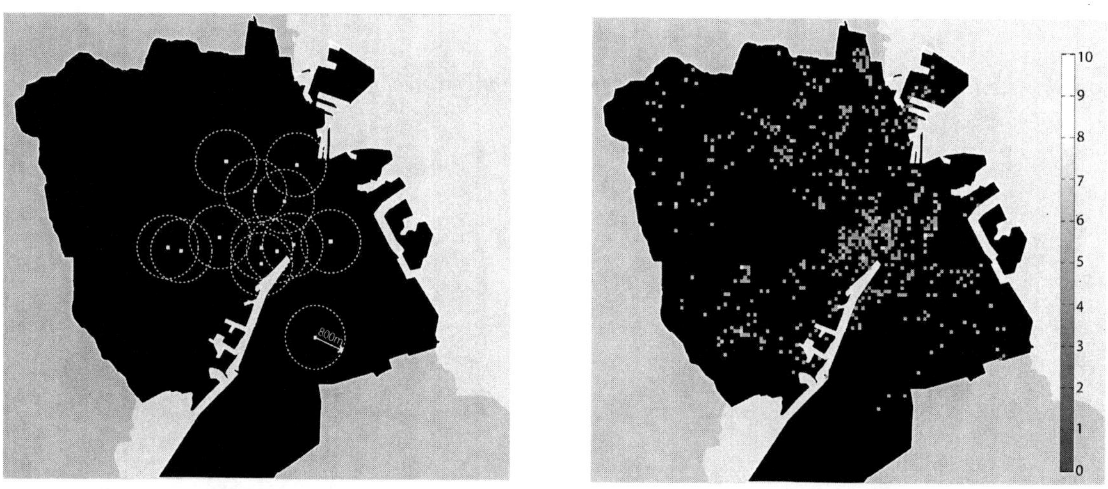

- A 'package' is equivalent to one item from any of the four goods categories
- Each person carries the same number ***p*** of packages
- All goods in the model are the ones available through the service
- All the carriers in the model are the ones enrolled in the service
- People request goods from the shop that is closest to them
- Carriers who are defined as available (based on their daily mobility patterns) will always accept a request to deliver goods

Defining the Variables

As mentioned above, to model the service effectively, the Copenhagen area was divided into a discrete number of cells, each of which represents a number (density) of residential properties. A generic cell is denoted as **d.** Time is indicated with **t,** categories of goods with **G,** and groups of people with **P**.

The model of the logistics system uses the following information, gathered through the data collection process described in the previous section.

- **Demand (G,d,t)**: Number of goods of type G requested by the location d, at time t - as generated through the weighted distribution explained above.
- **Distance (G,d)**: Distance of good of type G from the location d – based on the closest shop that sells this good type.
- **Availability (t):** Number of available people of category P, at time t The availability is the sum of the availabilities of different categories of people at the same time:

$$Availability(t) = Availability(P_1, t) + \ldots + Availability(P_{nP}, t)$$

- **Mobility (G,d,t):** Number of people that travel at time t from the location of a shop of good **G** to the location **d**. The mobility is the sum of the mobility of different categories of people

$$Mobility(G,d,t) = Mobility(P_1, G,d,t) + \ldots + Mobility(P_{nP}, G,d,t)$$

One of the goals of our proposed delivery service is to minimize the number of kilometers that people travel to obtain goods. We do this by utilizing the public to deliver goods when they are already traveling in the same direction. To measure the effectiveness of such a service, we define the BAU travel demand as:

$$TravelDemand(G,t) = \alpha \sum_{i=1}^{nD} Demand(G, d_i, t) Dis\tan ce(G, d_i)$$

Here, it can be seen that the travel demand for carrying a certain package under the BAU condition is modeled as a function of the distance between the residential cells and the shops, where α is the weight given to the distance. For instance α = 2 means that we are considering that the requesters have to travel from home to the shop and go back (2 times the distance), and that this trip does not overlap with other movements they already do, e.g. for work. As we recognize that people almost always coordinate their trips to get goods in the city, we assume a typical value for α of 1.5. The business as usual travel demand represents our benchmark from which to assess the effectiveness of the service we are proposing.

To evaluate the best-case scenario when the FriendFreight service is running, the FFTD must also be calculated. We can indicate with $n_{di}(G,t)$ the number of carriers traveling to location d_i, = **1,...,*nD***, at time *t* to carry goods of category ***G***. Each carrier will carry a total of ***p*** packages. It results that the FFTD is:

$$ServiceTravelDemand(G,t) = \beta\, p\sum_{i=1}^{nD} n_{d_i}(t) Dis\tan ce(G,d_i) + \\ + \alpha\sum_{i=1}^{nD}\left(Demand(G,d_i,t) - p\, n_{d_i}(t)\right) Dis\tan ce(G,d_i)$$

Here, the FFTD is the sum of two components: the demand that is the result of the travel the carriers who are enrolled in the service undertake and a remaining demand that is needed to deliver goods that can not be distributed through the service. The latter demand assumes the same form as the BAU condition while the former is defined as follows.

While the service is in operation, each trip of length **Distance (G,d)** generates a travel demand that is proportional to the distance traveled. It should be noted that the theoretical travel demand for carrying a certain package through using the service should be equal to zero because we assume that the carrier of the goods is already heading to the same location as where the package needs to be delivered. However, we recognize that this it is unlikely that the carrier will be going to the exact final destination of the good. Therefore we associate a small travel demand to every trip that is a function of the distance traveled. This average weight placed on the trip is represented by $\boldsymbol{\beta}$ where in general $\boldsymbol{\beta < \alpha}$. . An assumed typical value for $\boldsymbol{\beta}$ is 0.2.

To evaluate the theoretical maximum reduction in travel demand through using the FriendFreight service, the model needs to determine how goods are assigned to carriers. For instance, if we have 100 available cyclists in the service and have 150 requests for flowers, how does the model decide which flowers to assign to which carriers? Here, when a request is generated from a cell that represents a certain residential density, we know, from the assumptions made earlier that the item will be requested from the closest bookshop. We also know the number of potential carriers that are traveling in the direction of the bookstore/residence at this time. In the model, if two residences request books from the same bookstore, and a potential carrier is traveling in the direction of both residences, the farthest residence is assigned to the as this reduces travel demand by the greatest amount and thus creates a 'best-case scenario' that can help to evaluate whether the service is worth implementing.

Problem 1: Find the vector-valued function $S(t) = \{ n_{di}(G,t)\}_i$ that maximizes:

$$\sum_{i=1}^{nD} n_{d_i}(G,t) Dis\tan ce(G,d_i)$$

This problem operates under the following constraints:

1. The number of people assigned to goods must be less than or equal to the number of people available

$$\sum_{i=1}^{nD} n_{d_i}(G,t) \le Availability(t) \quad \forall t$$

2. The number of carriers that are a) heading in the right direction and b) are assigned to carry goods must be less than or equal to the number of people heading in the right direction and less than the total demand for goods.

$$0 \le n_{d_i}(G,t) \le \min\left\{Mobility(G,d_i,t),\ Demand(G,d_i,t)/p\right\} \quad \forall i = 1,\ldots,nD \quad \forall t$$

3. A carrier can only be defined as a whole number.

$$n_{d_i}(G,t) \in \aleph \quad \forall i = 1,\ldots,nD \quad \forall t$$

The solution to the optimization problem provides us, for every time step $_t$, with a vector $\boldsymbol{n_d(G,t)}$, which indicates how to optimally assign to the carriers the packages that need to be delivered, in order to minimize the service travel demand.

The solution can be easily computed using linear programming and in particular variations of the simplex algorithm for discrete variables. Note that the assumptions for the value of α and β do not influence the optimization problem. They simply allow us to compute a quantitative measure of performance in the proposed system.

TEST RESULTS

Having set up the parameters of the model with the information collected and described in the previous section, we can now evaluate the theoretical effectiveness of the service through a series of tests. For all tests, we consider:

- Only 10% of items required by households will be delivered by the service and therefore, the numbers of items required per day for flowers, books, dry-cleaning and medical items are: 273, 567, 311 and 617 respectively.
- Considering the typical size of these types of items we define that the number of packages each carrier can deliver is one.
- The area of Copenhagen is divided into cells each of which represents a density of residential households. The size of each cell is 500 x 500m.
- The assignment of goods to carriers is done by solving the optimization problem 1 above, using the Matlab Optimization Toolbox.

Test 1

Establish which category of goods can be most effectively delivered and the minimum number of carriers (available people) required in order to optimize the service travel demand.

For every good category, we can create a graph that establishes how the service travel demand decreases while the percentage of people available (combined number of students and workers) increases (for an example using dry cleaning see Figure 9). The travel demand in the case of business-as-usual is also compared to provide a measure of effectiveness of the service. It is intriguing to see that the service travel demand reaches saturation when the number of available carriers increases to a certain point. This essentially means that there is no perceivable benefit to increasing the total number of carriers available past

Figure 9. Total travel demand compared to service travel demand, for different percentages of people involved in the case of dry cleaning

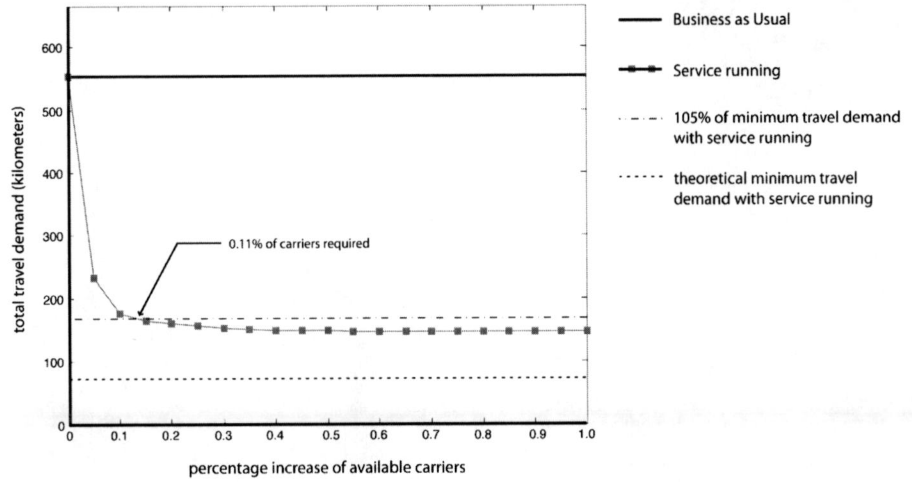

a certain threshold. The reason for this behavior is due to the spatial distribution of shops which does not always match the spatial availability of people, resulting in not having people close to a certain shop and heading in a certain direction. This implies that even if we increase the overall availability of people, a certain amount of goods still cannot be delivered. This aspect is shown in Figure 10.

To define the optimal number of carriers needed, we consider the availability point when the service travel demand reaches the settling point, with a tolerance of five percent. In other words we consider the minimum level of availability of people so that the travel demand reaches 105% of its optimal value (see dot-dashed line in Figure 9). In particular, for the case of dry-cleaning items, shown in Figure 9, the available people (number of carriers) at the settling point corresponds to 0.11% of the total combined number of workers and students (218 people). That is, when the service is running, we see a reduction in total travel demand of around 70% for this good type. Although not shown here, the optimal number of carriers for flowers, books and pharmacy items is respectively 0.09% (169 people), 0.2% (387 people) and 0.33% (438 people).

Once we have defined the optimum number of carriers required to minimize the FriendFreight travel demand (FFTD), we can examine how many requests there are for certain goods across a day and how effective the service is in distributing these goods spatially. In other words, are there areas where this service will not work because of the distinct distribution of shops? Additionally, how many goods out of the total can be delivered?

Figure 10. Spatial distribution of demand and met demand. Note that the spatial distribution of demand is equal for all types of goods.

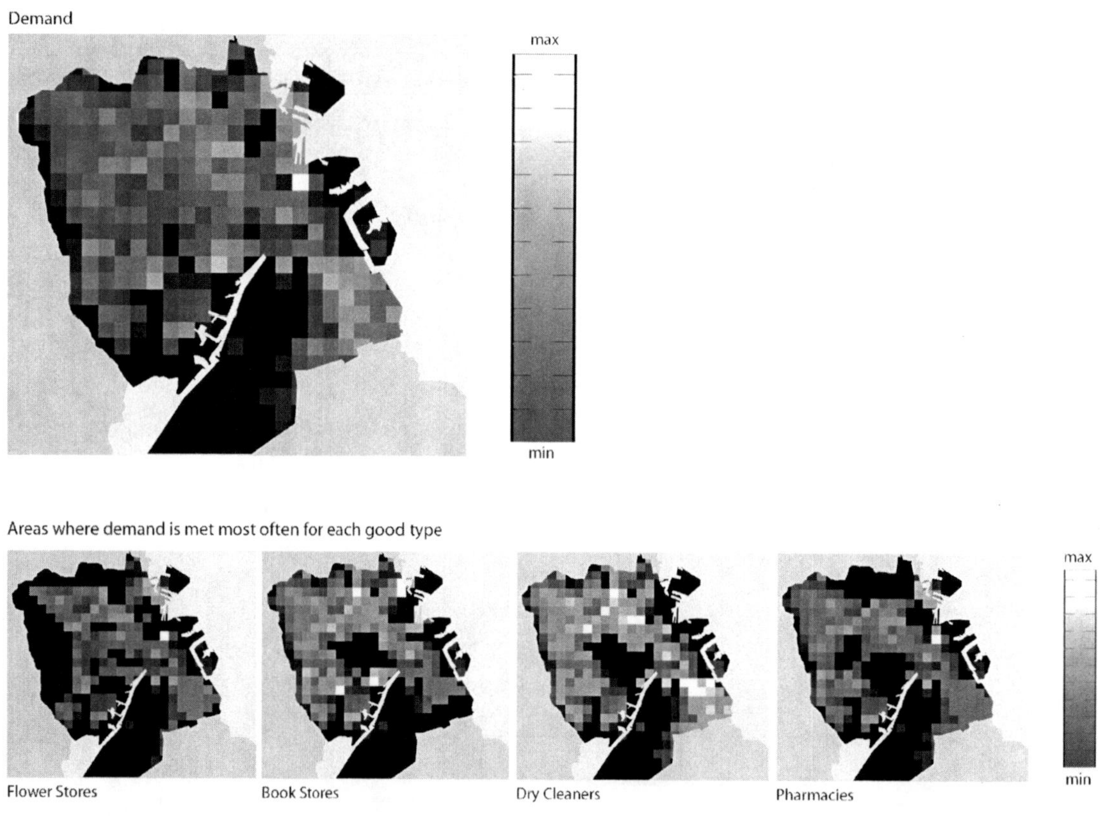

For flowers, books, dry-cleaning and medical items the following percentages of requests are met: 62%, 75%, 70% and 71% respectively. Figure 10, tells us the areas of the city where demand for goods can be met for most of the time using this service. Additionally, it is interesting to see that different levels of performance and ideal spatial coverage are seen for different goods categories. Where demand is not met, this is an indication that with the current spatial distribution of shops, the service would not work in these areas and delivery should be restricted to those areas where coverage is total.

As a result of the first test, we found that dry-cleaning items require a lower percentage of carriers than other items, but at the same time the spatial demand is met quite well with a spatial match of approximately 88%. We thus decided to isolate this category for further tests.

Test 2

Establish how the demand and availability of carriers over time influences the performance of the service.

Using information about the demand per hour for dry-cleaning as well as the percentage of available people per hour as calculated in the previous section, we rerun the model to find the results shown in Figure 11.

At different hours of the day, the performance of the service varies. However, since different categories of people have different temporal movement patterns and therefore availabilities, the hours in which dry-cleaning demand can be met is generally extended using a combination of different categories of people. In conclusion, we find that it is best to have a mix of categories of carriers enrolled in the service in order to increase the effectiveness of the service for a wider time range.

CONCLUSION, LIMITATIONS AND FURTHER WORK

The purpose of this chapter was to highlight that a rising access to location and movement information opens up exciting new possibilities for services that can alleviate some of the issues inner urban areas are facing and to test the possibility of implementing such a service within a real-world context. To do this, we proposed a new type of goods delivery service – FriendFreight – that harnesses the untapped resource of the freight capacity of bicycles. It was then discussed how the use of

Figure 11. Total travel demand compared to service travel demand, at different times in the day using workers, students or a combination of both.

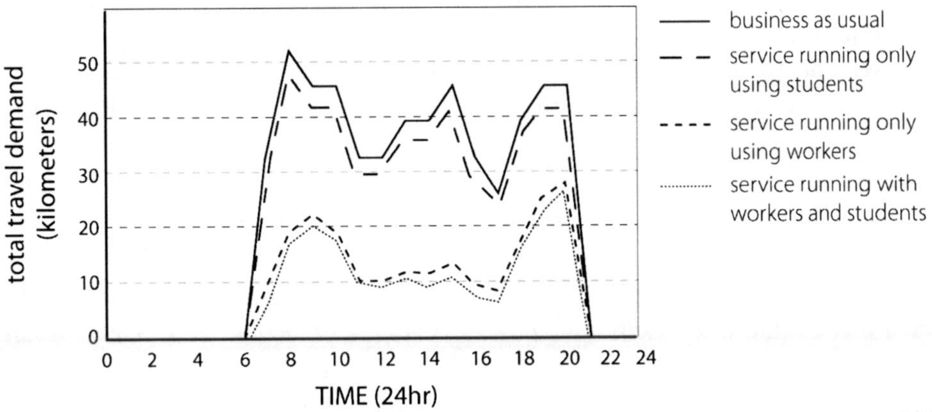

real-time information in this service contrasts with the work found in the field of City Logistics and results in a more bottom-up approach to reducing the negative affects of freight transportation. The social challenges associated with implementing this concept were outlined by briefly examining existing websites, such as eBay, Couch Surfing and LinkedIn. Here, it was asserted that in transactions that are not face-to-face but rely on the translation of digital information from a distance, an incentive for participation can either take the form of money, but can potentially also revolve around a common social goal. In addition, trust and reciprocity can be built through established mechanisms such as reputation systems, through the wording and ethos of the website itself, or through utilizing existing friendship and trust networks. When it comes to the privacy of location information, a strong common goal or incentive as well as clear indications for how the data will be used and disclosed may ensure that members do not see the collection and use of this information as a threat.

The final section of this chapter tested the feasibility of the service through creating a best-case scenario in terms of the greatest reduction in travel demand given the spatial layout of Copenhagen and the delivery of certain types of goods. Once it was established that the majority of bicycles in Copenhagen have enough freight capacity to carry goods, data collection was used to determine four likely categories of goods that could be delivered and their spatial and temporal demand. Two categories of potential carriers - university students and workers – were also selected and data about their mobility patterns and availability over time was collected.

Several tests were performed with the model using different conditions for the percentage of people involved in the service and the percentage of goods that need to be delivered. Of these tests, the three that are the most useful (should the service be implemented) are presented in this chapter.

What was found was that some items could be more effectively delivered than others – something that is dependent on the unique spatial distribution of shops and residences in Copenhagen as well as the availability of cyclists to carry goods. This seems to suggest that the service might be restricted to one set of goods rather than another and that the viability of the service and the type of goods that can be delivered through it are likely to vary from city to city.

The results also show that some areas of the city are better than others when it comes to having goods successfully delivered to residences, the majority of the time. In addition the spatial pattern of this met demand varies from good type to good type. This is interesting because delivery zones are often based on neighborhood boundaries and yet these become completely arbitrary in the successful running of the FriendFreight service.

Concerning carriers, it was found that it is best to have a mix of students and workers enrolled in the service as this allows an increase in the number of goods that can be delivered across the course of a day. Lastly, when delivering 10% of goods through the service, it was also found that travel demand could be significantly reduced. In the case of dry cleaning the theoretical reduction was up to 70%.

There are, of course, some obvious limitations with our model. The way it has been defined gives a base-case scenario in which it is known where all people are going and their availability. It is also assumed that if we define a carrier as available (based on their daily mobility patterns) that they will always accept a request to deliver goods. It is recognized that this type of global optimization only gives us a theoretical approximation of the real system when the service is in operation. However, as an initial test case, it can still provide some key recommendations and with additional data on the exact movement patterns of people and their willingness to be involved in the service, it is felt that the model could be further developed and would be a useful indication of whether or not it was worth implementing the service.

In conclusion, the FriendFreight service presents a novel way to combine the freight capacity that is inherent in the personal mobility vehicles of the general public with the movements of small goods in cities through harnessing the power of digital information that can be distributed at a distance. What should be remembered, however, is that while a rise in access to real-time location information allows these types of bottom-up services feasible (as proved in our computer model), issues such as trust, reciprocity and incentives for participation are equally important in having this service become viable.

Lastly, although we examined only one type of vehicle – bicycles – in this chapter, it is felt that more work could be done to assess the potential for a service like this to be used to deliver other types of goods with larger vehicles. This may have a greater impact on reducing the negative effects associated with travel demand and thus increase the potential for the service to be used to address the overall sustainability of goods movements in cities.

ACKNOWLEDGMENT

We would like to acknowledge the many people at the Massachusetts Institute of Technology who supported this research and gave us valuable feedback. In particular, thanks to Carlo Ratti, Assaf Biderman, Bill Mitchell, Liang Liu, Gautier Krings and Jon Reades. Additionally, this work would not have been possible without the City of Copenhagen and we would like to thank them for their support and provision of statistical information without which the modeling of the service would not be possible.

REFERENCES

Adams, P., Ashwell, G., & Baxter, R. (2003). Location Based Services – an overview of the standards. *BT Technology Journal*, *21*(1), 31–42. doi:10.1023/A:1022572210026

Beresford, A., & Stajano, F. (2003). Location Privacy in Pervasive Computing. *Pervasive Computing, IEEE*, *2*(1), 46–55. doi:10.1109/MPRV.2003.1186725

Brin, D. (1998). *The Transparent Society*. Reading, MA: Addison-Wesley.

Dablanc, L. (2007). Goods Transport in large European Cities: Difficult to Organize, Difficult to Modernize. *Transportation Research Part A, Policy and Practice*, *41*, 280–285. doi:10.1016/j.tra.2006.05.005

Dobson, J., Fisher, P. (2003, Spring). GeoSlavery. *IEEE Technology and Society Magazine*, (pp. 47-52).

Fusco, G., Tatarelli, L., & Valentini, M. (2003). Last-Mile: A procedure to set-up an Optimized Delivery Scheme. In *Proceedings of the 3rd International Conference on City Logistics*, (pp. 147-161). Madeira, Portugal: Elsevier.

Gedik, B., & Liu, L. (2005). Locations Privacy in Mobile Systems, A personalized Anonymization Model. In *Proceedings of the 25th IEEE International Conference on Distributed Computing Systems*. (pp. 620-629). Los Alamitos, CA: IEEE Press.

Ito, T., Hayano, M., Naito, T., Asakura, Y., Hato, E., & Wada, T. (2004). Empirical Analysis on hazardous material transportation using road traffic sensors and accident data. In *Proceedings of the 3rd International Conference on City Logistics*, (pp. 235-250). Madeira, Portugal: Elsevier.

Koehler, U. (2004). New ideas for the City Logistics Project in Kassel. In *Proceedings of the 3rd International Conference on City Logistics,* (pp. 321-332). Madeira, Portugal: Elsevier.

Kunze, O. (2004). A new interactive Approach on Route Planning with Tight Delivery Time Windows. In *Proceedings of the 3rd International Conference on City Logistics,* (pp. 83-96). Madeira, Portugal: Elsevier.

Lozano, A., Antun, J. P., Santos, C., Alarcon, R., Granados, F., & Hernandez, R. (2005). Basis for a policy for the development of Logistics Platforms in the Metropolitan Zone of Mexico City. In *Proceedings of the 4rd International Conference on City Logistics,* (pp. 147-161). Langkawi, Malaysia: Elsevier.

Miller, S. (1995). *Civilizing Cyberspace: Policy, Power and the Information Superhighway*. Reading, MA: Addison-Wesley.

Molz, J. (2007). Cosmopolitans on the Couch: Mobile Hospitality and the Internet . In Molz, J., & Gibson, S. (Eds.), *Mobilizing Hospitality, the ethics of social relations in a mobile world* (pp. 65–80). Burlington, VT: Ashgate Publishing Ltd.

Munuzuri, J., Larraneta, J., Onieva, L., & Cortes, P. (2004). Estimation of an origin destination matrix for urban freight transportation: Application to the city of Seville. In *Proceedings of the 3rd International Conference on City Logistics,* (pp. 67-82). Madeira, Portugal: Elsevier.

Nemoto, T. Browne, M. Visser, J. & Castro, J. (2005). Intermodal transport and City Logistics policies. In *Proceedings of the 4rd International Conference on City Logistics,* (pp. 15-29). Langkawi, Malaysia: Elsevier.

Pichler, R. (2000). *Trust and Reliance – Enforcement and Compliance: Enhancing Consumer Confidence in the Electronic Marketplace*. Unpublished doctoral dissertation, Stanford Law School, Palo Alto, CA.

Quack, H., Van Duin, R., & Visser, J. (2006). City Logistics over the Years. Lessons learnt, Research Directions and Interests. In *5th International Conference on City Logistics.*

Resnick, P., Zeckhauser, R., Swanson, J., & Lockwood, K. (2006). The value of reputation on eBay: A controlled experiment. *Experimental Economics, 9,* 79–101. doi:10.1007/s10683-006-4309-2

Statistikbanken. (n.d.). Retrieved from http://www.statistikbanken.dk/statbank5a/default.asp?w=1440

Taniguchi, E., Thompson, R., & Yamada, T. (2001). Recent advances in modeling city logistics . In Taniguchi, E., & Thompson, R. G. (Eds.), *City Logistics II* (pp. 3–34). Kyoto, Japan: Institute of Systems Science Research.

Thompson, R. G. (2004). Intelligent Vehicle Routing and Scheduling. In *Proceedings of the 3rd International Conference on City Logistics,* (pp. 97-110). Madeira, Portugal: Elsevier.

Yamada, T., Yoshimura, Y., & Mori, K. (2004). Road Network Reliability Analysis using Vehicle Routing and Scheduling Procedures. In *Proceedings of the 3rd International Conference on City Logistics,* (pp. 111-122). Madeira, Portugal: Elsevier.

Chapter 12
Attribute Hierarchies to Tally Fish

Andreas Mitrakas
European Network and Information Security Agency (ENISA)

ABSTRACT

This chapter presents a model that addresses some shortcomings associated with limitations in the way that electronic identities are used in an application environment. This model, Snowflake, allows the exchange of validation data in order to establish trust when identity attributes are used in applications. Snowflake leverages upon groups of validation authorities as electronic intermediaries to facilitate application interoperation. The application area of choice is the electronic reporting of fish catches by a vessel's master in line with the European Union legal framework. Grouping together the services of multiple validation authorities maximises the utilisation of resources available especially in environments that do not shy out in sharing resources and data such and the electronic fish logbook provides a good platform in this respect.

INTRODUCTION

Electronic identity management has emerged as the lynchpin of various eGovernment applications that depend on user authentication. Electronic identities based on electronic signatures allow the identifying, authenticating and ensuring the non-repudiation and integrity of a transaction. Electronic identity management also allows users to act under a role in environments where a multifunctional approach is necessary. Attributes can be leveraged upon to come up with identity services in areas where there is a need to link a user with a role rather than with its real life identity. The growing use of electronic identity technologies in disparate groups of users has also highlighted the need for interoperability at the application level in order to allow for better diffusion of the benefits associated with these transactions. As applications grow in what is often perceived as isolated contexts, it is often necessary to ensure efficient application management by leveraging upon data sets generated in these isolated

DOI: 10.4018/978-1-61520-769-5.ch012

contexts. Snowflake is an architecture that allows the exchange of information in order to establish trust in an electronic identity management based on a validation centric model. The remainder of this paper addresses the following areas: an overview of interoperability and identity, some considerations on an electronic identity model based on intermediaries and Snowflake, a model that enhances interoperability through a specific business process that relies on grouping together validation authorities.

The remainder of this Chapter describes the background of identity management upon which eGovernment services build. The use of attributes is described in the context of extending their functionality in a way that allows for carrying out performative transactions. To allow the interoperation of attribute hierarchies to interoperate, an interoperability architecture based on the validation of attribute certificates is discussed in a way that permits the interoperability of multiple validation authorities. This attribute validation model is set up against the transactions background of the electronic fish logbook, in line with the prevailing legal framework in the European Union. This Chapter claims that (a) attributes can replace fully fledged identities where it is not necessary or desirable to reveal fall or most identity elements in due consideration of privacy or the requirements of a transaction; (b) additionally a validation centric interoperability model can provide an alternative to interoperable certification authorities, the policy criteria of which might be a challenge to reconcile; (c) discreet policy environments and formal applications and transactions such as the electronic fish logbook can benefit from the functional interoperability allowed under the proposed certificate validation model in a way to facilitate European Union policies in terms of functional cooperation in discreet areas of public services.

INTEROPERABILITY IN MANAGING IDENTITIES

Interoperability is the ability of products, systems, or business processes to work together in order to accomplish a common task. While interoperability is a technical consideration it typically serves business needs by interlinking otherwise disparate systems that need to be leveraged together in order to deliver a common service. In eGovernment processes interoperability addresses such considerations as how information is used or exchanged in the various public administration applications, which parties have access to certain types of information and how information is presented to various types of end users that might belong to different contexts. Factors that contribute to determining the ability of a system to interoperate with others include the way a system is built, trade secrets and potential shortcomings in aligning with other peer systems. As a measure to contain such shortcomings user communities and government agencies have set up their shared interoperability structures like for example e-GIF in the UK, SAGA in Germany, IDABC at European Union level etc. Sometimes eGovernment services have been based on seemingly isolated and unrelated computerized applications that have overlapped in function and content within public organizations, thus resulting in relatively isolated islands of technology (Eymeri, 2001; Janssen & Wagenaar, 2004). This approach has clashed with the concept of information systems that had been viewed as an integral part of public services (Janssen & Wagenaar, 2004). In the specific case of electronic identification eGovernment, interoperability has emerged as a requirement for the provision of shared services in terms of authentication and non repudiation of public acts and systems use. Such shared services can eventually help organizations make available recurrent, pattern-based services to serve large populations of users. To facilitate large deployments service oriented architectures help the orchestration of eGovernment services and support cross-agency processes (Gortmaker, Janssen, & Wagenaar, 2007).

Electronic identities are often based on electronic signatures or more specifically the management of digital certificates that are supported by a public key infrastructure (PKI). While PKI offers an attractive approach to managing identities of end users that are hierarchically organised they encourage a silo approach that encourages islands of identity hierarchies. Electronic identities in eGovernment are likely to experience a drive towards interoperability due to the need for eGovernment services to serve large disparate sets of users that might currently duel in isolated contexts while being serviced through silo applications (Mitrakas, Hengenveld, Polemi, & Gamper, 2007). Cross certification emerges as a response to identity environments that seek to link with each other and allow leveraging on a PKI to carry out transactions across discreet application environments. In PKIs the lack of automated cross-certification services can be attributed to the inadequate standardization at technical and operational levels. The instruments to control PKI islands include certificate policies (CPs) and certificate practice statements (CPS) that often reflect legal divergence in the framework pertaining to the electronic certificates issued under such policies. PKI certificate profiles tend to follow diverging approaches challenging the relying party that has to positively compare them and establish trust and equivalence levels. While the CP structure is defined in current standards such as ETSI TS 101 456 (2004) (See also, RFC 3039, 2001) or ETSI TS 101 042 (2002), there is still a significant gap in the standardization of the way that the CP requirements are implemented by certification authorities. An example can be made out of the identification requirements that regard the roles of the subjects involved, certification and registration requirements, etc. Although the certification policy requirements have been broadly standardised across the board, there is no such attention just yet reserved to other significant aspects of certificate management such as the validation policy requirements, and the application policy requirements. The significance of the lack of sufficient standardisation levels can be mostly felt in a validation based framework that seeks to service application oriented goals. The lack of standardising policy requirements for certificate validation as such means that relying parties are likely to accept certificates on the terms of the issuing authority, which is part of a certification authority, alone while a more neutral validation centric approach remains out of reach. Along the same lines the application framework also calls for the setting of application level standards for application level policies that set out the conditions to access and deliver services therein

The apparently limited ability of eGovernment architectures to interoperate exacerbates concerns related to policy management. Policy in this regard is necessary to translate complex technical terms in warranties for the delivery of the service and the conditions for policy enforcement. Interoperability can be sought where eGovernment certification services need to be leveraged in a different context than the one that they originate from, e.g. a Belgian identity card issued by the city of Brussels is used to access a service in Tartu, Estonia. Public Administrations further enhance eGovernment services by keeping their service providers at arm's length through Service Level Agreements (SLA) and making available their services on the basis of policies that set out the conditions of accessing such services by stakeholders and users.

On the outset the divergence in views as what an interoperable electronic identity model should look like can be seen as a deterrent in progressing with interoperability in transactions. As there are various interoperability models, competition has led to diverging routes being adopted by eGovernment actors that little resemble the models adopted by the industry when faced with similar challenges. The root management approach followed by the PKI industry under the lead of Microsoft has led to industry led solutions that were corroborated by the accreditation scheme WebTrust for CAs promulgated by the American Institute of Chartered Accountants (AICPA). In this instance trusted roots of certification authorities that were de facto recognised by Microsoft in their associated applications, such as browsers were further validated with regard to the procedures that prevailed in safekeeping them and maintain a trusted status for them. AICPA made sure that the conditions were solidly put in place. Member State authorities such as the Ministry of Economics of Belgium recognised the WebTrust scheme for servicing their own purposes and in order to confer a "recognised" status to such service providers as GlobalSign when seeking to deliver services to the government of Belgium. Other ad hoc models such as the networks of trust of leading certification service providers have also aimed at providing solutions by extending the reach of a trust anchor over a subordinate community of stakeholders.

The stakeholders in electronic identity are reluctant to allow leaking or spilling over identities motivated by their inherent desire to control electronic identities much like they have been used to controlling physical identities of data subjects over time. The motivation is obviously based on concerns associated with the use of credentials in unauthorised environments, or beyond the closed user groups that have originally been thought that they belong etc. Such examples include customer relations in the private sector or the reluctance to collaborate at cross border level with regard to citizen identity management in eGovernment services (Cameron, 2005). The shifting landscape from isolated IT structures to web services facilitates the coupling of services and makes it necessary to add functionality to the electronic identity layer as an instrument to facilitate and control end users. The perspective of open grid services architecture and the shift towards cloud computing and shared environments raises the stakes to address electronic identity interoperability and quench the thirst of eGovernment services to interoperate with each other as well as with private sector services to the benefit of end users

Estevez (Estevez, Janowski, & Ojo, 2007) has suggested that "Most countries implementing e-government have developed national infrastructure projects that enable agencies to deliver local services more effectively. Infrastructure development involves many technical issues, such as: how the services are delivered, how the agencies connect to one another, and how they can tackle common problems (Authentication, secure transactions, etc.). Infrastructure development projects must be closely viewed with an eye on e-services opportunities. After identifying such opportunities, the common problems related to their development should be synthesized and analyzed." Identity management scores high on the list of pending issues for the successful implementation of eGovernment services mostly due to the horizontal requirement of government services to authentication the entitlement of the users to gain access to such services and benefit from them.

IDENTITY AND INTERMEDIARIES

To identify is a means to ensure that an entity can be distinguished from any and all remaining ones within a given context. The rule of excluded middle prevails in defining identity as everything is what it is and not something else, e.g. $x=y \vee x=\sim y$. This convention applies in physical identity of individuals, the role the play, attributes of their identities etc. As a result various features of identity have cropped up, as electronic identity variants, like for example:

- Identity of persons such as physical identity,
- Identity of roles such as role/attribute management,
- Identity in order to carry out a transaction while being in a performative mode.

Identity is a concept that remains discreet from authorisation as being who one claims to be (authentication) or performing an action that has the ability to change the status of things in a way that cannot be repudiated, does not necessarily mean that one is also authorised to carry out a specific act or omission. Long standing concerns regarding privacy, is an additional consideration when acting under a role in an electronic environment. To mitigate privacy risks when acting in a specific context such as a professional environment masking real life identities can have certain benefits as opposed to revealing the full fledged features of one's own identity; especially when the full identity might not necessarily be an essential element to carrying out the transaction. Validating a payment request would require having access to a properly issued certificate of a financial actor for the purpose of validating transactions and not necessarily a certificate issued to Mr. John Smith, Financial Assistant. In a business oriented context, that is free from lateral concerns on criminality, pseudonymity should be sufficient where physical identification is not required as such.

Non repudiation credentials can be issued in line with the qualified certificate requirements have been invoked according to Directive 1999/93/EC of electronic signatures and meet the policy requirements articulated in ETSI TS 101 456. Furthermore, strict identification of the individual signatory may not be necessary if the signatory can be identified within a specific application context anyway. In this regard a pseudonym could rather be used, under condition of course that the context or closed user group, remains firmly in control of the physical, real world identity of the signatory. Additional conditions to assert pseudonyms include revealing only portions of one's own identity as it might be required in order to carry out a specific transaction, which allows for the instantiation of attributes and role management in the form of issuing attribute certificates. It is understood that under the conditions of the ITU X.509v4 protocol, issued in 2000 new features were introduced to allow issuing attribute certificates in a way that makes it possible to use them for access control and authorization. Privacy requirements are observed if only essential data is revealed for clearly identified purposes.

In electronic identity much of the work done with regard to managing an electronic identity infrastructure is carried out through intermediaries. The principal, which in this case is a government agency, typically serves as registrar of requests for the issuance of electronic identity credentials. Other tasks such as the issuance proper, the validation, token management etc., are usually outsourced to several specialised service providers. The exacerbated role of electronic intermediaries is a feature in electronic transactions, which is a service-centred model. In electronic identity, intermediaries have gained a discreet role, thanks to the drive of Public Administrations to outsource large portions of the electronic identity management to service providers. Public Administrations have nevertheless kept core parts of activity for themselves only. In Sweden for example, the public sector also relies on identities that are issued by banks. In Belgium a fully outsourced model has been used while the Public Administration has kept for itself basically a role in the identification of citizens.

To date electronic identity intermediaries have provided discreet electronic identity related services that cover the issuance of credentials, time-stamping services, token management etc. Major functions of an electronic identity context that leverages upon intermediaries in electronic identity management include the application owner, the application service provider, the eGovernment link and services provider (Mitrakas et al., 2007).

THE ELECTRONIC FISH LOGBOOK

The approach described above can be particularly useful in environments where a role is desired rather than the real life identity of the user of the system. Such is the case of reporting systems for fishing catches in line with Member State quotas (Council Regulation (EC) No 1966/2006 of 21 December 2006 on electronic recording and reporting of fishing activities and on means of remote sensing), prevailing in the European Union. Electronic recording and reporting in fishing builds upon the widespread use of vessel monitoring systems by the principal flag or coastal states.

Currently, several satellite systems, (e.g Inmarsat, Iridium, and Argos satellite communications systems globally and the regional Euteltracs and Globalstar systems) have set the stage for a competitive market to track fishing vessels. The Global Positioning System has added new functionalities to positioning accuracy and the European Union Galileo system makes available its services too. Information and satellite technology combined have been the instruments available to the fishing industry to achieving sustainable harvesting of fish (OECD, 2005).

Since the introduction of the Common Fisheries Policy of the European Union, a requirement to recording fish catches has been established. In fishing services an electronic fishing quota logbook is compiled upon receipt of the catch reports by vessels' masters while at sea. These catch reports are data files that contain a list of what has been caught by a fishing vessel. Data fields include the date of catch, fish species, quantity and calibre. Although vessel monitoring is automatic, catch data is all but manual. A signature is required for all logbook recordings that are used to monitor the activity of a vessel with regard to fishing. Attributes can be as well used in order to ensure the role of the signatory, the vessel's master, who might otherwise be channelled to using systems that do not necessarily represent her actual needs.

SNOWFLAKE: AN ATTRIBUTE BASED MODEL FOR INTEROPERABLE IDENTITIES

In electronic identity attribute certificates can be used to deliver signature enhanced user authentication while acting under a role. Role management should be leveraged upon as appropriate in eGovernment where strict identification of the signatory is not mandatory by law or contract. Such is the case when a professional or an actor performing within a particular transaction context has to prove her ability to act under a certain capacity. This role acting should be kept discreet from acting as individual, which is likely to result in enhanced privacy concerns. For example an accountant might be required to prove her professional qualification through an attribute in order to sign a financial statement.

An interoperability model based on certificate validation is proposed to be used in delivering several services that require a permit within an eGovernment environment, such as for example the electronic fishing logbook etc. Snowflake allows the smooth exchange of critical information in order to establish trust in an electronic identity management environment that is based on a validation centric model through the exchanging of attribute related data. Snowflake builds on the concept of electronic intermediaries that act as shared e-government services to facilitate application management that may also include certification authority related services. Leveraging upon shared e-government services nodes can enhance the way eGovernment services evolve in terms of implementing identity management requirements. Snowflake may further be interlinked to a hub electronic identity management service architecture that allows for the interoperability of electronic identity data related to qualified certificates. The Snowflake framework calls for placing special emphasis on policy aspects as a means to provide an appropriate framework that meets technical, organisational and legal requirements.

Authorization policy statements can be based on authorization assertions that include rights delegation or capabilities, used in push model authorization systems and resource-centric access policy statements used in pull models. In an authorisation policy a proxy is used as a token to allow one to operate with the rights and privileges of the principal that granted the proxy. Neuman (1993) has described various kinds of proxy certificates such as full proxies, restricted proxies, bearer and delegated proxies to conclude that by extending an authentication system to support restricted proxies, it becomes possible to support distributed authorization and accounting mechanisms. In an application environment it suffices to delegate to a proxy the ability to carry out an act on behalf of a principal and limit the proxy's discretion by means of an access control list in order to contain its ability to further delegate its privileges. Clearly there is a need to extend the sheer technical approach to include transactions that have legal and policy consequences, while they touch upon the rights of large user communities, as it is the case in eGovernment; this is the objective of the remainder of this paper.

Snowflake is an attribute certificate interoperability model that tackles several shortcomings typically associated with PKI provided that conditions such as the following are met:

- Attributes based on electronic identity (that meet the issuance requirements for qualified certificate, albeit not for the identification of the signatory) for role management
- Only mission critical info is extracted from a qualified certificate issued to relate a role to an individual
- The quality of registration of attribute data is a key feature
- The archiving of the registration data is essential too.
- Privacy is observed because identity data is used on a need to have basis
- Attribute credentials ensure the identity of the actor
- Attribute certificates interoperability model to ensure scalability
- A validation centric model removes the typical shortcomings of PKI hierarchies
- User centric model that protects personal data
- Scales better to help "federation" of identities

Application centric solutions can enhance interoperability in specific operational environments. Especially in eGovernment this approach might bear fruit in terms of leveraging upon infrastructures that are already in placc. A policy framework can ensure that this approach becomes binding among its participants and that technical, organisational and legal constraints of each approach are respected.

Policy in this context makes reference primarily to a CPS. A certification authority typically uses policy in the form of a CPS to convey legally binding limitations to certificate users, being subscribers and relying parties. A CPS is a statement of the practices that a CA employs in issuing certificates (ABA, 1996). A CPS is a comprehensive treatment of how the CA makes its services available and delimiting the domain of providing electronic signature services to subscribers and relying parties. A certificate policy (CP) is sometimes used together with a CPS to address the certification objectives of the CA implementation. While the CPS is typically seen as answering "how" security objectives are met, the CP is the document that sets these objectives (ABA, 2001; Mitrakas, 2005b).

In a distributed model where each participating state can maintain its own policy autonomy, policy mapping could support the use of multiple policies in the application context of fishing quotas (Oduntan, 2004). The rating of trustworthiness based on certificate policies of the participating issuing authorities and validation authorities can be envisaged. Policy mapping can be based on trust criteria. An automated policy mapping tool can support the decision making process of the relying party to make informed decisions with regard to the trustworthiness of the issuing authorities against the criteria that relying parties attach importance to (Batarfi & Marshall, 2007; Backhouse, Hsu, Baptista, & Tseng, 2003).

In this model the validation authority is deemed to be the appropriate or designated entity upon which trust is leveraged for the purpose of enhancing interoperability. The validation authority may on its own turn introduce its own validation requirements that could be enshrined in a discreet policy framework. In this model this framework is dubbed as validation policy statement and it encompasses the validation content of standardised policy frameworks such as ETSI TS 101 456. A Validation Practice Statement could be used to describe the operational technical and legal constraints of the validation centric interoperability model (Mitrakas, 2008).

VALIDATION-CENTRIC INTEROPERABILITY

A validation centric interoperability model can be plotted along a set of overarching validation authorities. These overarching validation authorities meet the operational and legal requirements of their own discreet contexts and they adhere to the policy requirements that are prevalent in discreet application environments for the purpose of validation. Such requirements can be instantiated in a validation practice statement that is a policy document addressing the needs of the group status of these validation authorities. Such overarching validation authorities see to featuring a meta-directory that is typically based on validation data featured in the typical directory of a validation authority. It is necessary, to ensure that the policy and technology measures implemented by each of the participating overarching validation authorities allow for the accessing and broadcasting of certificate validation data across the board of relying parties. The participating overarching validation authorities may select their policy constraints by themselves in order to meet the functional requirements of their operational context. Attribute certificates issued, can meet the need to identify for example ea vessel's master for the purpose of signing data elements corresponding to the electronic fish logbook. The basic validation model based on an overarching validation authority is illustrated in Figure 1.

End entities such as citizens, businesses or other public administration organisations that are entitled to request services within the proposed infrastructure must register locally with the certificate service of their own administration. To facilitate the interactions among several public administrations, certificate services further need to trust each other by recognizing each other's top root so that they allow end entities to carry out end user transactions.

To validate end-entity certificates an overarching Validation Authority (oVA) makes available validation services by redirecting certificate validation requests to the appropriate validation service of a local issuing authority. An oVA provides the end entity with the technical capability to check the certificate status in order to facilitate an application that relies on validation data. Therefore the end entity in its role as relying party is facilitated to access validation data. Validation data can be transmitted using the prevailing protocols of online status protocol (OCSP) (RFC3280, 2002; RFC2560, 1999). Browsers and other applications have the ability to allow users to add other root certificates besides the ones included therein by default as needed. The end user can therefore update the default lists of CA roots to match their individual requirements.

A potential risk associated with the oVA is that by easing up root certificates that can take place in the context of end entity applications,

Figure 1. The basic validation model based on an overarching validation authority

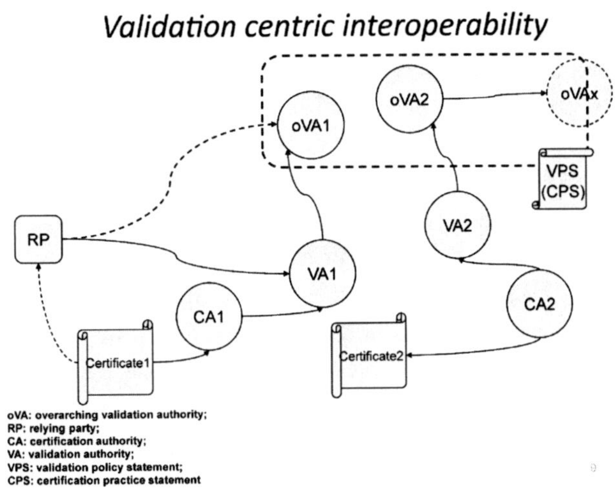

malicious actors could take advantage of them in order to add false root certificates. Such false roots can be further used to carry out fraudulent transactions. To overcome this problem, a list of trusted certification authorities is digitally signed by a dedicated Overarching Certification Authority (oCA). An oCA is a mere technical facilitator and is therefore entrusted with the competence to set up a policy and a technical framework to support the cross-border recognition of certificates issued by the various local certification authorities that belong to the trust chain. The group of overarching validation authorities leverages on the individual validation data that each oVA features to produce aggregates that are accessible to designated relying parties according to terms and conditions that are set out by each of the oVAs in their own operational contexts.

Policy mapping is a necessary condition for the oVAs to produce interoperable data aggregates. Besides policy mapping, setting up equivalence levels is a significant requirement that might also have to be introduced at a later stage in order to ensure compliance with prevailing rules and reconcile the differences among the approach towards voluntary accreditation conditions according to article 3.2 of Directive 99/93/EC on a Community framework on electronic signatures. Relying on intermediaries could result in collecting applicable policies in a repository as well as specifying audit conditions to provide advance notice concerning the legal status of the scheme or legal requirements under which a certificate has been issued or a root operates (Mitrakas & Bos, 1998). Policy mapping tools based on decision support systems could be developed in order to facilitate the management of disparate policies of participating certification authorities.

The way that operational requirements of an oCA are addressed, requires some attention as each time a new CA seeks recognition by the oCA its name and root certificate have to be added to the master trust list that is signed by the oCA and subsequently distributed to the implicated end entities. Distributing updated trust lists might face scalability shortcomings if there are more than a small number of certification authorities used to propagate these lists. This problem could be addressed by setting up a single reference directory for certification authorities that seek including the root in a given trust list. By accessing the directory of certification authorities' roots, an end entity that transacts with public administration entities enhances its ability to establish the trustworthiness of the certification authorities issuing the certificate in question because it can directly validate the certificate path online. The role of the oCA can be limited to providing the secure publication of the directory as well as ensuring the inclusion of a pointer to the directory by all trusted local CAs. Including certification authorities in a directory however is likely to require a classification mechanism for the certification authorities included therein (ETSI TS 102 231, 2005).

An additional area of future attention could address policy frameworks related with the application layer in eGovernment transactions. To respond to transparency requirements it is expected that online applications will increasingly become more demanding in explaining to the end user what they do and ultimately warranting the performance of the service. Relying on robust infrastructures for the delivery of end users services is likely to enhance the user experience and reliability of the electronic medium in a context such as the electronic fish logbooks. Through a combination of dedicated legislation and policy (e.g. CPS) this requirement has been partially addressed.

In eGovernment it is still required to use policies at the application layer level where the requirement for transparency in the transaction and accountability of the parties involved is higher than in electronic business (Mitrakas, 2005a). Finally, specifying policies further to meet the needs of particular application areas, groups or organisations are additional expectations, like for example in the area of the electronic fish logbook. Typical content for an application layer policy may include aspects associated with the kinds of underlying services that are involved, the profile of the providers that make these services available, architectural premises, disclosures with regard to data bases, personal data, validation services as well as liability or warranty that might be offered to the end user. The scope of application layer policy is to contribute to the transparency of the application platform that might ultimately lead to trust and legal safety (Mitrakas, 2005b).

While interoperability remains a constant requirement for information society, eGovernment applications need to also shift towards greater interoperability levels. A much needed requirement is to strive for a policy framework that meets the requirements of law and associated applications in due respect of the end users' rights, such as their expectation to privacy. In an example case, illustrated in the figure below the vessel's master is issued with an attribute certificate which is made available by an Attribute Authority (AA). The attribute certificate is dependable because it relies on data obtained by validating a qualified certificate that is available on electronic identity cards issued to citizens in several European Union Member States; purposely issued certificates can be considered in those member States that do not have a national electronic identity card program yet. As there is no real need to use the "real life" certificate to carry out a designated business transaction, such as reporting fishing quotas while on high seas, the vessel's master can use her attribute certificate. This certificate is subsequently validated and the submitted data signed with it is further admitted by the designated port authority alongside the data logs that it has been used to sign.

Interoperability in Snowflake is achieved by combining policy requirements in the context of the overarching validation authorities (oVA), the attribute validation authorities (AVA) and the overarching attribute authorities (oAA). It is important to consider sets of authorities in this case, in order to facilitate the priorities pursued by individual users that need to use certificates in their own contexts (Figure 2). Groups of overarching entities are more likely to meet the requirements of individual entities that could be industry groups, Member States, etc.

Sets of policy requirements with emphasis on the validations can be instantiated by means of a validation practice statement. Snowflake facilitates an application centric approach that ensures validation of identity data and can accommodate

Figure 2. Interoperability in Snowflake

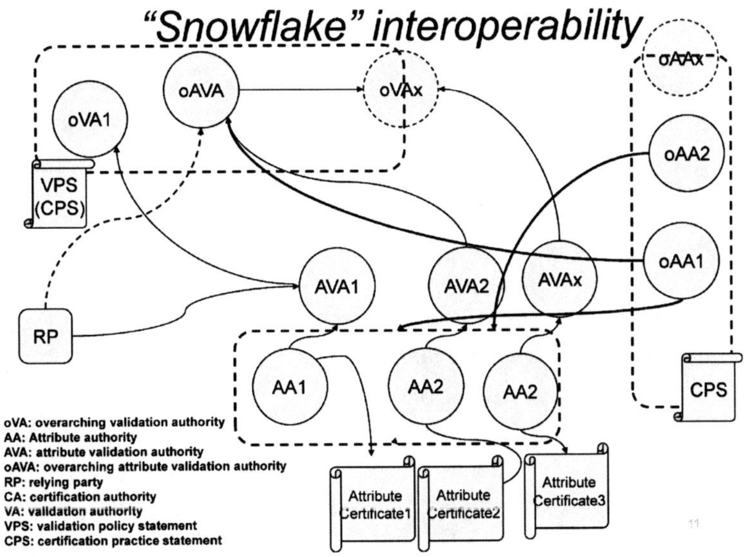

an array of applications. By simplifying the certificate validation process, Snowflake contributes to interoperability of multiple applications. To propagate the Snowflake model more than one electronic identity intermediaries could be set up in a way that facilitates discreet application contexts that share electronic identity infrastructures in eGovernment.

CONCLUSION

This paper claims that attributes can replace fully fledged identities where it is not necessary or desirable to reveal fall or most identity elements. This approach is desirable when specific identity features are not desirable to be revealed in due consideration of privacy or the functional requirements of a transaction, like it is the case when acting under a role. A validation centric interoperability model can provide an alternative to interoperable certification authorities, the policy criteria of which might be a challenge to reconcile. Certification authorities, especially those the public sector have been proven to be a difficult act to carry out when it was required to have them cross certified for example.

Shifting efforts instead, in enhancing the interoperability of validation authorities, which make services available on the basis of certificate validation lists, might be proved more rewarding a task. Snowflake is a validation centric model that enhances interoperability in an electronic identity environment by relying on grouping together validation authorities. Trusting validation authorities to deliver basic interoperability functions is acceptable because validation is a role that does not necessarily cause concern with regard to the issuance requirements that a certification authority adheres to. Therefore it is more likely to be accepted in various application contexts.

Grouping together various overarching validation authorities allows for multiple service providers to spell out their own service requirements and group them together in service contexts. Electronic intermediaries can be put in charge of bundling the various service requirements in a way that makes such services meaningful in distributed application contexts. Snowflake leverages upon the concept of shared environments that are facilitated through an electronic intermediary. Discreet policy environments and formal applications and transactions such as the electronic fish logbook can benefit from the functional interoperability allowed under the proposed certificate validation model in a way to facilitate European Union policies in terms of functional cooperation in discreet areas of public services. Further work on the policy aspects associated with certificate validation can enhance the Snowflake model and underpin its practical use in terms of sharing resources in the area of electronic fishing logbooks with regard to the signature requirements, based on a validation centric approach that observes functional and privacy requirements.

ACKNOWLEDGMENT

The views expressed in this paper are strictly personal and they do not reflect in any way whatsoever the views of the author's employers past or present. The background of this chapter associates with the author's involvement in the framework of the SHEEL (Secure and Harmonised European Electronic Logbook) project (contract no: 502153) that was co-funded by the European Commission. The author wishes to thank Mr. Stephane Mans for his valuable insight in this respect.

REFERENCES

American Bar Association. (1996). *Digital Signature Guidelines*. Chicago: Author.

American Bar Association. (2001). *PKI Assessment Guidelines*. Chicago: Author.

Backhouse, J., Hsu, C., Baptista, J., & Tseng, J. (2003). The key to trust? Signalling quality in the PKI market. In *Proceedings of the 11 European Conference on Information Systems* (ECIS 2003), Naples.

Batarfi, O., & Marshall, L. (2007). Conformance testing a set of criteria for assessing trust. *Journal of Computers*, *2*(1). doi:10.4304/jcp.2.1.44-49

Cameron, K. (2005). *The Laws of Identity*. Redmond, WA: Microsoft Corporation.

Estevez, E., Janowski, T., & Ojo, A. (2007). *Planning for e-Government – A Service-Oriented Agency Survey*. United Nations University, UNU-IIST Report No. 361.

ETSI TS 101 042 (2002). *Policy requirements for certification authorities issuing public key certificates*, Sophia Antipolis.

ETSI TS 101 456 V1.2.1 (2004). *Policy requirements for certification authorities issuing qualified certificates*, Sophia Antipolis.

ETSI TS 102 231 V1.2.1 (2005). *Provision of harmonized Trust Service Provider status information*, Sophia Antipolis.

Eymeri, J. (2001). The electronic identification of citizens and organisations in the European Union: State of Affairs. In *37th Meeting of the Directors-General of the Public Service of the Member States of the European Union*, 26-27 November 2001, Bruges.

Gortmaker, J., Janssen, M., & Wagenaar, R. (2007). Requirements on cross-agency processes in eGovernment: the need for a reference model . In Mitrakas, A., Hengenveld, A., Polemi, D., & Gamper, J. (Eds.), *Secure eGovernment web services*. Hershey, PA: IGI Publishing.

Janssen, M., & Wagenaar, R. W. (2004). Developing Generic Shared Services for e-Government. *Electronic . Journal of E-Government*, *2*(1), 31–38.

Mitrakas, A. (2005a). Policy Frameworks for Secure Electronic Business . In Khosrow-Pour, M. (Ed.), *Encyclopedia of Information Science and Technology* (*Vol. I-V*). Hershey, PA: Idea Group Publishing.

Mitrakas, A. (2005b). Policy-driven signing frameworks in open electronic transactions. In Soft Law constraints in eGovernment, BILETA 2005, (British Irish Law Education & Technology Association), Belfast, 7 April 2005.

Mitrakas, A. (2008). Heidi: an electronic identity intermediary model in eGovernment . In Bouwman, H., Bons, R., Hoogeweegen, M., Janssen, M., & Pronk, H. (Eds.), *Let a Thousand Flowers Bloom: Essays in Commemoration of Prof. Dr. René Wagenaar*. Amsterdam: IOS Press.

Mitrakas, A., & Bos, J. (1998). The ICC ETERMS Repository to support Public Key Infrastructure. *Jurimetrics*, *38*(3).

Mitrakas, A., Hengenveld, A., Polemi, D., & Gamper, J. (2007). Towards secure eGovernment. In Mitrakas, A., Hengenveld, A., Polemi, D., & Gamper, J. (Eds.), *Secure eGovernment web services*. Hershey, PA: IGI Publishing.

Neuman, B. C. (1993). Proxy-Based Authorization and Accounting for Distributed Systems. In *Proceedings of 13th International Conference on Distributed Computing Systems*.

Oduntan, G. (2004). The evidentiary issues arising from the proposed use of satellite based vehicle monitoring system and electronic logbooks in the FishCAM Project of the European Union. *International Journal of law and Information Technology, 12*(1).

OECD. (2005). *Fishing vessel monitoring systems: past present and future, OECD High seas task force*. Paris: Author.

RFC 2560. (1999). Internet X.509 Public Key Infrastructure: Online Certificate Status Protocol – OCSP.

RFC 3039. (2001). Internet X.509 Public Key Infrastructure - Qualified Certificates Profile.

RFC 3280. (2002). *Internet X.509 Public Key Infrastructure: Certificate and Certificate Revocation List (CRL) Profile,* (obsoletes RFC 2459).

Section 4
Selected Readings

Chapter 13
Video-Based Motion Capture for Measuring Human Movement

Chee Kwang Quah
Republic Polytechnic, Singapore

Michael Koh
Republic Polytechnic, Singapore

Alex Ong
Republic Polytechnic, Singapore

Hock Soon Seah
Nanyang Technological University, Singapore

Andre Gagalowicz
INRIA, Le Chesnay, France

ABSTRACT

Through the advancement of electronics technologies, human motion analysis applications span many domains. Existing commercially available magnetic, mechanical and optical systems for motion capture and analyses are far from being able to operate in natural scenarios and environments. The current shortcoming of requiring the subject to wear sensors and markers on the body has prompted development directed towards a marker-less setup using computer vision approaches. However, there are still many challenges and problems in computer vision methods such as inconsistency of illumination, occlusion and lack of understanding and representation of its operating scenario. The authors present a video-based marker-less motion capture method that has the potential to operate in natural scenarios such as occlusive and cluttered scenes. In specific applications in sports biomechanics and education, which are stimulated by the usage of interactive digital media and augmented reality, accurate and reliable capture of human motion are essential.

INTRODUCTION

Work on motion capture and analysis started as early as the 19th century, when Eadweard Muybridge began photographing horses to analyze their movement. During that period, a French physiologist Etienne-Jules Marey also embarked on his chronophotography work in studying the human performance filmed. Chronophotography is an application of the study of movement (science), and photography (art).The importance of motion capture (mocap in short), is motivated by its applications over a wide spectrum of areas. Tracking and following the movement of human joints over an image sequence, and recovering the 3D body posture and kinematics are especially useful for the study of the human locomotion and bio-mechanical applications (Corazza, 2006), such as gait analysis, injury prevention, rehabilitation and sports performance enhancement.

Since the mid-1970s, human motion analysis applications have made significant progress thanks largely to the tremendous advancement in digital electronics technologies. The measuring and analyzing of human movement have applications in many domains ranging from kinesiology, ergonomics, sports, 3D animation, 3D tele-presence, augmented reality, video surveillance, video data compression as well as medicine and clinical practice. Many, if not all, of these systems require the tracking of the motion trajectory of the subjects to yield kinematic information for further analysis.

There are many technologies developed to capture the human motion. They range from magnetic, mechanical and optical systems, for which the subject needs to wear sensors and markers on the body, to the non-intrusive one which is based purely on using video cameras. The kind of set-ups, methods and technologies that are used for motion capture are largely determined by their respective operational needs, that is, application specific. For example, it is a common practice in biomechanical studies for the subjects to wear reflective markers for movement analysis while operating in a very well-controlled environment.

In order to be suitable for human biomechanical applications, which is our main focus, the motion capture technique has to yield accurate and reliable quantitative information. To extend the applications, cluttered, outdoor and occlusive environments have to be considered. Also, the freedom of movement to the performer is an important consideration, which implies that wearing of sensors on the body has to be avoided. These issues are especially crucial if we are to consider a real scenario such as a sports tournament. Therefore, it has prompted many researchers to explore the use of video-based computer vision motion capture, which do not need any sensor or marker on the subject, for measuring the human movement.

Over the years, the use of video applications in the teaching of sports and physical activities have ranged from qualitative analyses to augment feedback in learning new sports skills (Koh & Anwari, 2004) to multimedia online environments with rich displays of animations and videos and embedded discussion forums to facilitate analysis and critique. These were designed to convey multiple representations of actual skilled sports performances and to facilitate the use of cognitive and social processes in learning and inquiry in a collaborative manner (Lim & Koh, 2006). However, efforts to use learning technologies are often hampered by prevailing attitudes and logistical issues. It is hoped that a marker-less motion capture system will circumvent the logistical issues and make for a seamless integration into novel pedagogical approaches in teaching sports and physical activities. Done well, this new form of learning technology has the potential to bring about a cultural change in the way we teach and learn.

Motivated by the enormous potential and widespread applications that lead to the pursuit of a video-based marker-less motion capture system, in this chapter we describe the application and technical advancements, future challenges and problems that lay ahead as we work towards a marker-less motion capture system. We will also present the work of our research team in the development of a marker-less motion capture system (see the following).

The outline of this chapter is as follows:

We will first review the existing technologies and methods. This is followed by a discussion on the challenges and outstanding problems. We will also present a video-based marker-less motion capture system proposed by our research team. The last few sections will illustrate the possible extension of our work with potential applications before the conclusion.

EXISTING TECHNOLOGY

Existing commercially available motion capture (mocap) devices fall into 3 main categories: magnetic, mechanical and optical systems. All these methods require the subject to wear some form of devices or markers on the body. Attaching markers to the subjects poses a hindrance to the true movement of the performers and in some instances, may render the movement unnatural. There is therefore an emerging need to develop a system to operate in a setup where the subjects wear natural clothing and move in the natural context as opposed to laboratory conditions. A possible solution to answer this need is to explore the use of computer vision-based and graphics algorithms.

Main Commercial Motion Capture Methods

Magnetic motion capture systems utilize sensors placed on the body to measure the low-frequency magnetic field generated by a transmitter source. The sensors and source are cabled to an electronic control unit that compiles their reported location within the measurement field. The electronic control unit is networked with a host computer that represents these positions and rotations in 3D space.

Mechanical motion capture systems require the performer to wear a mechanical armature fitted to their body. The fitting is usually done with elastic straps and belts, which hold plastic plates against the body of the performer. The sensors in a mechanical armature are usually variable resistance potentiometers or digital shaft encoders. These devices encode the rotation of a shaft as a varying voltage (potentiometer) or directly as digital value.

Existing commercial optical motion capture systems utilize reflective or pulse-LED (infra-red) markers attached to joints of the athletes' body. Multiple infra-red cameras are used to track the markers to obtain the movement of the subject. Currently, this kind of system is mainly used for biomechanical analysis. The main manufacturers of optical systems are Motion Analysis, Qualysis and Vicon.

In general, existing commercial mocap systems are only suitable for a very well controlled (mainly indoor) environment. Table 1 shows a comparison into the main features the different motion capture systems. Figure 1 shows an example of the motion capture setup using reflective markers carried out in our lab for analysis the walking pattern. This would be unrealistic and impossible to implement during a sport tournament for example. Furthermore, post-processing of motion capture data is tedious and time consuming e.g. re-establishing the correspondent markers across time due to the confusion of sensors in optical marker systems.

Computer Vision-Based Motion Capture

The inputs to all vision-based system are purely images acquired from electro-optic sensors such as video cameras or infra-red cameras. Electro-optic devices are popular largely owing to their

Table 1. A comparison of different motion capture systems

Type of Systems	Advantages	Disadvantages
Magnetic	No sensor occlusion issue Position and rotation are measured absolutely	Highly intrusive Cumbersome for the performers and movement constriction Sensors suffer from interference Operates mainly in indoor and well-controlled scenarios
Mechanical	No sensor occlusion issue Sensors do not suffer from interference	Highly intrusive Cumbersome for the performers and movement constriction Need frequent calibration of mechanism Operates mainly in indoor and well-controlled scenarios environment
Optical (with Marker)	Low intrusive Good realism of movements	Operates mainly in indoor and well-controlled scenarios Markers suffers from occlusion, confusion and undesirable swapping Highly priced (about US$ 200000 or more)
Computer vision-based	Non-intrusive Natural freedom of movement for performer Potential to perform in outdoor and cluttered environment	Technology is not fully matured, it needs more research and validation

Figure 1. Example of motion capture using reflective markers

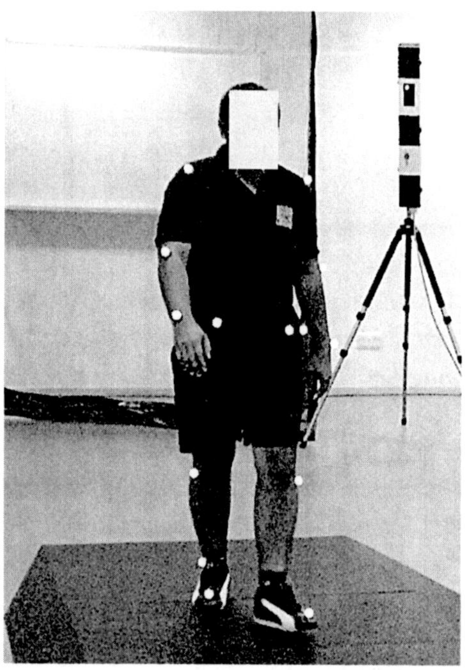

non-intrusive and passive nature. Also, they are easy to set up and their prices are cheap. In order to get a good coverage of the performance, multi-cameras are usually used. Computer vision-based approaches for motion capturing, with their potential to operate in natural environments, have started to catch the attention of industrial motion capture manufacturers.

Use of Computer Vision-Based Methods

Large numbers of work on the tracking and analysis of human motion using computer vision techniques had been proposed over the years. Reviews on computer vision-based motion capturing methods had been done in the years 1994 and 1999 (Aggarwal & Cai, 1994; Gavrila, 1999). In a more recent review, Moselund (2006) extended the earlier reviews of year 2001 (Moselund & Granum, 2001). The algorithms, technologies and equipment setups that have been used for motion capture are driven by the main application areas: surveillance, interaction/control, or analysis. In systems that require accuracy and reliability human models are usually used to facilitate the tracking (Carranza, 2003; Lerasle, 1999). Whereas in applications for visual surveillance and activities monitoring (McKenna, 2000), which require speed and robustness, such systems usually heavily involve low-level vision techniques such as image segmentation, thresholding, statistical and probabilistic formulations.

In the computer vision-based mocap survey by (Moselund & Granum, 2001), tracking is defined as establishing a coherent relation of subject and/or landmarks between frames. Pose estimation of subject is the process of identifying how the subject is configured in the scene e.g. posture of a human described by its kinematics at an instantaneous time. For 3D model-based motion capture approaches, the tracking and pose estimation are usually closely coupled with each other. The notion of motion tracking is used differently and loosely defined throughout the literature of visual analysis of human. In this article, since tracking and pose estimation are tightly coupled, and the aim is to quantify the human posture at each instantaneous time and follow it over a time sequence, the overall process refers to 3D human motion capture or 3D human motion acquisition.

Nearly every vision-based mocap follow the steps: (a) segmentation of subject from the rest of the image; (b) transformation of these segmented images into some kind of higher level representation to suit a particular tracking algorithm; and, (c) how the subject should be tracked from one frame to the next.

In these kinds of frameworks, many proposed algorithms have relied heavily on the image segmentation, which is a very crucial part of the system, and modeling of its result. In addition, some assumptions regarding the background scene had been used e.g. constant and low cluttered background. The information that is presented for segmentation can be either spatial or temporal image data.

Common Motion Tracking Approaches

The use of silhouette is one popular approach where 3D volume data of human is built from multiple 2D silhouettes to yield the human posture that is tracked at each instantaneous time sample (Bottino & Laurentini, 2001; Cheung, 2003a; Mikic, 2003). This kind of approach is commonly known as the shape-from-silhouette (also called visual hull) method. It relies on calibrated still cameras, and segmenting the subject from the background by assuming constant and low cluttered background. The silhouette approach is a very mature topic of research in the domain of computer vision and image analysis.

Various motion tracking algorithms, frameworks and steps were proposed based on the visual hull approach. In Cheung (2003a), they extended their temporal shape-from-silhouette technique (Cheung, 2003b) to track the articulated rigid segments through the alignment of multiple-views silhouette poses across time. Then in Kehl & van Gool (2006), the authors proposed a *stochastic meta descent* minimization algorithm to fit a human model made up of super-ellipsoid parts to the volumetric data. In Nobuhara & Matsuyama (2006), the authors used a deformable mesh model with 'repulsive force' to guide the inter-frame 3D volumetric deformation and at the same time detecting collision. Then in Mikic (2003), they estimate the human movement parameters by using the extended Kalman filter after utilizing the Bayesian network to estimate the body part sizes from 3D voxel data. Also in Carranza (2003), they adapt an articulated 3D human model to the multi-view silhouette via Powell's minimization.

The extraction of motion and body structure through optical flow related method had also been proposed (Bregler, 2004). Other methods by Planker & Fua (2001) and Grest (2005) propose a recovery of motion through the estimation of disparity and depth map. Also there are suggestions (Kim & Pavlovic, 2007; Rosales & Sclaroff, 2006) to perform motion capture through the learning or training via a human motion database collected a prior. However, such a database may be incomplete and when the general human model dynamics are to be learned, the amount of data, model complexity and computational resources that are needed may be impractical.

Estimation of Kinematics and Skeleton Structures

We need to acquire the true kinematics of the subject undergoing motion so that they can be useful for analysis. A human undergoing motion is mainly resulted from the contraction of muscles that drives the skeleton-bones, which can be represented via a kinematics chain. Thus, the visual appearance is an outcome of skin deformation caused by the skeleton movements and muscle actions.

Generally speaking, all the motion capture systems described before do not directly measure the movement of bones and skeletons. In the case of motion capture system that set up the subject to wear reflective markers, the reported markers location on the surface of the skin are used for estimating the kinematics and skeleton structures by using some form of mathematical least-square minimization techniques.

In medical applications, the human skeleton is obtained via X-ray. However its drawbacks are that X-rays are not easily available and it is impossible in the current technology that the human motion can be measured by X-ray. Lafortune (1992) attempted to explicitly measure human movement by inserting pins into the bones of the subject. Unfortunately, the procedure can result in the subject experiencing severe pain. Such a setup has very limited scope for application and is not practical in the general sport environment.

A common approach taken in video-based systems to estimate skeletons and joint locations is to use medial axis extraction algorithms based on mathematical morphology. Many techniques to compute 2D and 3D skeletons of object have been proposed over the years (Goh & Chan, 2007; Ma & Sonka, 1996). In Rosenhahn's (2005) work, skeletons were obtained from 2D images of the segmented subject. Wingbermuhle (2001) extracted the skeleton of the 3D model constructed by its visual hull. However, researchers have acknowledged that such approaches are very sensitive to noise and the uncertainties due to noise are not easily tractable. Other works attempt to improve the skeleton estimation by labeling body parts to kinematics chain of super-quadrics representing voxel data (Theobalt, 2004). It is also possible to derive a common kinematic structure through representing the shape of the subject in each time frame by using augmented Multiresolution Reeb Graph. However, we have to take note that the skeletons obtained mathematically are not the same as anatomic skeletons and do not take into consideration the skin deformations. Hence, numerical skeletons will not give the true anatomic joint locations e.g. the human's elbow joint is not at the centre of the surface skin.

CHALLENGES AND OUTSTANDING PROBLEMS IN VISION-BASED METHOD

Despite the numerous research works on vision-based motion capture being done and published, there are still many challenges and problems to be resolved before they are ready to be accepted for wide usage.

Computer Imaging Issues

In video images, their colour information is produced by many factors such as: (a) reflectance material of subject; (b) posture of subject; and, (c) light source. Hence, an image is a result of a highly dimensional combination, in which the lack of understanding and representation of its operating scenario is a main cause to the existing problems. Some of the researchers have termed this as the semantic gap (Dorai & Venkatesh, 2002). For example, the human posture undergoing motion is highly complex since they have many degree-of-freedoms (DOFs) and the skin is also deformable.

Most of the existing computer vision methods rely heavily on feature segmentation to be the entry point to their systems. Feature segmentation aims to extract the natural imaging features such as edges, highly textured points and corners from images. However, feature segmentation suffers severely from occlusion, lack of texture and foreshortening. There is no fool-proof method for image segmentation that can universally cater to all the environments. In addition, there is no consistent way to integrate different methods into a hybrid system.

Human undergoing motion will produce skin deformation, usually termed as skinning. The skinning issue has been studied in computer graphics; however it had not been properly reviewed in relation to any existing vision-based motion capture. Skinning is necessary for the geometric properties and color texture on the human skin to be warped correctly when undergoing motion. Some examples of human skin deformation include bending of elbow which causes the skin to stretch, the wrinkling of skin and the 180 degrees twisting of wrist which causes the skin at the wrist to twist about 180 degrees, while the skin near the elbow is twisted by only about 5 degrees. Other kind of deformations can arise from the bulging of muscle or the bending of the body. All these deformation factors are anatomically related. Therefore, the anatomical structure of 3D model must be able to mimic the subject closely.

One of the simplest and fastest ways to animate skinning is to consider a rigid skin model that is attached onto the articulation of the respective bones and to ignore the deformations near the joints. However, rigid skinning does not look realistic near the joints. The deformation of human skin is highly complex. By itself, it has attracted many research interests over the years, and mimicking the real human skin still remains a challenge.

When moving ahead towards a larger scale operating environment, there are issues such as the multiple interaction of people, the high occlusion and clutters as well as illumination uncertainties. These issues have yet to be resolved or properly tackled in the area computer imaging and artificial intelligence.

Hardware Issues

The progress of computer technology has enabled very fast processing speeds of up to GigaHertz on a single terminal, and even possibilities of putting together multiple processors and computers to perform the computation. However, there are several hardware issues that we have to address before the motion capture system can be fully optimized to its potential. Considering a future system setup of many cameras with true RGB colours at high definition (HD) of 1920×1280 pixels, the physical properties of the hardware have to be ready for this challenge.

Bandwidth for data communication between the hardware devices and processors has always been one of the bottlenecks of the overall computational performance. When dealing with video data, the amount of data transferred between the devices poses a very serious issue. Although many video compression methods have been proposed, the compression ratio is always restricted to the entropy of the signal when near loss-less compression is needed for image analysis by the mocap system.

The size and the storage speed of the storage devices (in relation to the data bandwidth) of the computer is another key issue. These storage devices including the CPU and GPU memories must be able to cope with the amount of data whilst maintaining a real-time application. Last but not least, the CPU and GPU computing speed and parallelism of computation are also areas that cannot be ignored, considering the amount of information which the processors have to analyse.

MOTION CAPTURE SYSTEM: A FRAMEWORK

This section describes a framework that we have proposed, and in which our research team is working on. The aim of our motion capture system is to create a framework and the realization towards a marker-less setup that can eventually be used in the cluttered and outdoor environments. The operating pipeline of our proposed framework is shown in Figure 2. The functionality of each block is modular and self-contained. We will also draw attention to some relevant methods that have been used in some other practices for each of the modules along the operation pipeline. And then, we will go on to show our results, which will illustrate the feasibility and potential of our approach.

Our methodology implements the model-based analysis-by-synthesis approach built on the concept of collaboration between computer vision and computer graphics to tackle our problems. The main concept has been proposed as early as 1990 (Gagalowicz, 1990), and has gained popularity (Terzopolus, 1999). The computer graphics will provide the knowledge-base for the 3D modeling, animation and synthesis, whereas, computer vision will supply the learning and analysis capabilities.

Camera System

The advancement in imaging technologies over the past 10 years has seen video technology move from the interlacing sensor scanning of the NTSC or PAL to the progressive scanning in most of the current digital video devices, and towards a common high definition (HD) in the not far away future. To acquire fast movement, a high camera frame rate and fast shutter speed is needed. Our cameras run at a frame rate of 30Hz and shutter speed of at least 1/250 second.

The cameras are connected to a host computer. In our system, the data interface between the digital cameras and the computer is connected through the IEEE 1394 fireware for the ease of interfacing with computers. The IEEE 1394b interface is able to provide data bandwidth of up to 800Mbits per second at its current state, and with up to 6Gbits per second of data rate in the next few forthcoming years. Data repeaters could also be used to increase the distance of data transmission.

In a multi-camera system, all the cameras have to be synchronized so that all the images are acquired at the same time step. One common approach is through the use of an external triggering source, which is sent to all the cameras. Alternatively, the IEEE 1394 interface™ can be used to synchronize the cameras that are connected to the same PCI interface card on the computer.

Camera calibration is a software process that initializes the focal length, orientation and position of the cameras. This process also performs the correction in the images caused by lens distortion. The calibration of these parameters of a multiple camera system is to enable the recovery of 3D geometrical information through computing

Figure 2. Operation pipeline for model-based video motion capture

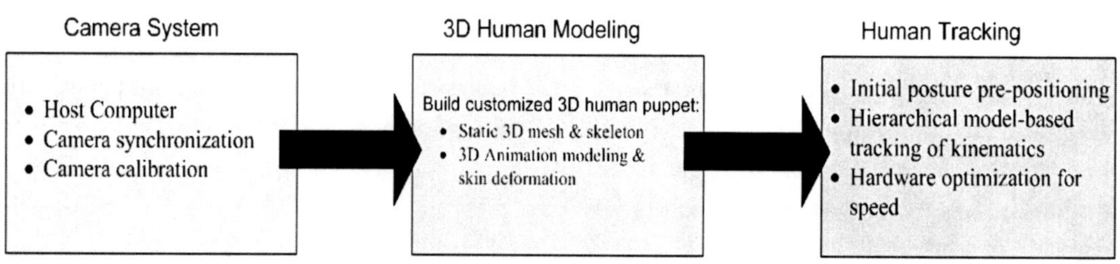

methods such as triangulation and back-projection of imaging rays.

Over the years, many camera calibration techniques have already been developed. Every camera calibration algorithm assumes that the 3D object information and their corresponding features in the 2D images are already established. The main approaches are through: closed-form methods, numerical solutions, or relating linear camera model to the perspective model by iterative solutions. Readers who are interested in the technical algorithm in camera calibration may refer to the work of Bouguet, Dementhon & Davis (1995) or Zhang (2000).

3D Human Modeling

To have a 3D geometrical human model that closely resembles the size and shape of the subject, a very crucial component in the mocap system that will perform this function is the accuracy and reliability of tracking the subject in motion. Many studies (Gerard & Gagalowicz, 2003; Goncalves, 1995; Kakadiaris & Metaxas, 1998) have already indicated that the process of tracking is very sensitive to the shape parameters used. Thus, it is inappropriate to use, for example, a generic "averaging human" model for accurate and precise tracking of human with a different shape and size.

Building the 3D Human Model

In order to make sure that the 3D human model can be used properly later on in the tracking process, a substantial amount of work has been spent on the research and development of the algorithm for 3D reconstruction of human model. Also, the software system for building this 3D human should allow only minimal to moderate amount of human interaction such as clicking on a few feature points and drawing a few curves. This is to ensure that the whole 3D reconstruction process does not take more than several minutes.

Representation of the customized 3D human model comprises the following steps:

i. The surface skin made up of regular triangle mesh.
ii. Bones made up of triangle mesh as well. Alternative, for simplicity, they can be 3D curves/lines formed by the medial axes of these bones. The bones of different body parts are also linked through a forward kinematics chain, the root of which starts from the pelvis. During 3D animation, the bones will drive the surface skin by manipulating its kinematics, which deform the surface.
iii. Given also the calibrated cameras parameters and real images of the subject, the texture coordinate of each vertex on the 3D model can be computed by back-projecting its 3D position through the camera parameters onto the 2D image coordinate i.e. (u, v) coordinate. This texture coordinate is used for shading and rendering of the 3D model during the motion animation.

This scheme of representing our 3D model using triangle mesh and texture were chosen since they are easy representations for image synthesis and are well supported in many existing graphics hardware.

There are many systems and methods that have been developed to construct the 3D human model. They usually construct the surface skin of the model, and their methods fall into two categories: (1) 3D scanner-based system, and (2) passive multi-camera system.

The scanner-based and active systems are available from many commercial companies such as Cyberware®, Hamamatsu and (TC)^{2}®. Typically, they use the triangulation principle, and with laser light or pattern projection method and a CCD camera. These systems usually capture the shape of the entire human body in about 15 to 20 seconds. The output of such a system is a set of highly dense point cloud representing the surface of the model. Before this model can be used, the point cloud needs to be furnished into a regular surface 3D triangular mesh by using 3D geometrical software with some manual cleaning up of possible spurious data points. Drawbacks of a full-body scanner-based system include its high pricing and the requirement for the subject to stay still and rigid for the whole duration of scanning, about 15 seconds for full body coverage, which is quite constrictive and not very practical in certain applications.

The passive multi-camera approach is much cheaper; also video cameras are more easily available and their set-ups are much simpler. Once the synchronized cameras are calibrated, we are able to reconstruct the 3D model of the subject when the features points and curves of the subject seen in the images are properly defined.

A Proposed Method for 3D Human Reconstruction

In our project, we used the passive multi-camera approach since it is cheaper and easier to set up. Here, we will describe our general setup and approach for modeling the 3D human. Readers who are interested in our 3D human reconstruction method can refer to Gagalowicz & Mathieu (2006) or Quah (2005). Other works related to 3D human reconstruction can be found in Matsuyama (2004) or Starck & Hilton (2003).

The method that we proposed uses multi-cameras to build the external surface skin of the subject, and at the same time estimate its underlying skeleton. We make use of a 3D generic human model with skeleton (top of Figure 3) and adapt it to the real subject seen in the multiple images, thus yielding the customized 3D human model of the subject with its skeleton automatically positioned inside the external skin.

Initialization of the 3D human reconstruction begins by establishing the 3D anatomically related characteristic points on the surface of the generic model and their 2D correspondents in the respective images (bottom of Figure 3). This is achieved by using an interactive point-matching tool that we have developed. Next, our algorithm builds the 3D human model through the use of camera

Figure 3. (top) Generic surface model with skeleton, and (bottom) example of characteristic points selection on an image of a real golf player that is corresponded to points on surface of the generic model

calibration and 3D ray triangulation. The feature points reconstruction yields the global geometry of the subject (Figure 4). However, these are insufficient to give the local body geometry. The local body contours of the subject are refined through the matching of silhouette curves (Figure 5). The skeleton of the subject is estimated based on the information of changes from generic model to the customized model. Figure 6 shows the result of 3D human of our subject with its estimated skeleton.

Human Tracking

We propose a model-based analysis-by-synthesis framework for human motion capturing (see block diagram on Figure 7). The concept of this work is built on the work by Gagalowicz & Gerard (2000) for tracking rigid objects. Prior to this framework, a textured 3D puppet model that closely resembles the subject was built (from the previous section) to facilitate the motion tracking. Our method does not use explicit or standard image background

Figure 4. 3D reconstruction via feature points that yield the global geometry of the subject, but inadequate for the local body contours

Figure 5. Matching of silhouette curves for refinement via local contour

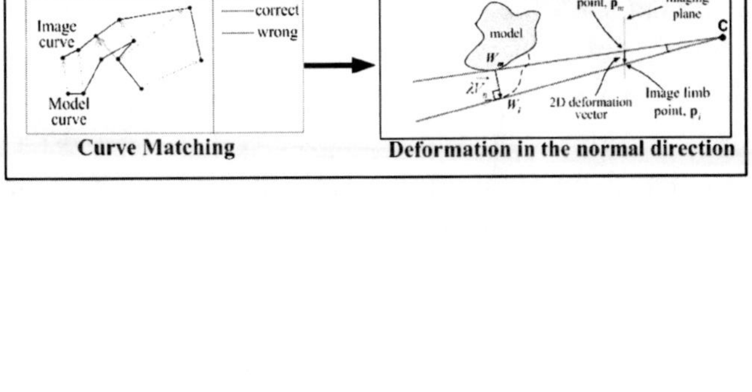

Figure 6. Final result of a close-up of the surface skin with its skeleton for the subject superimposed onto the image

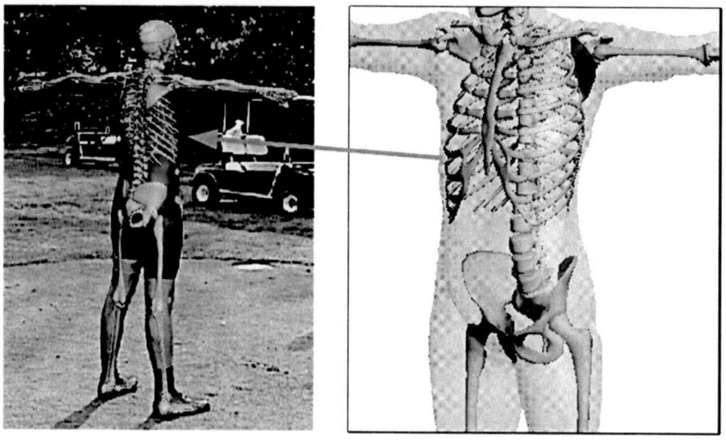

Figure 7. Model-based human motion tracking framework

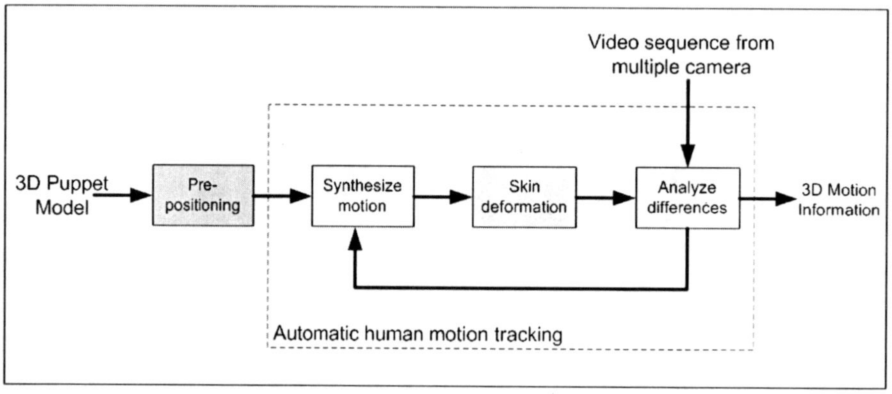

segmentation or *skeletonization*, which is very sensitive to noise. A more detailed explanation of our algorithm could be found in Quah (2007).

Our algorithm starts with automatic realization of the colour/texture onto the puppet model from its initial pre-positioned posture. The pre-positioning was done using a 3D modeling software just for the initializing to the first image. Next, our computation synthesizes the 3D puppet movements such that they minimize the differences between the synthesized movements and real athlete's motion in the subsequent images. This is achieved through the use of the simulated-annealing algorithm to search through the various probable degree-of-freedoms of the joint kinematics. The joint kinematics then drives the skin of the model puppet. The solution to the joint kinematics is taken when the skin puppet posture yields the smallest error when superimposed onto the video containing the subject. The colour texturing onto the puppet is analyzed and updated once every few frames so that the synthesis is not influenced by the changing articulated posture and illumination variations. In this project, we implemented the rigid skin deformation. The rendering for motion synthesis is the most computational intensive module and it is sped up by using the graphics processor unit (GPU).

Figure 8. Results of tracking in clutter and self-occlusion setting

Results & Discussions

In this section, we present some of our typical tracking results. The meshes of the upper-arm and forearm are superimposed onto sequences of images to visualize the tracking results. The top images show the results for the camera view that we used for tracking. The bottom images show the results on another camera from the same videoing scene that we used for verification and visualization. From the results, we can see that our algorithm is able to track the human arms in environments with clutter and occlusion (Figure 8). The 3D meshes of the arms are superimposed onto the sequence of images to visualize the tracking quality. They are overlaid onto 2 different camera views to validate their 3D depth validities. The left images are from camera 1 and the images on the right are from camera 2.

FUTURE WORKS

This section describes the future direction of our research team.

Human Skin Deformation and Shading

In practice, the human undergoing motion will cause the skin to deform in non-rigid manners. The accuracy and reliability of human tracking can be improved if these non-rigid deformations can be closely modelled. In computer graphics animation terms, the surface geometries of the animated characters could have the attributes of: twisting, bulging, bending, stretching and squashing. The scientific principles of these attributes could be used to derive and synthesize the proper surface geometries for use in motion tracking.

The shading for the synthesized image must also not be overlooked. Although very realistic images had been synthesized before (Boivin & Gagalowicz, 2001), to put it to use in practice together with proper skin deformation, different lighting condition and computationally efficient is still a big challenge.

Automatic Posture Pre-Positioning

In this article, pre-positioning had been done interactively by using 3D graphical tools. To initiate the tracking automatically, the initial posture of the subject needs to be recognized by the machine vision system. The 3D geometries of the postured model and subject images have to be properly registered. We also have to take into account the skin deformation while manipulating the kinematics of 3D model at the stanza posture to the initialized posture. One suggestion to do this is through machine learning of example postures from multiple subjects, and then synthesizes them during an automatic pre-positioning operation.

Full-Body and Multiple People Tracking

A natural extension to the work from this chapter is to perform full-body and eventually multiple people tracking in cluttered environments. For full-body tracking of a single subject, the complete 3D body model of the subject will be used, thus, more degrees-of-freedom and possibly more accurate skinning function have to be used. For extension to multiple people tracking, we have to consider occlusion of subjects and detecting collision and interactions between subjects. A higher level data topological relationship may also be needed if there are many subjects in the scene.

Computational Efficiency

Computational speed is a factor of consideration if we are to add in more sophisticated operations involving the likes of skin deformation, automatic posture pre-position and multiple people tracking. Specific dedication is needed to make sure that the algorithms that manipulate the model geometries, data synthesis and matching take place efficiently so that the whole tracking process can yield the desired results in a realistic time-frame. Efficient data streaming is also required, since the large amount of data transfer within the module of the tracking system can be a bottleneck.

IMPACT OF VISION-BASED MOTION CAPTURE AND PRACTICAL APPLICATION FOR THE FUTURE

This section describes the impact of a purely vision-based motion capture and its implication toward the domain of sports kinesiology. We also give a few practical applications that uses this system for the future.

The innovation of these technologies will potentially benefit the biomechanics community through the utilization of marker-less motion capture methods by minimizing skin-markers artifacts and equipment cost (as such configurations require only video-based cameras). Additionally such technologies may enable user-friendly setups accessible by the masses, particularly for the sports science, sports coaching, sports retailing, animation arts industry and main-stream education communities. With a wide spectrum of users, the push for the advancement of such technologies will be imminent.

There are many potential applications of Interactive Digital Media (IDM) that require capturing and analysis of human movements in the fields of sports coaching, sports science, education, medicine and animation domains. We have chosen two examples to illustrate the permeability of its application.

Educationally Contextualized Interactive Digital Media

In typical main stream schools, the learning of academic subjects like physics, mathematics, biology, physical education and the coaching of sports skills are usually taught predominantly within each discipline's domain. There are common inter-related concepts like mechanics in physics, mensuration in mathematics, kinesiology in biology and simple technique analysis in physical education or sports coaching which can be learnt within a common educational framework. The relevant connectedness between these academic and sports coaching domains are usually not perceived by the learner. To develop the holistic thinking student, one of the strategies is to have the connectedness between these domains, which forms a significant part of the formal main stream school curriculum, located within the learner. The learning of interdisciplinary academic concepts incorporating experiential physical activities in realistic conditions provides a potential platform for establishing the connectedness. Traditional classroom or laboratory strategies have their limitations in that the concepts taught are usually not fully contextualized to real situations involving actual activities. Therefore, this may not facilitate deep learning. Interactive digital Media (IDM) technology offers a very attractive tool to address this problem by providing contextualized augmented feedback, infused with inter-disciplinary information, in activities performed in real settings. Such situations and activities are best facilitated using imaging technologies especially in three dimensions (3D). Objective and augmented feedback through interactive 3D video-based approaches using Augmented Reality (AR) technologies can, thus, be used to enhance the teaching, learning and coaching processes. We will illustrate three pedagogical areas that our project may have potential contributions.

Inter-Disciplinary & Holistic Learning

Using IDM in learning within a common theme may involve various inter-related academic disciplines. For example, in the example of a golf swing, the angles, speed, time concepts in mathematics can be explored together with physics principles like kinetic energy, potential energy and momentum under the domain of skills acquisition or physiology focusing on the musculoskeletal forces and moment arms. Also, these concepts can be taught experientially whilst the students actually perform the swing in actual realistic environment. Benefits of experiential and interdisciplinary` learning environment in education are well documented (Badenoch, 2005; Redman, 2006).

Additionally, as computing hardware—both wired and wireless—approaches ubiquity, new opportunities emerge to use technology to enrich individuals' experiences of objects and places. Because all areas of academic inquiry potentially benefit from background and context, AR has the possibility of enhancing education across the curriculum. By holistically exposing students to an experiential, explorative, and authentic model of learning early in their higher education experience, AR has the potential to help shift modes of learning from students' simply being recipients of content to their taking an active role in gathering and processing information, thereby creating effective knowledge.

3D Spatial Understanding

In order to solve three dimensional mathematical and especially geometrical problems, spatial abilities are an important prerequisite (Brown & Wheatley, 1989; Fennema & Sherman, 1977; Guay & McDaniel, 1977). Many students have difficulties solving tasks that require spatial visualization skills and spatial thinking. They may use strategies such as learning construction steps by heart without fully understanding spatial problems and their solutions in 3D space. Spatial abilities can be improved by well-designed training (e.g. Souvignier, 2001) and a number of training studies have shown the usefulness of AR in training spatial ability (Durlach *et al.*, 2000; Larson & Van Rooyen, 1998).

Future Research: Extension of Developed Technology into Related Applications

Exploiting the developed new IDM technologies that aim to be faster and more accurate, there is great potential to use them, in future research, to develop 'Immersive Virtual Environments' and 'Education Games' applications where the participants' movement can be detected more accurately and faster and hence, more realistically contributing to a more enhanced experience. In sports coaching, automatic augmented objective feedback may be provided to both coaches and athletes where immediate feedback on speed, angle and distance information amongst others can be presented contributing to an effective and engaging coaching session. This will significantly facilitate the application of sports biomechanics for sporting excellence in line with the ethos of the Committee on Sporting Singapore (CoSS) (2000) encompassing various activities like SEA games, Asian games, Olympics and normal physical activates.

Also, there is potential for future research in the area of 'Digital Literacies and Modalities of Representation' where this project will lend itself to be a potential case study for understanding the issues related to pedagogy, learning, design and persuasion using multi-modal forms of electronic representations and media (e.g. interactive art, digital textbooks) in IDM environments. Also, in the 'Research on Impact of IDM use among Students' domain, this project will provide a framework for understanding, in future studies, the impact of widespread use of IDM on students' social, emotional and cultural development in a longitudinal context. The research activities and products (including achieved aims) in this project, potentially, provide the catalyst for more research using IDM technologies within learning environments through journal publications, conference/poster presentations and various professional sharing sessions. These will potentially generate more questions requiring further research.

Sports Analysis

In sports research, sports biomechanics typically provide the scientific basis for movement intervention by means of analyses and review of movement performance data. The use of such technology with a suitable user interface will empower the coach to gain insights into existing performance levels more readily and conveniently. This in turn can enable coaches to be better able to identify key movement attributes that are impeding performance. Feedback to learners can be near real time and contextualized so that new learning interventions introduced can be assessed for its efficacy and appropriateness in terms of the kinematics of the movement or that of the performance outcomes. Certainly, such a platform to augment feedback can provide a further impetus to motivate learners to improve themselves and to optimize their training time. Current approaches in sports coaching that involve the use of video technology range from qualitative analysis of sports techniques to more complicated and time consuming controlled environments for empirical movement analysis. Both methods are limited in their application either through a lack of sophistication or detail in the case of qualitative analyses or cost and availability in the latter environment. The advent of a marker-less system would help overcome this limitation and promote new pedagogical approaches to coaching sports.

CONCLUSION

AR technologies are technological tools that must be developed and contextualized before their benefits can be fully manifested. Currently, the Republic Polytechnic (Singapore) and Nanyang Technological University (Singapore) are examples of two institutions that have been awarded government research grants for research and development in this area. In this chapter, we reviewed the existing systems for human motion capture, which have many limitations. Looking ahead to the future and challenges, we present a model-based approach for marker-less motion capture that will widely extend its scope of applications in many domains of practices. From the results, we demonstrate its feasibilities and potentials to operate in challenging scenarios. We also illustrated that our system is particularly important in the sports and educational domain, with examples of ongoing development through the use of video and interactive digital media.

In closing, we are only limited by our imagination while we attempt to advance technology. Technology will not replace us but we may be replaced by those who can exploit technologies in the right context. We will continually advance knowledge; the more we acquire, the more is our desire; the more we learn, the more we are capable of learning.

ACKNOWLEDGMENT

This project was funded by the National Research Foundation of Singapore grant NRF2007IDM-IDM003-052. We would like to thank the INRIA golf stream project team for the filming of golfers. We would also like to thank Mr. Ta Huynh Duy Nguyen for assisting in the image acquisition. Also we would like to thank Saharuddin Omar and Ian Harris Sujae for the optical mocap setups.

REFERENCES

Aggarwal, J. K., & Cai, Q. (1994). Human Motion Analysis: A Review. *Proceedings of the IEEE Workshop on Motion of Non-Rigid and Articulated Objects,* (pp. 2-14), Austin, Texas: IEEE.

Badenoch, D. (2005). *Interdisciplinary Curriculum Design and Teaching for Integrative Learning and Development of Graduate Qualities: A case study of reconceptualising Physical Education: University of South Australia.* Paper presented to the Division of Education, Arts & Social Sciences' Teaching & Learning Colloquium, University of South Australia.

Boivin, S., & Gagalowicz, A. (2001). Image-based Rendering of Diffuse, Specular and Glossy Surface from a Single Image. *SIGGRAPH,* (pp. 107-116).

Bottino, A., & Laurentini, A. (2001). A Silhouette Based Technique for the Reconstruction of Human Movement. *Computer Vision and Image Understanding, 83,* 75-95.

Bouguet, J-Y. (nd). *Camera calibration toolkit for Matlab and in OpenCV.* Retrieved May 12, 2008, from http://www.vision.caltech.edu/bouguetj/calib_doc/

Bregler, C., Malik, J., & Pullen, K. (2004). Twist Based Acquisition and Tracking of Animal and Human kinematics. *International Journal of Computer Vision, 56,* 179-194.

Brown, D.L., & Wheatley, G.H. (1989).Relationship between spatial Knowledge. In C. Maher, G. Goldin, & R. Davis (Eds.), *Proceedings of the 11th Annual Meeting, North America Chapter of the International Group for the Psychology of Mathematic Education* (pp. 143-148). New Brunswick, NJ: International Group for the Psychology of Mathematics Education.

Carranza, J., Theobalt, C., Magnor, M., & Seidel, H-P. (2003). Free-Viewpoint Video of Human Actors. *SIGGRAPH*, (pp. 569-577).

Cheung, G., Baker, S., & Kanade, T. (2003a). Shape-from-silhouette for articulated objects and its use for human body kinematics estimation and motion capture. *Proceedings of the IEEE Conference on Computer Vision and Pattern Recognition*, (pp. 77-84), Madison, Wisconsin: IEEE.

Cheung, G., Baker, S., & Kanade, T. (2003b). Visual Hull Alignment and Refinement Across Time: A 3D Reconstruction Algorithm Combining Silhouette with Stereo. *Proceedings of the IEEE Conference on Computer Vision and Pattern Recognition*, (pp. 375-382), Madison, Wisconsin: IEEE.

Committee on Sporting Singapore (CoSS) (2000). Retrieved May 12, 2008, from http://app.mcys.gov.sg/web/sprt_towards_committeesporting.asp

Corazza, S., Mundermann, L., & Andriacchi, T. (2006). The Evolution of Methods for the Capture of Human Movement Leading to Markerless Motion Capture for Biomechanical Applications. *Journal of Neuroengineering and Rehabilitation*, *3*, 6.

Cyberware (n.d.). Retrieved May 12, 2008, from http://www.cyberware.com

Dementhon, D.F., & Davis, L. (1995). Model-based Object Pose in 25 Lines of Code. *Intl. Journal of Computer Vision*, *15*, 123-141.

Dorai, C., & Venkatesh, S. (2002). Media Computing: Computational Media Aesthetics (The International Series in Video Computing). *Springer*, Edition 1.

Durlach, N., Allen, G., Darken, R., Garnett, R.L., Loomis, J., Templeman, J., & von Wiegand, T. E. (2000). Virtual environments and the enhancement of spatial behavior: Towards a comprehensive research agenda. *PRESENCE-Teleoperators and Virtual Environments*, *9*, 593-615.

Fennema, E., & Sherman, J. A. (1977). Sex-related differences in mathematics achievement, spatial visualization, and affective factors. *American Educational Research Journal*, *23*(1), 51-71.

Gagalowicz, A. (1990). Collaboration between Computer Graphics and Computer Vision. *Proceedings of the International Conference on Computer Vision*, (pp. 733-737), Osaka, Japan: IEEE.

Gagalowicz, A., & Gerard, P. (2000). Three Dimensional Object Tracking using Analysis/Synthesis Techniques. In A. Leonardis, F. Solina, R. Bajcsy, & F. Solina (Eds.), *Confluence of Computer Vision and Computer Graphics* (pp. 307-330), Dordrecht, NL: Kluwer.

Gagalowicz, A., & Mathieu, M. (2006). Modeling 3D Humans from Uncalibrated Wide Baseline Views. Paper presented at *SAE Digital Human Modeling Conference*, Lyon, France.

Gavrila, D.M. (1999). The Visual Analysis of Human Movement: A Survey. *Computer Vision and Image Understanding*, *73*, 82-98.

Gerard, P., & Gagalowicz, A. (2003). Human Body Tracking using a 3D Generic Model Applied to Golf Swing Analysis. Paper presented at *MIRAGE 2003Conference*, Rocquanecort, France.

Goh, W.B., & Chan, K.Y. (2007). The Multiresolution Gradient Vector Field Skeleton. *Pattern Recognition*, *40*, 1255-1269.

Goncalves, L., Di Bernardom, E., Ursella, E., & Perona, P. (1995). Monocular Tracking of the Human Arm in 3D. *Proceedings of the International Conference on Computer Vision*, (pp. 764-770), Cambridge, Massachusetts: IEEE.

Grest, D., Woetzel, J., & Koch, R. (2005). Nonlinear Body Pose Estimation from Depth Images. In W. Kropatsch (Ed.), *Proceedings of the DAGM - German Association for Pattern Recognition Conference*, (pp. 285-292). Vienna, Austria: DAGM.

Guay, R.B., & McDaniel, E. (1977). The relationship between mathematics achievement and spatial abilities among elementary school children. *Journal for Research in Mathematics Education, 8*(2), 211-215.

Hamamatsu (n.d.). Retrieved May 12, 2008, from http://usa.hamamatsu.com

Kakadiaris, I.A., & Metaxas, D. (1998). 3D Human Body Acquisition from Multiple Views. *International Journal of Computer Vision, 30*, 191-218.

Kehl, R., & van Gool, L. (2006). Markerless Tracking of Complex Human Motions from Multiple Views. *Computer Vision and Image Understanding, 104*, 190-209.

Koh, M., & Anwari, K. (2004). Integrating video and computer technology in teaching – an example in gymnastics initial PE teacher-training in Singapore. *British Journal of Teaching Physical Education*, 35(3), 43 -46.

Lafortune, M.A., Cavanagh, P.R., Sommer, H.J., & Kalenak, A. (1992). Three-Dimensional Kinematics of the Human Knee during Walking. *Journal of Biomechanics, 25*, 347-357.

Larson, P., & Van Rooyen, A. (1998). The Virtual Reality Mental Rotation Spatial Skills Project. *CyberPsychology and Behavior, 1*, 113-120.

Lerasle, F., Rives, G., & Dhome, M. (1999). Tracking of Human Limbs by Multicular Vision. *Computer Vision and Image Understanding, 75*, 229-246.

Lim, W.Y., & Koh, M. (2006). Effectiveness of Learning Technologies in the Teaching and Learning of Gymnastics. *Pacific Asian Education, 18*(2), 69–77.

Ma, C. M., & Sonka, M. (1996). A Fully Parallel 3D Thinning Algorithm and Its Applications. *Computer Vision and Image Understanding*, 64, 420-433.

Matsuyama, T., Wu, X., Takai, T., & Nobuhara, S. (2004). Real-time 3D Shape Reconstruction, Dynamic 3D Mesh Deformation, and High Fidelity Visualization for 3D Video. *Computer Vision and Image Understanding, 96*, 393-434.

McKenna, S. J., Jabri, S., Duric, Z., Rosenfeld, A., & Wechsler, H. (2000). Tracking Groups of People. *Computer Vision and Image Understanding, 80*, 42-56.

Mikic, I., Trivedi, M., Hunter, E., & Cosman, P. (2003). Human Body Model Acquisition and Tracking using Voxel Data. *International Journal of Computer Vision, 53*, 199-223.

Moeslund, T.B., & Granum, E. (2001). A Survey of Computer Vision-based Human Motion Capture. *Computer Vision and Image Understanding, 81*, 231-268.

Moselund, T.B., Hilton, A., & Kruger, V. (2006). A Survey of Advances in Vision-based Human Motion Capture and Analysis. *Computer Vision and Image Understanding, 104*, 90-126.

Motion Analysis (nd). Retrieved May 12, 2008, from http://www.motionanalysis.com

Nobuhara, S., & Matsuyama, T. (2006). Deformable Mesh Model for Complex Multi-Object 3D Motion Estimation from Multi-Viewpoint Video. In M. Pollefeys (Ed.), *Proceedings of the 3rd Intl. Symposium on 3D Processing, Visualization and Transmission (3DPVT 2006)*, (pp. 264-271). Chapel Hill: University of North Carolina.

Peak Performance. (nd). Retrieved May 12, 2008, from http://www.peakperform.com

Planker, R., & Fua, P. (2001). Tracking and Modeling People in Video Sequences. *Computer Vision and Image Understanding, 81*, 285-302.

Quah, C. K., Gagalowicz, A., Roussel, R., & Seah, H S. (2005) 3D modeling of humans with skeletons from uncalibrated wide baseline views. In A.Gagalowicz & W. Philips (Eds.), *Proceed-*

ings of the International Conference on Computer Analysis of Images and Patterns, (pp. 379-389), Versailles, France: Springer.

Quah, C.K., Gagalowicz, A., & Seah, H.S. Markerless 3D Video Motion Capture in Cluttered Environments. In F.K. Fuss (Ed.), *Proceedings of the Asia-Pacific Congress on Sports Technology*, (pp. 121-126). Singapore: Nanyang Technological University.

Qualysis. (nd). Retrieved May 12, 2008, from http://www.qualysis.com

Redman, R.W. (2006). The challenge of interdisciplinary teams. *Research and Theory for Nursing Practice, 20*(2), 105-107.

Rosales, R., & Sclaroff, S. (2006). Combining Generative and Discriminative Models in a Framework for Articulated Pose Estimation. *International Journal of Computer Vision, 63*, 251-276.

Rosenhahn, B., He, L., & Klette, R. (2005). Automatic Human Model Generation. In A.Gagalowicz & W. Philips (Eds.), *Proceedings of the International Conference on Computer Analysis of Images and Patterns,* (pp. 41-48), Versailles, France: Springer.

Souvignier, E. (2001). Training räumlicher Fähigkeiten. [Training spatial abilities.]. In K.J. Klauer (Ed.), *Handbuch Kognitives Training* (pp. 293-319). Göttingen: Hogrefe.

Starck, J., & Hilton, A. (2003). Model-based Multiple View Reconstruction of People. *Proceedings of the International Conference on Computer Vision,* (pp. 915-922), Nice, France: IEEE.

(TC)^2. (nd). Retrieved May 12, 2008, from http://www.tc2.com

Terzopolus, D. (1999). Visual Modeling for Computer Animation: Graphics with a Vision. *Computer Graphics, 33*, 42-45.

Theobalt, C., Aguiar, E., Magnor, M., Theisel, H., & Seidel H-P. (2004). Marker-free Kinematic Skeleton Esimation from Sequence of Voume Data. In R. Lad & G. Baciu (Eds.), *Proceedings of the ACM Virtual Reality Software and Technology Conference,* (pp. 57-64), Hong Kong: ACM.

Vicon. (nd). Retrieved May 12, 2008, from http://www.vicon.com

Wingbermuhle, J., Liedtke, C-E., & Solodenko, J. (2001). Automated Acquistion of Lifelike 3D Human Models from Multiple Posture Data. In W. Skarbek (Ed.), *Proceedings of the International Conference on Computer Analysis of Images and Patterns,* (pp. 400-409), Warsaw, Poland: Springer.

Zhang, Z. (2000). A Flexible New Technique for Camera Calibration. *IEEE Trans. Pattern Analysis and Machine Intelligence, 22*, 1330-1334.

This work was previously published in Digital Sport for Performance Enhancement and Competitive Evolution: Intelligent Gaming Technologies, edited by N. K. L. Pope, K. L. Kuhn, and J. J. H. Forster, pp. 81-100, copyright 2009 by Information Science Reference (an imprint of IGI Global).

Chapter 14
Navigating a Speckled World: Interacting with Wireless Sensor Networks

Matthew Leach
Napier University, UK

David Benyon
Napier University, UK

ABSTRACT

The Speckled Computing project is a large multisite research project based in Scotland, UK. The aim of the project is to investigate, prototype, and produce tiny (1mm^3) computational devices, called Specks, that can be configured into wireless sensor networks, called SpeckNets. Our particular interest is in how people might interact in such environments, what interaction tools they require, and what characteristics are required to be provided by the operating system of the Specks. Interaction in these environments places the human physically inside an information space. At one time, the human may be interacting with one Speck, at another with a hundred, and at another with several thousand. Moreover, the Specks themselves have no input method, apart from their sensors, and no output display. We explore these issues through taking some theories of distributed information spaces, some design principles from information visualization, and report on some empirical studies of prototypes and simulations that have been developed.

INTRODUCTION

We are interested in human interaction with a new type of information space, one that involves computer miniaturisation, sensors, wireless communication, and networking. Devices are on the verge of reaching a critical point of size and affordability that will allow them to be embedded in our environment in their thousands, sensing their surroundings and opening up a broad range of new applications. They will also be embedded in clothing and jewellery, truly becoming part of the fabric of the world. It is a vision that was foreseen by Mark Weiser (2002), who coined the term "ubiquitous computing"; but it is likely that even he would be astounded at the miniaturisation of technology that has made his prediction possible, in projects such as SmartDust (Hoffman, 2003), and more recently the Speckled Computing Consortium (2006).

The vision of Speckled Computing is the development of miniature 1mm^3 computational devices (Specks), which will be simple, small, and cheap enough to be distributed in their thousands (Arvind, 2005). They will combine wirelessly to form a unique class of wireless sensor networks (WSN), called SpeckNets. Some SpeckNets will be embedded in the fabric of buildings, but others will be spontaneously created by scattering Specks over an area or even spraying them onto surfaces. The combined storage and processing of the microchip revolutionised computing in the late 20th century. Specks, offering combined processing, storage, sensing, and communications, are expected to revolutionise computing in the 21st.

The characteristics of SpeckNets give rise to some specific requirements for human-SpeckNet interaction. First, the invisibility of the Specks and the lack of any perceptible output mechanism means the human-Speck interface must be provided through some other mechanism. Second, people may literally be immersed in the network surrounded by Specks, of various types in various configurations, in three dimensions; they are not interacting with a device, they are inside an information space. Third, the SpeckNet may not know about the physical characteristics of the real world. In stable SpeckNets, it would be possible for the network to develop a model of the world and relate this to its own configuration. However, in recently created or in movable SpeckNets, the human will have to supply information about the physical world. Finally, there is the issue of scaling interaction from individual Specks to potentially thousands.

Previous work on WSNs and human interaction with these networks identifies a number of different types, depending on the method of deployment of the nodes, their size, the communication mechanism, and the network's topology (Romer & Mattern, 2004). There have been many WSN applications, each with its own interaction methods and techniques. For example, a sensor network has been embedded within a vineyard. The system would automatically trigger an event, such as turning on sprinklers when soil moisture is low or firing air cannons when birds were detected (Burrell, Broke, & Beckwith, 2004). Another example is the self-healing minefield (SHM), a minefield that can reorganise itself (through mobile mines) to cover gaps that appear (Meriall, Newberg, Sohrabi, Kaiser, & Pottie, 2003). ARGO is a global network with an intended 3,000 sensors that will monitor salinity, temperature, fresh water storage, and so forth, of the upper layers of the oceans, and transmit results via satellite. Deployment began in 2000, and as of February 2006, 2,385 floats were in operation (Argo Project Office, 2006). In most of these applications, data was sent from the network to a remote database.

In contrast, we are interested in WSNs where the person is inside and interacting with the network directly; SpeckNets. We describe a generic tool-kit for humans interacting with SpeckNets, focusing specifically on the unique characteristics that separate them from other WSNs. In the next section, we elaborate on the requirements that this interaction demands. This is followed by some background theories that have helped to shape the tools. Some studies of alternative designs are described that help to evaluate the effectiveness of both theory and tools. The chapter closes with some conclusions and indications concerning the future of interaction in WSNs.

REQUIREMENTS

The aim of our work in the Speckled computing consortium is to explore the requirements of human-SpeckNet interaction (HSI). One reason for doing this is to inform the design of the operating system that will be developed for the Specks. Another is to elaborate on possible scenarios and applications.

A typical scenario of HSI could be a surveyor performing an inspection on a domestic building. The surveyor would arrive at the house and be alerted that he or she has entered a SpeckNet. As part of the initial detection, the interaction tool will have initiated a harvest of top-level information, such as the types of sensors in the network, date of installation, and so forth. The surveyor would be presented with this information and then make a decision as to whether or not this information is valuable.

The surveyor may be interested in the SpeckNet's ability to supply moisture information, and initiate a harvest for values from moisture sensors over a certain threshold. This information would be retrieved and presented along with metadata (e.g., time of last reading) and new options (e.g., hourly readings available for last three weeks).

Having identified an area in the network that is of interest, the surveyor would be guided to the appropriate room to view the information. At this point the surveyor may interact with the representation to gain a better understanding (e.g., rate of spread, severity, etc.). This investigation may initiate a new search for more information.

This scenario captures many of the characteristics of HSI. The network will initially require detection. The user will not necessarily know the location extent or other characteristics of the network. Detection is concerned with ensuring people know that they are in an information space, and how they know what is connected to what.

The user will also be concerned with finding out what information is available from the information space, how it is distributed, and how to control the harvesting of the content. As a real-time system, it would be infeasible to work with all of the data. Instead, the user will control the nature of the data harvested, making it dynamic and user-driven.

The user needs to know what types of Specks there are in the SpeckNet, what attributes they have, and what the values of those attributes are and can be. Whilst this data is of a limited range (since Specks have a limited number of sensors), the data may have other attributes, such as when it was collected.

One consideration is the best way to present information. A visualisation is one way, but auralisation is another. Since the user may be fully immersed in the network in three dimensions, the presentation needs to make them aware of the scope of the network. Presentation of the details of the individual Specks and groups of Specks is also needed. The relative position of the user to any particular piece of data is required so people can find their way to a specific physical location.

As they are so small, Specks cannot have their own display other than perhaps a light. This is clearly inadequate to provide details of the sensor states and readings over a period of time, and so we must find an alternative.

Increasingly developers of WSN applications are beginning to understand the benefit of displaying sensor readings in a real-world context. Although due to the tradition of viewing data remotely, a virtual environment is often used to represent the real world with the sensor data overlaid. An early example was the fire alert system reported by Boone (2004), where sensors attached to trees in a forest could detect fires, and display the location in a 3-D GIS-based virtual environment. The benefit of the system was the ability to use the map to coordinate fire crews tackling the fire. A more recent system combines real-time sensor data (including meteorological readings as well as live audio and video) with the Google Earth application to allow an environment to be monitored remotely by anyone with an Internet connection (University of California, 2006). The ARGO project now also allows the tracking of its sensor buoys using Google Earth, but they are yet to integrate the sensor readings into the application.

The intended miniature dimensions and large numbers of Specks mean that they will blend into our surroundings to the point where they are indistinguishable from it; floors will store data on human traffic, machines will store data on their operational performance, buildings will store data on conditions, such as dampness within their walls. Our physical environment will be coated with a layer of data, with no fixed point of interaction; instead, a person will interact directly with the object they want to gain information about.

The combined requirement for possibly complex information tied to devices too small and numerous for individual interfaces makes the use of augmented reality (AR) a natural choice for HSI. A live video stream can be enhanced with computer-generated objects (rendered so that they appear to be within the actual scene) and then presented to the user. We have not explored immersive displays in our work, preferring to use non-immersive displays, as have been tried elsewhere, including computer monitors (Belcher, Bilinghurst, Hayes, & Stiles, 2003), PDAs (Wagner & Schmalstieg, 2003), and mobile phones (Moehring, Lessig, & Bimber, 2004).

The single largest technical issue remaining in augmented reality is that of accurately aligning the real and virtual environments, a process called registration (Azuma, Baillot, Behringer, Feiner, Julier, & MacIntyre, 2001). A number of systems allow the technology used for performing this registration to also offer 3-D input. A notable example is the ARToolkit (HitLab-Washington, 2006); a software library that includes facilities required to optically track images placed in the real world, and align computer-generated graphics based on their position and orientation.

Figure 1 shows how ARToolkit has been used in our previous work to both allow the alignment of 3-D geometry with real objects (in this case playing cards), as well as tools for interacting with the AR representations (in this case a proximity selection paddle). Like many researchers, we continue to use ARToolkit to allow us to focus on issues of interaction, rather than the technical issues of implementing a working AR system.

NAVIGATING INFORMATION SPACE

There are a number of conceptualisations of human-computer interaction that foreground the distributed nature of information spaces. Benyon (2001; 2005) highlights how information is distributed both physically and conceptually, and this leads to the need for people to navigate the information space. These ideas have been applied to Web site design (Benyon, 2006). Wright, Fields, and Harrison (2000) present a model of distributed information spaces, called the resources model, in which they focus on information structures and interaction strategies. Distributed cognition (DC) similarly argues that people make use of resources in the world as a central part of their cognitive activities (Perry, 2003), as does situated cognition (Lave, 1988). There are also similarities with Pirolli's "information scent," where people

Figure 1. Paddle selection in Yu-Gi-AR

are seen as "informavores"; utilising our evolved food-foraging mechanisms for information gathering (Pirolli, 2003).

In information foraging theory, people shape their behaviours according to the information ecology. Information foraging theory shows how cognitive and perceptual abilities lead to information finding. Information scent (e.g., good labelling) is concerned with proximal cues by which to judge distant information sources. Optimal foraging is the trade-off between finding, choosing, and handling information

DC foregrounds the importance of cognitive artefacts such as checklists, books, and diaries, and how they are distributed across an environment and are fundamental to cognition (Perry, 2003). DC is concerned with knowledge in the head and knowledge in the world (Zhang, 1997). The resources model (Wright et al., 2000) identifies some specific type of knowledge that designers can place in the world such as plans, histories, and action prompts.

In the navigation of information space approach (Benyon, 2001; Benyon, 2005; Benyon, 2006; Benyon & Höök, 1997), lessons from environmental psychology and urban design are used to inform the design of information spaces. Techniques, such as developing maps, guides, signposts, and landmarks, help people to develop both survey knowledge of the space and route knowledge to help them move through the space.

Three activities are highlighted by the approach. Wayfinding is concerned with how people reach a known destination. Exploration is concerned with finding out about the size and topology of a space. It is concerned with establishing a horizon, and how one part of the environment relates to others

A third activity that is undertaken in information space is object identification. Here people are less interested in the location of objects, or finding a path, or reaching a goal. Object identification is concerned with finding categories and clusters of objects spread across environments, with finding interesting configurations of objects, and finding out details about the objects.

The human interacting with a SpeckNet is physically inside both an information space and a physical space. People have to physically move through the physical space in order to select the part that interests them. They must also understand the information that is available in the SpeckNet and how it is distributed. We frame our discussion in terms of object identification, exploration, and wayfinding in information spaces.

Figure 2 shows the sequence of these activities in the context of human-SpeckNet interaction. The process begins at the exploration stage, gaining an understanding of the data distributed through the SpeckNet. Having chosen sensor readings of interest, the user moves (wayfinding) to reach the real-world location where they were generated. At the object identification stage, the user attempts to make sense of the data in the context of the environment. If satisfied, then the user can move onto other tasks; but if more information is required, then the process is repeated, leading to a spiralling pattern as they drill-down into the information space.

The actual tasks at each stage of the model could range widely, from complex to trivial. Exploration may require the comparison of thousands

Figure 2. Model of Human-SpeckNet Interaction

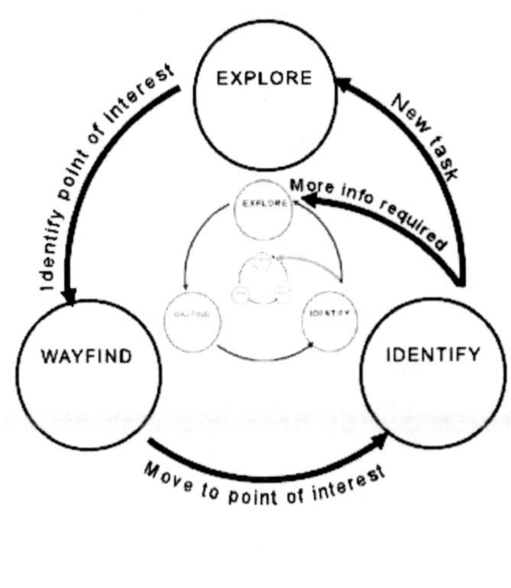

of data values, or a single one; wayfinding could require crossing a city, or simply turning around to deal with data behind you; object identification could require a user to cross-reference information from several different regions, taking into account the context of the environment and decades of professional experience, or simply being shown which switch to flick. However, despite the huge range of scenarios, the sequence is seen as fundamental to human-SpeckNet interaction.

Three main studies have been undertaken that explored principles of supporting users in HSI. The *Specktator* investigation explored user interaction issues within a small-scale SpeckNet, using hardware prototypes. A group of users carried custom-built Speck devices (a *Specktator*) that monitored their level of activity (through a pedometer) and visits to designated locations (via Specks distributed in the environment). Two groups of four people took part in the investigation, carrying the Specktator devices for a period of 3 hours. At the end of each trial, the participants regrouped and were presented with visualisations of the data collected.

The *AR HomeFinder* investigation explored information visualisation techniques in an augmented reality setting. A tablet PC (in essence, a 10.4" flat panel screen with a computer built into it) was used in conjunction with a Web cam to produce a handheld AR *window*. This allowed a virtual representation of houses to appear on a tabletop, which 20 participants (12 male, 8 female) then explored in three tasks. The focus of this work was on the selection of objects via a pen on the screen, or by targeting the camera (gaze method) and on adjusting parameters again via the pen and then by using the motion of the camera to interact. The study also explored providing an overview of the data through glyphs.

The *virtual warehouse* investigation explored the requirements for wayfinding tools (as well as the requirements for a SpeckNet to support them), using a virtual environment. Two wayfinding tools were compared (using waypoint markers and using relative positioning), and presented to the participants in a random order.

OBJECT IDENTIFICATION

In an information space, people need to find out what objects exist, their attributes, and what the values of these attributes are. Our experience with the object identification issues in SpeckNets comes from the Specktator and AR HomeFinder investigations.

For the Specktator, the chosen scenario was for each Speck device to function as a virtual pet, where its appearance would reflect the underlying data (see Figure 3). A semantic relationship was chosen where, for example, encounters with a Speck placed in a kitchen area would affect the appearance of the character's torso (from thin to fat), walking activity would affect limb size, and visits to a print room would affect head size (which assumed printed documents would be of an intelligent nature).

The decision to convey context via a feature and representation of a data value via the appearance of that feature is based on the work of Chernhoff, who used faces to represent chemical samples (Chernoff, 1973), and Spence, who used glyphs for both house and ship representations (Spence, 2001). The aim is to present a recognisable link between appearance and underlying data.

Figure 3. Specktator character representation

Although the study was small in scale, the results were promising, in that the semantic relationship proved easily understandable. Although perceived accuracy of the representations was low, and indeed it would have been impossible to view a character and determine, for example, how many trips had been made to the kitchen, the purpose was for comparison and for this, the visualisation seemed appropriate.

The AR HomeFinder investigation also explored the use of glyphs. Two types of glyphs were used as shown in Figure 4; simulating the potential for different types of Specks. House glyphs represented three values (number of bedrooms, size of garden, and cost of property), and school glyphs represented two values (number of students and pass rate). In the representation, there were 18 houses and 4 schools shown in an area approximately 25cm x 25cm.

Part of a questionnaire, presented during the trial, asked participants what they expected the components of each glyph to represent. Results show that a literal mapping, such as tree size conveying size of garden, could be intrinsically understood (chosen by 19 out of 20); but an abstract mapping, such as platform height for price, was less obvious (correctly chosen by 12 out of 20).

Participants were capable of identifying extreme values (e.g., most expensive house) almost immediately; even if a subsequent search was performed to confirm their initial choice. However, identifying actual values was still problematic, and so it is clear that some method of obtaining detailed values must accompany a glyph system; the most obvious of which is to allow selection of a glyph and present its values.

The selection task in the AR HomeFinder application required participants to find a house of a given price from the 18 presented. Two types of interaction were explored. A gaze interaction required the user to centre the camera on an object. After a few moments, the camera would zoom in to provide details. A pen interface required the user to tap on the object to initiate the zoom function.

The task was performed twice using a pen (or stylus) to select objects on the tablet PC, and then twice more using the camera to target the AR houses (termed gaze selection). Figure 5 shows the task completion times for the four repetitions, and it is clear that although initially poor performance (Gaze1), the gaze selection method quickly improved. However, despite promising results, the gaze interaction was not well received, being reported as both harder to select buildings, and harder to remember which buildings had been checked. In addition, 14 participants preferred the pen method against 4 that preferred gazing.

Figure 4. Example glyphs from AR HomeFinder

Figure 5. Comparison of selection methods

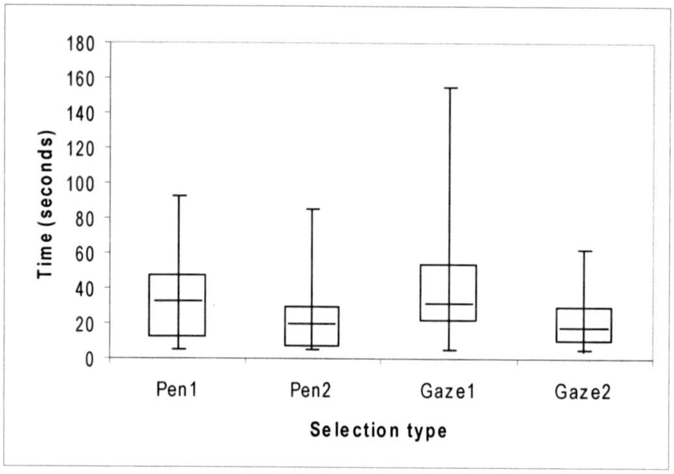

EXPLORATION

In some SpeckNet applications requiring human intervention, the data interpretation phase may be trivial, in that the network need only identify the point of intervention (e.g., directing firefighters to a fire). In other applications, the user may need to perform some exploratory analysis (e.g., the ALife system described by Holmquist, Gellersen, Kortuem, Schmidt, Strohbach, Antifakos, Michahelles, Schiele, Beigl, & Mazé, 2004, where the user must prioritise trapped avalanche victims). While the results reported for object identification show how individual Speck values can be interpreted; they do not address the large potential scale of data that may be presented.

The problem is, in essence, one of applying information visualisation principles to an AR environment. Specifically, the principles of *overview, zoom, and filter*, which were proposed as fundamental by Shneiderman (Card, MacKinlay, & Shneiderman, 1999). The second part of Shneiderman's mantra, *details on demand,* is covered by our category of object identification. These issues were considered in the second and third stages of the AR Homefinder investigation.

Figure 6. Clustering map

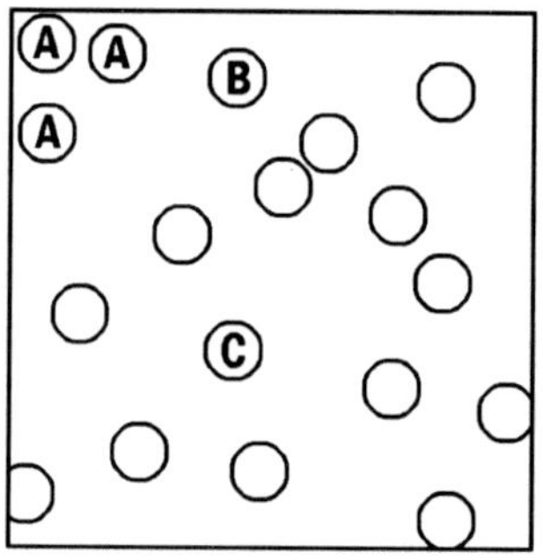

Creating a semantic relationship between the underlying data and the glyph representation is particularly important in enabling users to develop an overview of the data. Lee, Reilly, and Butavicius (2003) report that glyphs perform poorly when compared to spatial visualisations, which they attribute to the glyphs being read sequentially (that is, each consulted in turn until the required one is found). Our contention is that a semantic link creates visual cues or, following the metaphor of information scent, an *aroma* that can be recognised much more easily.

Some support for this position was presented in the object identification section, where we reported that participants in the third stage of the AR

Figure 7. AR HomeFinder tilt interface

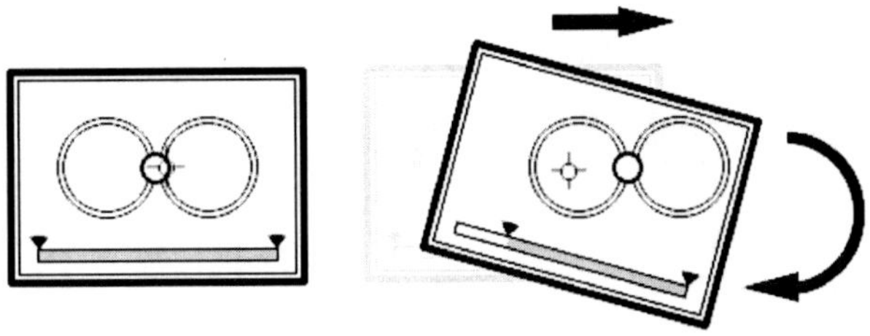

HomeFinder investigation could identify extreme values almost immediately (i.e., not sequentially searching to find them). In addition, participants were given additional tasks of identifying trends (similarities) and exceptions to those trends, and again it was observed that the majority were initially drawn to the correct objects.

An interesting observation relating to identifying trends is that, for example, the question "Identify a group of 1 bedroom houses" is open to interpretation as to what constitutes a *group*. Figure 6 shows the layout of houses on the map used, and circles with letters are one-bedroom houses. Five participants only identified the A's as constituting a group, while the remainder included B, with the exception of one individual who marked all.

The second stage in the AR HomeFinder study explored another tool in exploration, the use of dynamic filtering to focus attention. Participants were given a range of prices and asked to report the number of houses that were within this range. First, they used a standard dynamic filter scroll bar using the pen, and then a novel method of using the camera to interact. The inspiration for the second method came from Beardsley, Van Baar, Raskar, and Forlines (2005), who used the tilting of a projector to control scaling of an image.

The principle of the interface is shown in Figure 7, and consists of two components on the tablet PC screen; the selection rings and dynamic filter range bar (left). The image on the right shows the interface in use. The selection rings are fixed in 3-D space, and so, movement of the camera allows the user to left or right rings, to adjust lower or upper boundaries respectively. Actual adjustment of the value is achieved by tilting the tablet PC to the left or right, moving the parameter range to the left or right in turn. Tilting to a more extreme angle causes the parameter to adjust at a more extreme rate.

Participants were comfortable using the point-click method and were both fast and accurate in completing tasks (only one participant gave an incorrect answer, which was due to mishearing the range values). In contrast, the tilt interaction method performed very poorly, taking over twice as long to use, had three non-completions and three incorrect responses. Many design lessons were learnt that could improve the tilt interaction method, but it is unlikely to match the ease of use found from point-click interaction.

WAYFINDING

While the previous sections lay out a foundation for interacting with SpeckNet data from a region directly in front of the user, they may not always begin adjacent to it. Since the points of interest are generated from sensor readings, they could occur anywhere within the SpeckNet; and since we are interested in individuals interacting with these points directly, additional tools are required to guide them to the locations. The majority of our understanding in this area comes from the Virtual Warehouse investigation (shown in Figure 8).

Figure 8. Screenshot from Virtual Warehouse

Figure 9. Guidance methods (left: waypoint; right: relative)

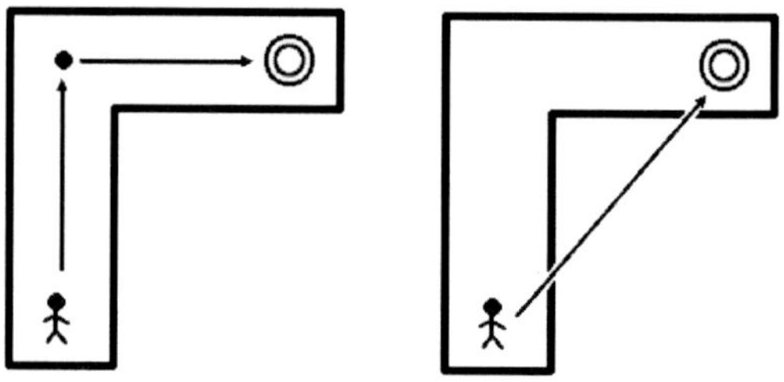

Directing someone to a particular area of interest within the SpeckNet is problematic since we cannot assume that they have previous knowledge of the environment. As such, some sort of navigational aide is required to direct them towards the destination. In contrast to the majority of existing guidance applications (e.g., GPS road guides), a SpeckNet is not expected to easily have access to global environment information. It is expected that the network will be able to generate knowledge of its own distribution, but will have limited knowledge of non-Speckled entities, and also limited knowledge of an environment's permeability to humans (i.e., the paths by which a person can move through the environment).

The first guidance tool tested was the familiar waypoint system used in the majority of navigational systems (in one form or another), where the user is given a sequence of points to pass through. The second method provided the user solely with relative information of direction and distance, which is the minimum information expected to be available from the SpeckNet (see Figure 9).

Figure 9 shows a comparison of the wayfinding times (i.e., from starting point to goal) for each of the collection conditions (W for waypoint, R for relative). The six paths are arranged from the most simplistic on the left (a short distance with one wall obstructing), through to the most complex on the right (longer distance with several intervening walls and different vertical level). In each pair, the result from using the relative system is shaded to aid comprehension.

The most noticeable result is that use of the relative system always resulted in a wider range of collection times than the waypoint system. In runs 1, 2, and 3, for example, the times were found to not be significantly different statistically, but it is clear that some individuals performed significantly worse using the relative system. And indeed, the average standard deviation for the relative system over the six runs was twice that of the waypoint system.

Runs 2 and 3 included unexpected blockages, which led to the increased collection times, and also less difference between the waypoint and relative systems. This is to be expected, since the waypoints did not take into account the blocked paths (simulating the intrusion of an object that the SpeckNet is not aware of), participants were forced to deviate from the path and find their own route, as with the relative system. This shows that although substantially superior, the benefits of the waypoint system could easily be compromised if the paths that it believes exist become blocked.

Table 1 shows the results of T-Tests comparing the waypoint and relative collection times for each run. Note that generally speaking, the difference becomes more significant (closer to zero) as the paths become more complex. The notable exceptions are runs 2 and 3, which as mentioned previ-

Table 1. Statistical comparison of waypoint and relative system collection times. A value of less than 0.05 is considered to be significant

Run	T-Test probability
1	0.2434
2	0.6003
3	0.6576
4	0.0675
5	0.0012
6	0.0005

ously, is attributed to the inclusion of blockages removing the benefits of waypoints.

However, despite the better performance of the waypoint system, we should not forget that the participants did still manage to reach the goal using the relative system. Therefore, the main conclusion from these results should be that it is possible to navigate using only relative information, but with penalty increasing over distance.

CONCLUSION

Our work on human-SpeckNet interaction has answered several specific questions about this particular type of HCI, and has raised some more for future forms of interaction. Wayfinding in SpeckNets is assisted by providing waypoints. Exploration is supported by providing an overview of the whole space, and the distribution of different types of Specks and different values for the attributes of Specks. Object identification is supported by having clear, semantic representation of attributes and their values.

The recommendation for waypoints in SpeckNet applications is not necessarily straightforward. Ideally, all applications would use them because of the proven benefits they offer, but there is the issue that their implementation has a cost. Either the SpeckNet itself must generate paths or they must be set manually by an operator, which has processing or manpower costs, respectively. As shown in the Virtual Warehouse study, if there are unexpected barriers, then this cost could be paid with no benefit. As such, the recommendation is dependent on the application. If there are substantial horizontal and vertical barriers (such as in a building) then waypoints are almost certainly required, but if not (as an extreme example, a SpeckNet embedded in a sports field to monitor the turf) then a relative system could suffice.

The huge difference for split levels suggests that performance gains could be achieved by using

Figure 11. Collection times in Virtual Warehouse

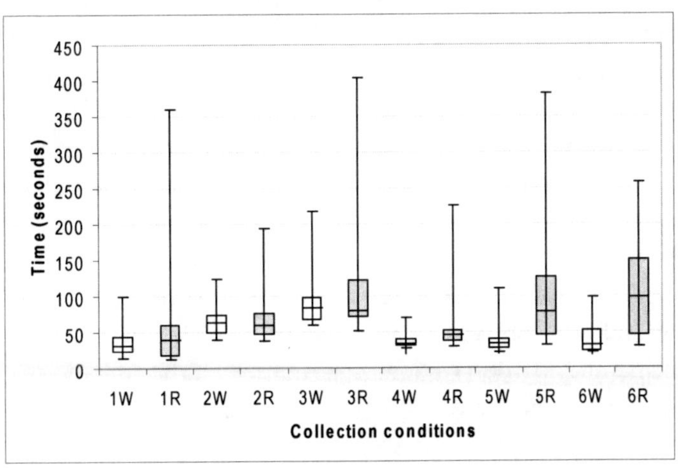

floor level indicators. Obviously, this would be an added feature, requiring additional resources, but offering rough location information could be a compromise between the two tested systems.

The results also confirmed the use of glyph representations as an ideal solution for SpeckNet data (encoding both data type and value, easy interpretation through use of semantic links, etc.), and so the general tools for querying an environmentally distributed database were identified. However, some shortcomings were also identified, namely, that data outside of the field of view was often overlooked.

It is likely that the poor feedback on the gaze selection was due to participant's overall perceived performance and unfamiliarity. The results indicate that it is a promising method, and it is hoped that users would become more comfortable with it through use.

In contrast, attempts to adjust parameter bounds using camera interaction were largely unsuccessful. Gazing may have performed better because it is a natural Human activity in the real world; looking at something you are interested in. Adjusting values is a less natural task, and so better suited to other form of interaction (either screen based or physical widgets).

Of course, there will be many novel devices, visualisations, and other displays designed to optimise performance in particular types of environment. However, we are seeking something more generic; the types of tools and widgets needed to support basic human-SpeckNet interaction. Primarily, this will be implemented on some personal device that mediates interaction with the SpeckNet, but the design principles also need to be supported by the operating system and data structure of Specks themselves.

A number of features of Specks and SpeckNets have been identified. Specks need to be able to identify their structure and attribute values in some accessible way, such as an XML description. Specks need some distance measure so that the SpeckNet can generate an overview of the mesh and indicate the network's horizon. Specks need a way of propagating the relationships between two locations in the SpeckNet, dependent on specific attribute values, so that the person can navigate from one location to another.

Interacting with SpeckNets provides an insight into the range of interaction issues that will arise in the future of pervasive and ubiquitous computing. Firstly, unlike most other forms of HCI, interacting with SpeckNets separates device and display. This issue generalises to other areas, such as context aware applications and other forms of ambient computing, where the smooth transition of an interaction from one device to another is a key requirement.

SpeckNets are also examples of interaction with multiple processors, another departure from traditional HCI. There are potentially multiple levels of interaction as the user interacts with different numbers of processors. We think that our too -kit will generalise to operate up and down through levels of abstraction. We have begun to investigate issues of immersion in the 3-D of the SpeckNet, but more has to be done.

There are also issues concerned with probing the SpeckNet in different ways. Currently, we just retrieve data from the SpeckNet, zoom, and filter based on attribute values. It would be interesting to explore other methods of probing that depend on the sensors the Specks have. For example, in a Specknet with light sensors, one could probe using a torch, but it is not clear what the functionality of such interactions might be.

Human-SpeckNet interaction is just one of the many new forms of interaction that will become familiar over the coming decades. There are clearly some significant new challenges for HCI, but it is also encouraging to see some of the principles that have evolved over the last 30 years do generalise to the new forms. More natural forms of interaction, such as the gaze zoom, will be developed that emphasise an appropriate "human scale" to the interaction.

REFERENCES

Argo. (2006). *Five years of progress, five decades of potential.* Retrieved September 26, 2006, from http://www.argo.ucsd.edu/FrAbout_Articles.html

Arvind, D. K., (2005). Speckled computing. In *Proc. Nanotech 2005, 3*, 351-354.

Azuma, R., Baillot, Y., Behringer, R., Feiner, S., Julier, S., & MacIntyre, B. (2001). Recent advances in augmented reality. *IEEE Computer Graphics & Applications, 21*(6), 34-47.

Beardsley, P., Van Baar, J., Raskar, R., & Forlines, C. (2005). Interaction using a handheld projector. *IEEE Computer Graphics and Applications, 25*(1), 39-43.

Belcher, D., Bilinghurst, M., Hayes, S. E., & Stiles, R. (2003). Using augmented reality for visualising complex graphs in three dimensions. In *Proceedings of the International Symposium on Mixed and Augmented Reality* (ISMAR 2003) (pp. 84-92).

Benyon, D., & Höök, K. (1997). Navigation in information spaces: Supporting the individual. In S. Howard, J. Hammond, & G. Lindgaard (Eds.), *Human-computer interaction: INTERACT'97*, (pp. 39-46). UK: Chapman & Hall.

Benyon, D. R. (2001). The new HCI? Navigation of information space. *Knowledge-based Systems, 14*(8) 425-430.

Benyon, D. R. (2005). Information space. In C. Ghaoul (Ed.), *The encyclopedia of human-computer interaction.* Hershey, PA: Idea Group Reference.

Benyon, D. R. (2006). Navigating information space: Web site design and lessons from the built environment. *Psychnology, 4*(1), 7-24.

Boone, G. (2004). Reality mining: Browsing reality with sensor networks. *Sensors Mag, 21*(9). Retrieved September 25, 2006, from http://www.sensorsmag.com/articles/0904/14/main.shtml

Burrell, J., Broke, T., & Beckwith, R. (2004). Vineyard computing: Sensor networks in agricultural production. *IEE Pervasive Computing, 3*(1), 38-45.

Card, S. K., MacKinlay, J. D., & Shneiderman, B. (1999). *Readings in information visualisation: Using vision to think.* San Diego, CA: Academic Press.

Chernoff, H. (1973). The use of faces to represent points in k-dimensional space graphically. *Journal of the American Statistical Association, 68*(342), 361-368.

HitLab-Washington. (2006). *ARToolkit Homepage.* Retrieved September 24, 2006, from http://www.hitl.washington.edu/artoolkit/

Hoffman, T. (2003). Smart dust: Mighty motes for medicine, manufacturing, the military and more. *Computer World: Mobile/Wireless.* Retrieved September 24, 2007, from http://www.computerworld.com/mobiletopics/mobile/story/0,10801,79572,00.html

Holmquist, L. E., Gellersen, H-W, Kortuem, G., Schmidt, A., Strohbach, M., Antifakos, S., Michahelles, F., Schiele, B., Beigl, M., & Mazé, R. (2004). Building intelligent environments with Smart-Its. *IEEE Computer Graphics & Applications, 24*(1), 56-64.

Lave, J. (1988). *Cognition in practice, mind, mathematics and culture in everyday life.* Cambridge, UK: CUP

Lee, M. D., Reilly, R. E., & Butavicius, M. A. (2003). An empirical evaluation of Chernoff faces, start glyphs, and spatial visualizations for binary data. In *Proceedings of the Asia-Pacific symposium on Information Visualisation, 24*, 1-10.

Meriall, W. M., Newberg, F., Sohrabi, K., Kaiser, W., & Pottie, G. (2003). Collaborative networking requirements for unattended ground sensor systems. In *Proc. IEEE Aerospace Conference*, 5.

Moehring, M., Lessig, C., & Bimber, O. (2004). Optical tracking and video see-rhrough AR on consumer cell phones. In *Proceedings of Workshop on Virtual and Augmented Reality of the GI-Fachgruppe AR/VR* (pp. 193-204).

Perry, M. (2003). Distributed cognition. In J. Carroll (Ed.), *HCI models, theories and frameworks*. Boston, MA: Morgan Kaufman.

Pirolli, P. (2003). Exploring and finding information. In J. Carroll (Ed.), *HCI models, theories and frameworks*. Boston, MA: Morgan Kaufman,

Romer, K., & Mattern, F. (2004). The design space of wireless sensor networks. *IEEE Wireless Communications, 11*(6), 54-61.

Schmidt, A. (2005). Interactive context-aware systems interacting with ambient intelligence. In G. Riva, F. Vatalaro, F. Davide, & M. Alcañiz (Eds.), *Ambient intelligence: The evolution of technology, communication and cognition towards the future of human-computer interaction*. IOS Press.

Shneiderman, (1988). *Designing the user interface* (3rd ed.). Reading, MA: Addison-Wesley.

SpeckNet (2006). *Speckled Computing Consortium*. Retrieved April 28, 2006, from http://www.specknet.org/

Spence, R. (2001). *Information visualization*. Essex: Pearson Education Limited.

University of California. (2006). *James Reserve: Data management systems*. Retrieved September 21, 2006, from http://dms.jamesreserve.edu/

Wagner, D., & Schmalstieg, D. (2003). First steps towards handheld augmented reality. In *Proceedings of the 7th International Conference on Wearable Computers* (p 127).

Weiser, M. (2002). Mark Weiser (1952-1999), The founder of Ubiquitous Computing. *IEEE Pervasive Computing, 1*(1).

Wright, P. Fields, R., & Harrison, M. (2000). Aanalyzing human-computer interaction as distributed cognition: the resources model. *Human-Computer Interaction, 15*(1), 1 – 4.

Zhang, J. (1997). The nature of external representation in problem solving. *Cognitive Science, 18*, 271 – 295.

This work was previously published in Exploration of Space, Technology, and Spatiality: Interdisciplinary Perspectives, edited by P. Turner, S. Turner, and E. Davenport, pp. 26-40, copyright 2009 by Information Science Reference (an imprint of IGI Global).

Chapter 15
Modelling Pedestrian Movement to Measure On-Street Crime Risk

Spencer Chainey
UCL Jill Dando Institute of Crime Science, UK

Jake Desyllas
Intelligent Space Partnership, UK

ABSTRACT

This chapter presents results for the first large-scale analysis of street crime rates that utilizes accurate on-street pedestrian population estimates. Pedestrian counts were generated at the street segment level for an area in central London (UK) using a modeling process that utilized key indicators of pedestrian movement and sample observations. Geocoded street crime positioned on street segments then allowed for street crime rates to be calculated for the entire central London study area's street network. These street crime rate measures were then compared against street crime volume patterns (e.g., hotspot maps of street crime density) and street crime rate statistics and maps that were generated from using the residential population as the denominator. The research demonstrates the utility of pedestrian modeling for generating better and more realistic measures for street crime rates, suggesting that if the residential population is used as a denominator for local level street crime analysis it may only misinform and mislead the interpretation and understanding of on- to pedestrians. The research also highlights the importance of crime rate analysis for understanding and explaining crime patterns, and suggests that with accurate analysis of crime rates, policing, and crime prevention initiatives can be improved.

INTRODUCTION

Measures and approaches for analysing patterns of street crime (i.e., robbery from the person and thefts from persons) are typically performed using volume statistics presented in tables and graphs, and as hotspot maps to identify the volumetric density patterns of street crime. With the proliferation of geographical information systems (GIS) into police and crime reduction agencies, hotspot analysis is seen as a crucial first step in developing intelligence to help identify and explain crime problems (Chainey & Ratcliffe, 2005; Eck, Chainey, Cameron, Leitner, & Wilson, 2005; Home Office, 2005). This analysis can then help inform the focused targeting of patrol and crime reduction resources to specific areas to help tackle the identified crime issues.

Hotspot maps certainly have a purpose, but can hide the relative levels of risk from being a victim of crime that people may experience. For example, hotspot maps show areas where crime is high but may purely be a product of the volume of people that frequent (or targets that exist in) the areas of high crime concentration. Geographic analysis that considers the spatial distribution of crime rates can provide added value to hotspot maps by considering some underlying population. This type of analysis may also make it easier to identify certain underlying causes of crime by classifying differences between hotspots (Clarke & Eck, 2003).

The analysis of crime rates alongside crime volume is regularly applied to burglary patterns (for examples see Chainey & Ratcliffe, 2005; Harries, 1999; Home Office, 2001a). Hotspot maps showing high concentrations of burglary may only reveal where there is a large amount of housing stock, hence the calculation of burglary rates against the underlying distribution of housing stock can provide an added dimension to the geographical analysis of burglary patterns by revealing where residents are at most risk of being a victim of this type of crime. Burglary rate maps are straight-forward to construct because most developed countries possess census data at fine geographic resolution that describe the number of residential households in each census geographic unit, although, the choice and source of other denominators for other crime types is not as straight-forward (Chainey & Ratcliffe, 2005).

Rate maps and statistics calculated for crime rates for other crime types often make use of the resident population. For example, in England and Wales, published crime statistics include crime rates for vehicle crime and robbery to the person by using the resident population as a denominator. As a general-purpose measure to compare differences between areas it does have utility, however, in some cases it can greatly mislead (Chainey & Ratcliffe, 2005). Table 1 lists UK Home Office published robbery statistics for April 2002 to March 2003 for a sample of seven police force command areas (Home Office, 2003). The table also shows that the average robbery rate in England and Wales was 2.1 crimes per 1000 population per annum.

Table 1 demonstrates the large variation in robbery rates between areas in England and Wales. This can partly be explained due to the genuine differences in robbery between these areas. For example, inner city areas such as Lambeth in central London do tend to experience more problems with robbery than a provincial town such as Stratford-upon-Avon. These differences being explained due to the different socioeconomic and demographic characteristics between the two areas. There is though a large variation in the resident population between these six areas. The robbery rate statistics suggest that the likelihood of being a victim of robbery in the City of London is three times that of Newcastle, but half of that when compared to the neighbouring London borough of Lambeth. This may be useful as a general relative measure between areas, but may also be inherently inaccurate, as it does not consider the daytime population that frequent these areas. For example, the City of London is the financial

Table 1. Robbery rates in eight police command unit areas in England and Wales, shown with the average rate for England and Wales and each area's resident population. Source: Home Office, 2003.

Police force area	Robbery rate per 1000 population per annum	Resident population
Birmingham	6.1	976000
City of London	6.3	7200
Lambeth	15.1	267000
Manchester	12	393000
Newcastle-upon-Tyne	1.8	260000
Stratford-upon-Avon	0.2	112000
Westminster	9.6	182000
England and Wales	2.1	52 million

centre of the UK, with land use mainly consisting of offices. Many thousands of people travel into the area each day for work, but very few people live in this area of central London. Similarly, the London Borough of Westminster is a centre for business, education, shopping, and tourism, and attracts over one million visitors each day.

The impression given by crime statistics in this form is that the areas of Lambeth, Westminster, and the City of London are some of the most unsafe areas in the country for robbery crimes committed against persons on the street. For all the areas shown in Table 1 the robbery rate only considers the residential population and does not take into account how the levels of daytime population may influence the robbery levels experienced between areas. For example, the City of London witnessed only 45 robberies between April 2002 and March 2003 whereas Birmingham experienced 5,976 robberies. The impression given by the crime rate statistics is that Birmingham is marginally safer than the City of London.

Table 1 highlights the exaggerated effect that resident population may have as a denominator for calculating crime rates. This denominator offers some utility for comparisons between large areas, but could particularly exaggerate and mislead the identification of crime problems when considered at finer levels of geographic resolution (e.g., census blocks and census output areas).

The problems with applying inappropriate denominators for calculating and interpreting crime patterns were highlighted by Clarke and Mayhew (1998). While their analysis was on vehicle crime, the principles of applying suitable denominators to calculate crime rates are applicable to all crime types. The British Crime Survey suggests that two-thirds of vehicle thefts occur when a vehicle is parked in the vicinity of the owners' home (half of this crime occurring when the car is parked on the street rather than a driveway or garage) and 17% occurs in public car parks. In Clarke and Mayhew's research they used exposure (the time the car was parked at a particular location) as the denominator revealing that the risk of a car being stolen from a public car park was actually over four times that of the risk of the vehicle being stolen while parked on the street outside the owner's home. This research was fundamental in identifying the differences between types of localities in risks to being a victim of crime and demonstrated that although car parks may not necessarily appear as hotspots on a crime density map, the risk of having a vehicle stolen from a car park is much higher than any other locality. The impact of this research was the call for specific crime prevention initiatives to be targeted to these high-risk areas. The similar problem of inappropriate use of denominators for street crime may also mean that certain patterns are hidden in the data, and opportunities are being missed to think more considerately about effective crime prevention for tackling street crime.

THE MEASURE AND ANALYSIS OF STREET CRIME

Street crime is usually only analyzed at the local municipal and neighborhood level by exploring patterns in its volume. Attempts that are made to analyze street crime rates use resident population statistics. These figures can have little bearing on daytime street population, particularly in city centres and areas where there are shopping malls and entertainment centres where street crime tends to be most highly concentrated (Home Office, 2001b). If resident population values are used as denominators in these areas then it is likely that the rate will be highly exaggerated and misleads the interpretation of crime patterns. Similarly, in areas where the volume of street crime could be low, the street crime rate could be underestimated, as the area is a locale of high resident population.

Street crime volume statistics and hotspot density maps do have their purpose. They can be used as important tools for helping target resources to relatively small geographic areas where they could have a positive impact in reducing crime. However, volume analysis may miss two important features that can aid intelligence development. Firstly it could fail to identify those places where pedestrians on the street are at most risk of being a victim of street crime. Instead, resources may only be targeted to those places where the numbers are high, but the chances of someone being a victim could be relatively low. Secondly, analysis of crime rate patterns provides an important dimension to explaining why crime may persist in certain places. By identifying areas where the crime rate is high can be more useful in revealing the underlying causes to the crime, rather than identifying the problem as a product of the volume of people that frequent that location. For example, analysis of crime density patterns and crime rates could distinguish between areas acting as a crime generator or a crime attractor (Brantingham & Brantingham, 1995; Clarke & Eck, 2003).

Accurate analysis of street crime rates requires measurement of the on-street population. This type of measure has been utilized for better understanding crime rates in shopping malls (Brantingham & Brantingham, 1995) and for analysis of populations available to stop and search (MVA, 2000), but has yet to be applied in larger scale studies. The on-street population measure would also need to be geographically precise because large differences in pedestrian traffic can exist within very short distances (Duxburry & Desyllas, 2000, 2001). Developments in pedestrian modeling make it possible for accurate estimates of the on-street population to be calibrated, for specific time periods, and at fine levels of geographic resolution. With the use of street crime data that can similarly be geocoded to these same high levels of precision it is possible to analyze geographic patterns of street crime rate.

AIMS OF THIS RESEARCH

This research had five main aims. First it aims to demonstrate how on-street pedestrian volumes can be generated using pedestrian modeling, and secondly how data derived from the model can be linked with crime data, in a large-scale study area, to compare patterns of. The third and fourth targets were to investigate if geographic patterns of were different to street crime volume patterns (i.e., hotspots), and if maps generated using the resident population as a denominator were different to patterns revealed in rate maps generated using on-street population values. Finally, the research aimed to generate better information for understanding street crime, proving if there was value in street crime rate analysis alongside analysis of volumetric patterns.

STUDY AREA AND DATA

Study Area

The study area for this research was within the London Borough of Westminster in central London (see Figure 1). The area includes busy pedestrian areas such as Oxford Street, Soho, Piccadilly, Trafalgar Square, Leicester Square, Parliament Square and Covent Garden, and 19 London Underground stations. It also includes residential areas such as Pimlico and Westminster.

This study area was selected because extensive pedestrian modeling had already been performed in the area for the purposes of monitoring changes in pedestrian levels resulting from the introduction of the central London congestion charging zone (CCZ)—the CCZ was introduced in February 2002 and aims to reduce traffic congestion by requiring drivers to pay a daily fee if they drive into the zone. It is also an area that experiences high levels of street crime, but attracts many millions of visitors for work, education, shopping, and tourism—within the study area are several major attractions such as Buckingham Palace, Nelson's Column, Big Ben, 10 Downing Street, the Houses of Parliament and Westminster Abbey. The study area is referred to as the "West End study area" throughout this chapter.

Crime Data

Street crime data for the period January 2003 to December 2003 was sourced from the Metropolitan Police via the Westminster Crime and Disorder Reduction Partnership. Street crime was defined as robbery from the person of personal property and theft from the person. Robbery involves some form of violence or the threat of violence, while theft from the person involves property being

Figure 1. The West End study area in Central London

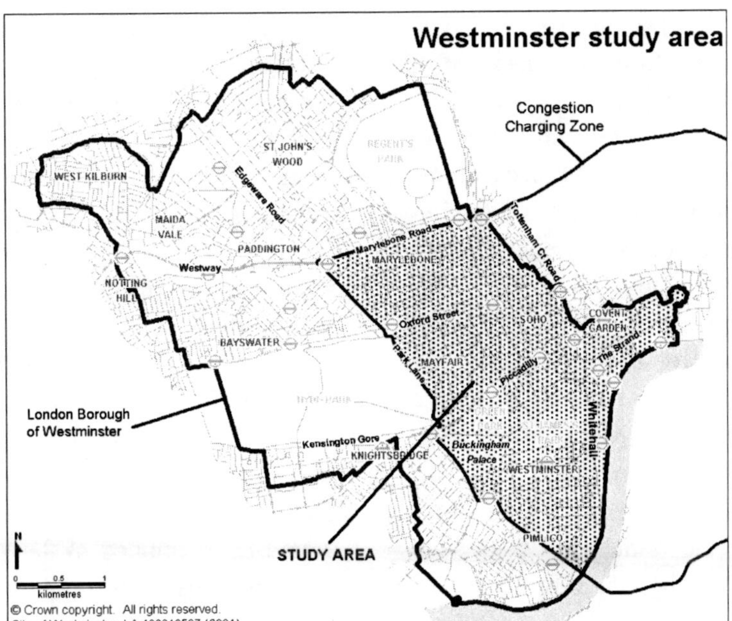

stolen from the person without the person being violently attacked or threatened with violence. It may though include some degree of force being applied, for example, the struggle for a bag that has been snatched by the offender from the victim. It does not include pick pocketing.

The street crime data was geocoded by Westminster Crime and Disorder Reduction Partnership. Precise geocoding of street crime data can present more of a challenge compared to other types of crime because of the ambiguous entry of location information that can be recorded in a street crime record. For example, if a person was robbed on Oxford Street in central London they may not know if it happened outside number 5 Oxford Street or number 105 Oxford Street. Oxford Street is nearly 2 kilometres in length, so difficulties will result in pinpointing the crime location if only "Oxford Street, London" is the location entry in the crime record. Data input can include an organization name (e.g., Selfridges) or "location text" entry (e.g., "junction with Regent's Street") that can be manipulated and used to precisely geocode the crime record (Chainey & Ratcliffe, 2005). However, nearly 20% of records may remain where the only address or location entry in the crime record is the street name. This may result in these crimes being positioned to the centroid or some other "common" location on the street. For short streets this does not cause a problem for most analytical applications. Indeed for this study, as the spatial resolution of interest was the street segment, short streets that were not dissected by street junctions means that this geocoding problem for these places was insignificant. For long streets (e.g., Oxford Street) the positioning of crime events to a common point can result in creating a "false hotspot".

A function that the Westminster CDRP's geocoding software performs is one that successfully handles the problem of uncertainty in precisely determining where the crime took place. The function works in the following way:

- It identifies those crime events that have been successfully and precisely geocoded on the street in question.
- The location of these precise records determines the weights for positioning the affected records. The greater the number of incidents that have been precisely located to a position on the street the greater the weight and draw to that location for points that require this calculated positioning.
- The weighting function also includes an element of random distribution along the street so that no particular location can overly dominate.
- The function is then applied to the affected records, grid coordinates are determined, and this calculated position is recorded in the record's audit code so it can be identified from those that did not require this treatment (Chainey & Ratcliffe, 2005).

The geocoded street crime data were then validated by the researchers conducting this study. This was performed using a methodology for geocoding accuracy analysis as reported in Chainey and Ratcliffe (2005, p 61-63). This revealed the street crime data to be more than 95% accurate and fit for purpose for this research.

Geocoding of crime records in the UK most often makes use of the Ordnance Survey ADDRESS-POINT dataset. This address list can be added to with other points of interest and nonaddressable locations (e.g., public lavatories, parks, car parks) for the purposes of geocoding. This approach is applied in Westminster. However, it does mean that a street crime is most typically geocoded to the position of the building of the address or location where the offence occurred, rather than on the street and outside of a premise. For example, a robbery that may have occurred outside Selfridges Department Store on Oxford Street would be positioned to the ADDRESS-POINT seed point that is geographically referenced to a point at the approximate centre of the building.

A process has to be applied to this geocoded data to position it on the street segment where the crime occurred. This process is explained in a following section.

Pedestrian Data

As part of a study into the impact of the central London congestion charging zone (CCZ), Transport for London (TfL) and the Central London Partnership commissioned Intelligent Space to develop a pedestrian model of the total walking volumes for each segment of the street network in central London. The model was calibrated using geographic information on key indicators that influence pedestrian movement with observed pedestrian flow data from 107 sample locations. The model has been tested and validated independently by Transport for London, who has approved its accuracy in generating on-street pedestrian estimates. The model estimated flow values for 25 square kilometres of street length and over 7,500 street segments. A description of the pedestrian modeling method is provided in a following section.

The origins of the pedestrian model were for comparing changes in pedestrian movement as a result of the introduction of the CCZ. The CCZ only applies between the hours of 07.00 and 19.00. Therefore a limitation of the model was that it had only been calibrated to generate estimates of pedestrian movement between these hours (we refer to this period as "daytime" hours from this point forward).

Figure 2 shows the output of the model for each street in the West End study area. A thematic range was used to indicate those areas of high pedestrian flows, from areas with low flows. These are also shown with photographs for each thematic class to help visualize the level of pedestrian traffic experienced on the street.

Crime and Pedestrian Features of the Study Area

The West End area of London within Westminster provides an ideal study area for exploring patterns of on s. Published statistics on street crime levels in Westminster often appear to exaggerate the actual on street risk as these statistics are calculated

Figure 2. Modeled daytime pedestrian counts per street segment in the West End of London, shown thematically with a photographic representation of the on-street population. Source: Chainey and Desyllas (2004).

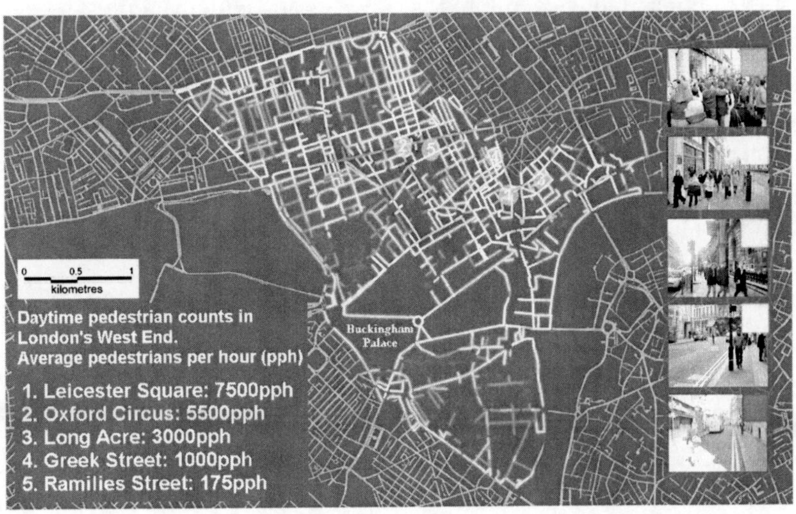

against the relatively small resident population that lives in this area, and not the millions of tourists, shoppers, and commuters that visit or pass through the area.

Table 2 lists the volume of robbery and theft from the person in the study area for the period January 2003 to December 2003. As the West End study area's pedestrian model was calibrated for observing daytime pedestrian movement it was only possible to consider street crime events that had occurred during this same daytime period. However, this did not influence the aims that were originally set out in the research, the main focus being to prove the concept of a large scale study of investigating patterns, and comparing it against hotspot density patterns and crime rates generated using the resident population as a denominator. Table 2 lists the sample of crime data that was available for daytime hours that could be linked to street segments, from which street crime rates could be calculated.

The West End study area contains some of the highest flows within Central London and the core of the pedestrian movement network. For example, the average pedestrian daytime population per year (ppa) on Oxford Street was 33 million—the arterial route that runs from East to West through Oxford Circus (see Figure 2). One of the key characteristics of pedestrian flows is the large difference in the volume of flows on streets within close proximity of each other. This is exhibited in the study area where there is a great diversity in levels of pedestrian flow. For example, some of the side streets leading off Oxford Street have an estimated 0.7 Million ppa—these streets joining directly onto sections of Oxford Street with a ppa of 25 Million.

METHODOLOGY

Pedestrian Modeling

Pedestrian modeling uses a flexible and testable framework of multiple regression analysis whereby key indicators that influence pedestrian movement are calibrated against observed flows from the study area. This modeling process is shown in Figure 3. The modeling process allows for continuous retrials for new areas or new policy levers, and independent validation by users of the model outputs—in Figure 3 the validation of the central London pedestrian model was performed by Transport for London (TfL). With every extension of the model, the importance of all factors can be retested to ensure an optimal balance between specific site accuracy and the wider application of the model design.

Observation Samples

Three studies were conducted in the central London study area to collect sample observation data. Two of the studies were conducted between 08:00 and 19:00, with the third taking place between 12:00 to 18:00. The dates of each study were March 2000, July 2001 and August 1999. The methodology adopted to collect observation samples was similar to that of Benham and Patel (1976). This involved the selection of "study zones"

Table 2. Street crime in the West End study area (January 2003 – December 2003)

	Number of crimes	Number of crimes during daytime hours
Robbery	487	125
Theft from the person	651	277
Total - street crime	1138	402

that were representative areas for capturing the variations in pedestrian traffic in the study area. Pedestrian flows were sampled for each street in these study zones.

Observation points were located on both pavements at the mid point of each street segment. These pavement midpoints were then organized into circuits for observers to carry out sampling. This involved each observer walking a circuit, taking pedestrian flow counts at each pavement midpoint for five minutes, so that each mid-point was covered once for each hour of the survey. If the observer finished a circuit before the end of the hour they would return to the circuit starting point and wait until the end of the hour before commencing the next count. Flow rates per hour for each street segment in the study zones could be derived by multiplying the five-minute counts by 12. The sampling and gathering of pedestrian counts can also be performed by utilizing automated counting technologies such as passive infrared, or by performing observations using street closed circuit television cameras (UTMC, 2000).

A key characteristic of sample data that the model has to capture is the variations in counts that may result with direction: there were street junctions in the study area that experienced flows averaging over 3,000 people per hour on one street and zero people per hour on the perpendicular street. This natural fine scale anisotropic character of pedestrian movement network can be masked by aggregating the data and presents issues associated with "ecological fallacy" (Robinson, 1950). For this reason, the modeling was performed at the street segment level, meaning that the unit of the model was the flow along any street, including both pavements if the street was not pedestrianised. No attempt was made to average flows in any other way, such as along a line of sight (Hillier, Penn, Hanson, Grajewski, & Xu, 1993).

Another key characteristic of observed movement the model was required to replicate was the uneven distribution of flows, with a small number of streets exhibiting very large higher flows compared to most others. Pedestrian flows are logarithmically distributed (Duxburry & Desyllas, 2000, 2001). The pedestrian model therefore has to be able to identify the primacy of these key streets in order to represent the pattern of pedestrian movement successfully.

The sample size observation sites for the pedestrian model was 231 street segments. These observations were not just for the streets with the West End study area, but the central London area that the pedestrian model was originally calibrated for.

Statistical tests were then used to determine the key indicators that significantly impact pedestrian

Figure 3. The pedestrian modeling process (Source: Chainey & Desyllas, 2004)

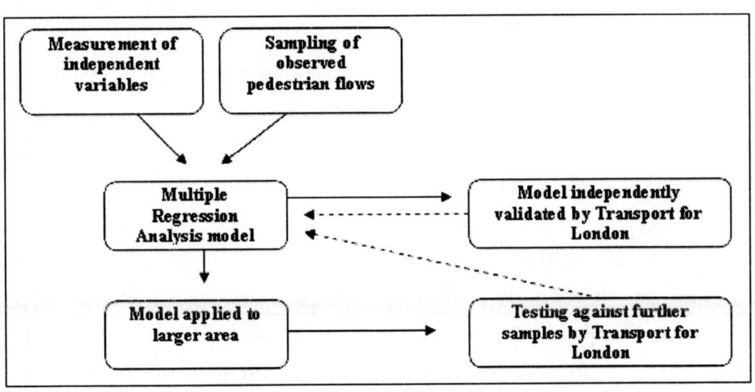

flows within urban landscapes. The key indicators that were identified were

- The pattern of land use
- The accessibility of public transport facilities
- The capacity of pavements or footways
- The visibility or "desire lines" for movement within the layout of the street grid

Each indicator is explained next.

Land Use

The mix of land uses has a great influence on pedestrian movement because short trips are only possible where local destinations for pedestrians are available. This was demonstrated in several of the first pedestrian models developed in the 1970's following the tradition of land use transportation models by Lowry and McLughlin (Willumsen & Ortuzar, 1994) and large scale urban models of the 1960's (Lee, 1973). Pushkarev and Zupan (1975), and Benham and Patel (1976) also demonstrated a strong correlation between independent land use variables and pedestrian movement, resulting in this indicator as being an input variable that would be tested for its suitability in a modern central London pedestrian model.

The pattern of land use in central London was analyzed for the pedestrian model in order to test the influence that different building use had as a "trip generator" for pedestrian flows. Statistical tests from this analysis showed that the key land uses which influenced the pattern of pedestrian movement were retail, food, and drinking establishments. This partly relates to the importance of shopping and leisure trips within overall pedestrian movement—these generate 24% of overall movement within the UK, whereas commuting account for up 16% of trips.

The fact that a shop, restaurant, bar, or pub may exist on a street can be inconsequential if the pedestrian cannot see this form of land use. A separate model was therefore generated to analyze how many premises of this type a pedestrian would be able to see as they walked around the central London street network. This output revealed a large number of highly visible shops and food/drink establishments were clustered in certain places in the West End study area. The regions in view of retail and food/drink land uses in the West End were put forward as a key indicator for the pedestrian model.

Pedestrian Capacity

The infrastructure provision for walking (in terms of the space capacity available for pedestrians or footway width) is an important factor as it can either support flows or discourage walking. The influence of footway capacity on pedestrian flows has been well established since Fruin formalized the level of service (LOS) concept for walkways (Fruin, 1971), showing that reduced capacity leads to congestion and thereby reduced flows. Previous pedestrian modeling studies such as Pushkarev and Zupan (1975) found walkway area to be a significant independent variable. Capacity is important because low quality pavements of inadequate width are an active deterrent to walking.

The capacity of central London's streets was analyzed using a computer program to calculate the average total pavement width of each street segment. This was possible due to the rich base mapping data that is available in Great Britain that is produced by the ordnance survey—Britain's national mapping agency.

Street Configuration

Street configuration (the urban design or layout of streets) has an important influence on pedestrian flows. Street systems are complex and irregular, and pedestrians have to navigate through these networks in order to go anywhere. Pedestrians are sensitive to the complexity of routes and tend to choose simple, more direct options for most trips over complicated, indirect routes (Hillier, 1996). This means that the urban design of an area can have an important effect on supporting or hindering pedestrian movement. The layout of streets can either provide clear, simple routes that are easy to use for pedestrians or it can provide more complex, confusing routes which are much harder to use.

Street configuration was measured by calculating lines of site from every location across the West End study area. This revealed how certain streets offer very clear pedestrian visibility in comparison to others within the dense labyrinthine street network of parts of central London where visibility is poor.

Access to Public Transport

Transport facilities are a key factor in pedestrian flows because pedestrian movement is limited to short trips and so other modes are critical to generating pedestrian activity within an area. There are over 8.6 million trips on the public transport network of London each day (Transport for London, 2003), most forming part of a multimodal trip that includes walking.

Within the study area of central London, the importance of transport for pedestrian movement is underlined by the majority of daytime pedestrian space users within central London who arrive into the area using mechanized transport of some form. The pedestrian mode is often the first and last stage of the overall trip, so pedestrian movement in the study area often originates from the last point of mechanized transport. As transport nodes are important in delivering pedestrians to the area that they will walk in, the pattern of the flows at the local level is influenced by how walkers distribute away from and towards these nodes. Those streets offering clear distribution channels around transport nodes will be more heavily used than those that do not lead towards transport facilities.

It is possible to capture these differences at a large scale, for local areas, using a computer model. For the central London pedestrian model a measure of accessibility to London underground stations within the street network was developed using visibility graph analysis. The output from this model was a calculation of the path complexity from each sample point to the nearest Underground station entrance.

Model Testing

Standard quality control procedures were used in the statistical modeling, such as checking the frequency distributions of each of the variables to ensure that all logarithmic distributions were transformed, checking that all coefficient signs were as expected and testing for autocorrelation of independent variables. Stepwise regression was used to confirm that the four key indicators had a statistically significant influence on observed pedestrian movement. A linear regression of the expected number of pedestrians (calibrated from the model) was performed against observed flows, and this produced an r^2 of 0.82. This is a far higher correlation than any of the previous published urban pedestrian models; Benham and Patel's (1976) correlation of 0.764 was the closest, but with a much smaller sample of 20 cases. The published correlations of the Pushkarev and Zupan (1975) and Hillier, et al. (1993) studies were 0.61 and 0.55 respectively.

Multiple regression analysis of the key indicators revealed that visibility in the central London study areas was the most important variable. This suggested that given that pedestrian trips in central London tend to be short, visibility could be the most appropriate spatial measure of the street network for capturing the aggregation of short journeys. Another assumption would be that simpler routes tend to use the more visible spaces and pedestrians economise on route complexity, not distance. Visibility may also be considered as the most direct aid to wayfinding, especially as pedestrians have imperfect knowledge of the street network, with pedestrians choosing the most obvious routes, which are also the most visible.

The pedestrian model was then independently verified by TfL and a program of testing against additional observation data was carried out. This included a recalibration of the model using the values from two areas to estimate the pedestrian levels for a third area that was not included in the original models calibration. When the estimated values were compared to the new third area in central London it produced an r^2 of 0.74. This demonstrated that the model held up well to scrutiny against data that was not used to calibrate the model.

Linking Crime Data to Pedestrian Flows

Geocoded crime data were linked to pedestrian flow data by following a dual-combinational approach of assigning the crime to its relevant street segment. This was performed in a GIS using a database of street segments. The process involved referencing the street name that was listed in the crime record against the street name recorded in the street segment database, and positioning the crime on the nearest adjacent street segment as long as the street names between the two files matched. This process was particularly important for locations near junctions or where a street crime was geographically positioned at some distance from the street it occurred on (e.g., this occurred for some large buildings where the ADDRESS-POINT seed was several metres from the street where the crime occurred and on occasion was more adjacent to a different street). Validation of this process assured the linking of crime data to street segments had been performed accurately.

Table 3. Street crime rates (per 1000 population per annum) in the West End study area between January 2003 - December 2003

	Rate (per 1000 population per annum)
Robbery	3.1
Theft from the person	6.8
Street crime rate	9.9

Calculating Street Crime Density and Street Crime Rates

Street crime density maps were generated using kernel density estimation (see Eck et al., 2005; Chainey & Ratcliffe, 2005 for more details on this technique). These maps were generated in MapInfo professional using default settings in Hotspot detective (Ratcliffe, 2004).

Rate measures can be calculated by comparing a numerator (i.e., crime) against a denominator (i.e., population) and presented as normalized value (e.g., the number of crimes per 1,000 population). Resident population is commonly used to generate street crime rates. 2001 Census data on resident population for geographic output areas (OAs) was sourced from the UK's online Neighbourhood Statistics Service and linked to geographic polygons of OAs in MapInfo GIS. A count of the number of street crimes in each OA was calculated in MapInfo from which a street crime rate (crimes per 1,000 population) for each OA was calculated. These street crime rate patterns at the OA level were then mapped.

Street crime data that had been linked to street segments was aggregated to form a count of crimes per street segment. These values were then calculated as a street crime rate against the daytime pedestrian population for each street segment and were mapped.

Street Crime in the West End Study Area

For the period January 2003 to December 2003, 1,138 street crimes were recorded in the West End study area. The street crime rate per 1,000 populations for daytime hours in the study area was 9.9 per 1,000 population per annum. This rate was calculated using a 2001 Census derived population of 40756 residents. Table 3 shows rates for robbery and theft from the person for the daytime period.

Figure 4 shows the pattern of street crime across the study area during daytime hours. This shows that the major robbery density hotspot areas (4a) were in Soho and in pockets along Oxford Street that were centred around Tottenham Court Road, Oxford Circus, and Bond Street Underground stations. The rate map (4b), drawn using Census Output Areas, suggests that the area between Piccadilly and Oxford Street, and the eastern section of Oxford Street were the highest risk areas. The OAs with the highest rates were greater than 45 per 1000 population per annum, with street crime peaking in an OA with a rate of 58.8 per 1000 population per annum.

Figure 4c shows theft of the person hotspots (during daytime hours) in the study area. Hotspots were apparent along the middle section of Oxford Street around Oxford Circus, the area near Marble Arch at the west end of Oxford Street, the area extending from Piccadilly and across Leicester Square and into Covent Garden, and the area due north of Piccadilly in Soho. The rate map (4d)

Figure 4. Street crime in the West End study area: (a) robbery hotspots, (b) robbery rates by Output Areas, (c) theft from the person hotspots, (d) theft from the person rates by Output Areas

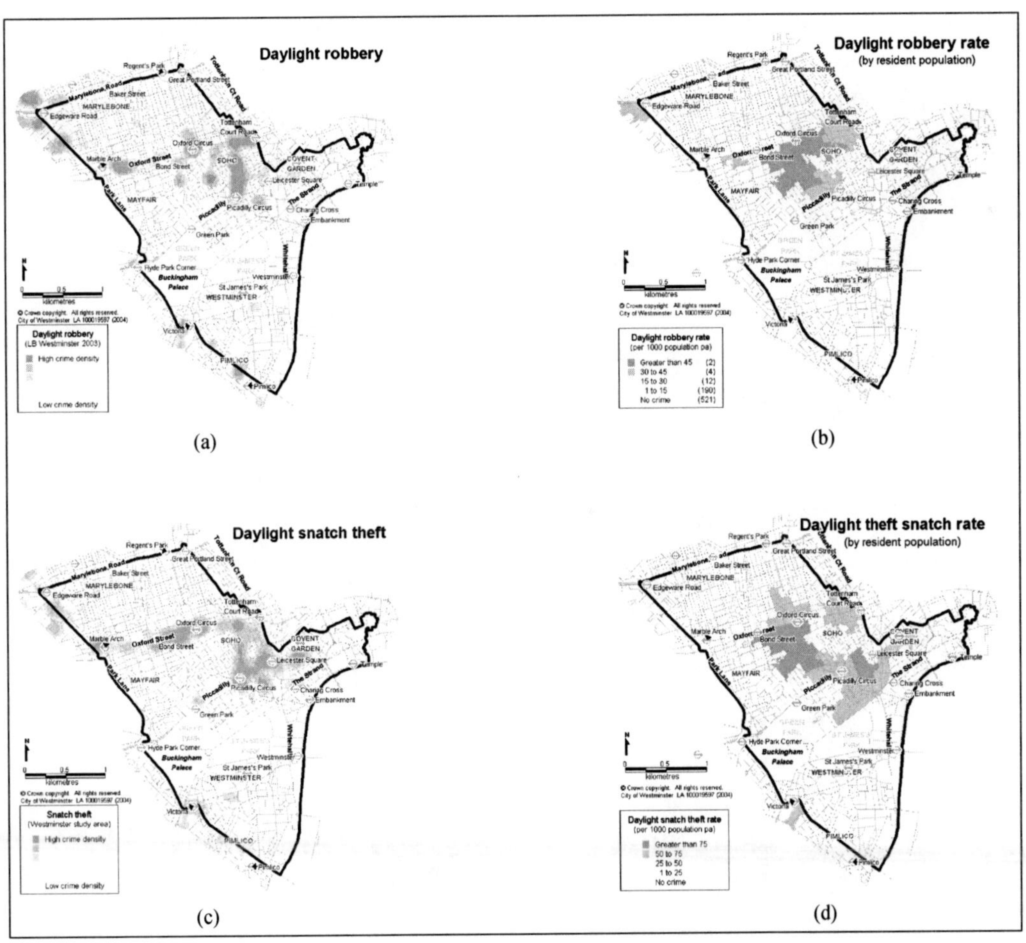

suggests the area between Piccadilly and Oxford Circus (and to the west of Regent Street) as one the most vulnerable areas for this type of street crime, along with the area surrounding Leicester Square. These areas recorded rates that were greater than 75 thefts per 1,000 population per annum, with the worst OA surrounding Oxford Circus, recording a rate of 133.7 thefts from the person per 1,000 population per annum. This could be interpreted as over 1 in 10 residents in the area being a victim of thefts from the person.

PEDESTRIAN

Street crime rates were calculated for each street segment using the modeled pedestrian counts as the denominator. The first most striking finding was how significant the difference was between street crime rates calculated with on-street populations and street crime rates calculated using the resident population. The street crime rates in the areas that recorded the highest levels, using resident population, exceeded 45 per 1000 population per annum for robbery and 75 per 1000 population per annum for theft from the person. Using pedestrian counts as the denominator for calculating, the areas with the highest street crime rates recorded levels of:

- **Robbery:** 0.0019 per 1000 pedestrians per annum
- **Theft from the person:** 0.0051 per 1000 pedestrians per annum

In comparison to the rates reported using the resident population the difference in magnitude is extremely significant. The pedestrian risk statistics for robbery equate to requiring 500,000 people to walk along the highest risk segment of street in the West End study area before a pedestrian becomes a victim of this crime. Similarly, these calculations suggest that it requires 200,000

Figure 5. Robbery rate patterns in the West End study area derived from modeled pedestrian counts

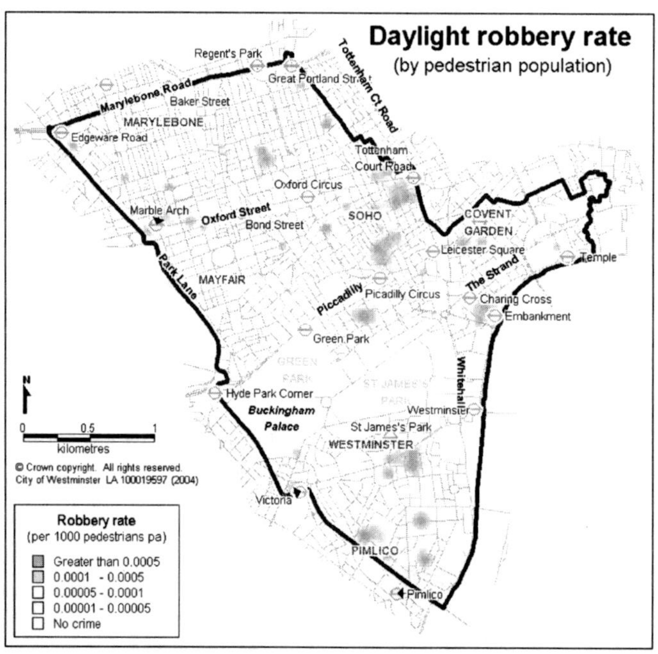

people to walk along the segment of street with the highest calculated rate of theft from the person in the West End study area before a pedestrian becomes a victim of this crime.

As well as differences in magnitude, there are also significant differences in the geographic pattern of street crime rates when resident populations and on-street populations (as denominators to street crime) are compared. Figure 5 shows geographic patterns of robbery rates derived from modeled pedestrian counts. This reveals different areas with the highest on-street risk of street crime to pedestrians in comparison to the rate map shown in Figure 4b. This demonstrates that the attention drawn to the robbery rates in areas along Oxford Street, to the west of Regents Street and north of Piccadilly into Soho appear to mislead when the residential population was used as the denominator. Certain areas in Soho do remain as high-risk areas, but other areas in the north and south of the study area are revealed that had not appeared previously as exhibiting high robbery rates.

Figure 6 shows the geographic patterns of theft from the person crime rates derived from modeled pedestrian counts. The different patterns shown by this rate map in comparison to the residential population derived rate map are again significant. The central section of Oxford Street, the area west of Regents Street and Leicester Square were the areas where street crime rates appeared to be highest in the resident population based rate maps. In comparison, the street crime rate map derived from pedestrian counts exhibited many other areas as the worst affected areas for pedestrians suffering from this type of crime. These areas include the area immediately south of Tottenham Court Road Underground Station and into Soho (e.g., an area known as Soho Square), Covent Garden, and small pockets situated to the north of Oxford Street, the area to the south of Charing Cross Underground station, the top of The Mall (north of St. James's Park), and the area between Victoria Station and

Figure 6. Theft from the person crime patterns in the West End study area derived from modeled pedestrian counts

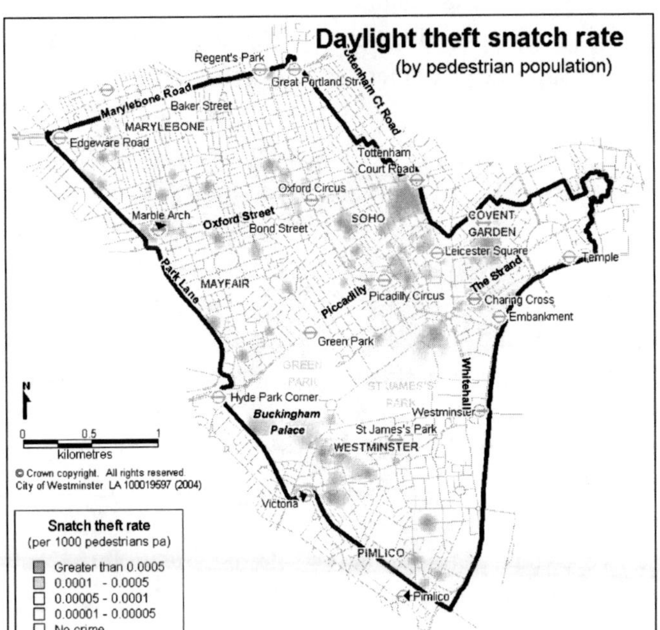

Buckingham Palace. Neither Leicester Square, the central section of Oxford Street, nor the area to the west of Regent Street showed up as areas of high risk to this type of crime.

These results suggest that the rates calculated for street crime that use the resident population as a denominator offer very little in helping identify and understand relative patterns of street crime risk, and only confuse and misinform. Instead, street crime rates calculated using street segment modeled pedestrian counts offer a more reliable basis for identifying and interpreting patterns of risk for this type of crime.

DISCUSSION

This research was the first ever study that used a large-scale pedestrian model to generate precise street crime rates. In many ways the research has served as a proof of concept exercise to demonstrate how the use of modeled pedestrian count data can act as an accurate denominator for understanding measures and patterns of . In this section we discuss the results in terms of the street crime rate statistics that are generated as a result of using pedestrian counts as the denominator to street crime volume and geographic patterns of street crime rates, and the application of these measures.

Street Crime Rate Statistics

This research has shown the clear differences in street crime rate statistics that can be generated from using the resident population and pedestrian population as denominators to the volume of street crime. By using the pedestrian population as a denominator it has been possible to produce new statistics that offer a more accurate (and realistic) measure of the likelihood of a person becoming a victim of street crime. We argue that from this research the resident population value does nothing but misinform measures of risk. Pedestrian-based measures were different by several levels of magnitude to the resident population measures that are routinely used. For example, in the worst affected street crime areas the was not 133 per 1,000 population per annum (as suggested by using resident population data), but was 0.005 per 1,000 population per annum.

Reassuring the public about the actual levels of crime is an agenda that many national and state governments are active to promote to help better address the public's fears, worries and perceptions of crime. To do this it requires the appropriate presentation of statistical facts that can truly reflect the very rare chance of someone becoming a victim of street crime. Policing and Crime and Disorder Reduction Partnership activity in Westminster is very much focused towards reducing street crime and the fear that this crime can present, and by using measures that describe that in its worst areas it takes half a million pedestrian journeys along this street before a member of the public may become a victim of robbery can only help to reassure those that are resident or that visit Westminster of the likely risks and the rarity of being a victim of a robbery or having an item stolen from them.

Street Crime Rate Geographic Patterns

This research has shown the clear degree of difference between the geographic patterns of street crime rates generated from modeled pedestrian counts and those generated for hotspot density maps and street crime rate maps generated using the resident population at the OA level. Volume hotspots are often used as the only geographical analysis tool for targeting police operations and crime reduction initiatives on street crime. These hotspot maps do have a purpose as they help to focus crime reduction and policing initiatives to the areas where crime does concentrate. However, this research has revealed that there are other areas where the likelihood of being a victim from a street crime is different to that of the volume hotspots. This has demonstrated that several of the areas that experience a high volume of crime often only exist due to the large number of pedestrians (and larger number of opportunities) that travel through these areas. For example, an area well-known in central London is Oxford Circus. The street segment to the east of Oxford Circus, outside a large shop called Top Shop is a theft from the person hotspot (there were 8 thefts at this location in 2003) and therefore very often receives targeted police attention and is covered by closed circuit television (CCTV) surveillance. However, in any one daylight hour over 5,000 people on average walk this section of street. It therefore requires 2,500,000 pedestrian journeys along this section of street before a member of the public may become a victim of this type of crime.

Other streets in the study area experienced low volumes of street crime, but their pedestrian traffic was also observed as being relatively low. Calculation of the number of street crimes against pedestrian levels for these areas revealed several of these streets to experience high levels of street crime rate. Yet, because they did not appear in volume hotspot statistics they were places that typically receive very little, if any, policing or crime reduction attention. Other areas were identified where the volume of crime was high and the rate was also high, but their attention for targeted policing and for crime reduction was most often considered with the same solution that was used for the high volume, low risk areas.

Areas that have a street crime problem may require different treatment due to differences in the reasons why crime occurs at these locations. In the analysis of street crime hotspots, each hotspot is often assumed to display similar qualities that explain why crime concentrates at these places. More considered analysis that can classify each hotspot in terms of the spatial differences in volumes and rates can help to unpack why crime may persist at these particular places.

Hotspots can be broadly classified into three types; crime generators, crime attractors, and crime enablers (Brantingham & Brantingham, 1995; Clarke & Eck, 2003). Each can be typified by differences in the volume of crime and the crime rates that each experience (see Table 4). For example, crime generators are places that appear on hotspot maps due to the high numbers of crime that are experienced at these types of locations. However, they are also places where the relative crime rate is low, due to the large numbers of targets and opportunities that are a product of these types of areas. For example, a crime hotspot at a busy shopping centre can be classified as a crime generator. Other areas that appear on a density map as hotspots could also be crime generators, but some, where the crime rate is also high can be classified as crime attractors—these areas tending to be places where specific crime opportunities exist (e.g., places where people can by drugs) or where offenders know there to be a good reputation for committing crime of a certain type at these places (e.g.,

Table 4. Classifications of hotspots, determined by differences in the volume of crime and the rates that each experience. Adapted from Clarke and Eck (2003).

	Number of crimes	Rate
Crime generators (e.g., shopping centres)	HIGH	LOW
Crime attractors (e.g., areas around bars)	HIGH	HIGH
Crime enabler (e.g., park pavilion)	LOW	HIGH

robbery that may occur late at night around the vicinity of bars and nightclubs). Further analysis of a crime rate map may also reveal places that exist as crime enablers—places where the volume of crime is low but where the rate is high. For example, these could be places where there is an environmental or situational lack of control, and little regulation of behavior—such as a pavilion in the middle of a park, a public lavatory, or a bus shelter, which is prone to repeated incidents of vandalism and criminal damage. The number of crimes committed against these targets may be relatively low, but their isolation can make them particularly vulnerable to criminal activity.

Examples of these three definitional differences between types of hotspots can be seen in the study area when comparing Figures 5 and Figure 4:

- **Crime generator:** Oxford Circus, where the volume of robbery is high but the relative rate is low.
- **Crime attractor:** The area immediately north of Piccadilly Circus and into Soho where the volume of robbery is high and the relative rate is also high.
- **Crime enabler:** An example of this type of hotspot can be seen in figure 5 in the area immediately south of Piccadilly Circus where the relative robbery rate is high, but the volume of crime (as shown in Figure 4) is low.

Classifying hotspots in this way helps to diagnose differences between them. Analysis of crime rates alongside analysis of volumetric patterns is therefore important in intelligence development for helping to understand and explain why crime may persist at particular locations. To address these issues it therefore calls for different types of police and crime prevention responses, rather than a single response that is considered as appropriate for each location. For example, a targeted police presence may help address one type of hotspot, but in others an effective and sustained reduction of crime may require a situational crime prevention measure to address the problems.

Spatial and Temporal Analysis of Street Crime Rates Using Pedestrian Modeling

This research has concentrated on the use of pedestrian modeling for analyzing spatial patterns of street crime rates, with the model's output being limited to pedestrian movement during daytime hours. Time also plays an important role in crime analysis, and pedestrian traffic varies across geographical space over time. Extending the pedestrian model to cover night-time hours would be an obvious next step, allowing for a complete 24 hour picture of street crime to be analyzed. The pedestrian model could also be calibrated to be precise to hours in the day, therefore providing the opportunity to explore in spatial and temporal terms. As the use of pedestrian modeling is considered by a larger audience, including police and crime reduction professionals, it is hoped that extending the model to cover these times will be possible.

CONCLUSION

This research has used a pioneering approach for modeling on-street pedestrian movements in central London. The pedestrian model combines variables that influence pedestrian movement (such as land use, pavement width capacity, and on-street natural surveillance) with samples of on-street pedestrian counts collected at strategic points within the study area to generate accurate and precise measures for pedestrian movement. These pedestrian measures have then been used to calculate street crime rates at the street segment level to determine how these rates compare and differ to hotspot density maps and street crime rate maps generated using the residential population (at the census output area level) as a denominator.

The research has found the following:

- The difference in measures of street crime risk between using resident population and pedestrian population as denominators were significant. For example, in the worst areas for street crime that was 133 per 1,000 population per annum when the resident population was used as a denominator, and was 0.005 per 1,000 population per annum when the on-street pedestrian count estimate was used. The research has therefore produced new statistics that offer a more accurate (and realistic) measure of the likelihood of a person becoming a victim of street crime, and argue that the resident population value does nothing but misinform measures of risk
- The degree in difference between the geographic patterns was also significant. Volume hotspots are often used as the only mapping analysis tool to target efforts for street crime. These hotspot maps do have a purpose as they help to focus crime reduction and policing initiatives to the areas where crime does concentrate. However, this research has revealed that there are other areas where the likelihood of being a victim from a street crime is different to that of the volume hotspots, and that the high crime volume areas often only emerge due to the large number of pedestrians that travel through these areas. Other streets in the study area experienced lower volumes of crime, but their pedestrian traffic was also relatively low. Street crime rate measures for these areas were higher than several of the crime volume hotspot areas, yet because they did not appear in volume hotspot statistics they typically receive very little attention for targeted policing or crime prevention. Other areas were identified as both high volume and high-risk areas, but their attention from policing and for crime reduction was typically treated with the same solution that was used for the high volume, low risk areas.

This research provides a starting agenda for discussing improvements to street crime reduction initiatives that consider not only crime volume but also risk to pedestrians—by starting with a measure and a map of the real risk faced by people on foot, stakeholders can begin to target initiatives for risk reduction. The significance in the results generated from this study call for the need for further work that explores the risk generators in high crime rate areas, and how attention to these will help support sustainable crime reduction programmes in street crime and improvements to pedestrian personal safety.

FUTURE RESEARCH DIRECTIONS

Pedestrian modeling in London has been designed to allow for other variables to be included once they become available and prove to contribute to further improvements. The most promising variable to include would be a measure of urban density, with previous research showing that there is a clear, log-linear negative relationship between urban density and private transport energy use per capita (Newman & Kenworthy 1999). This suggests that higher urban densities result in higher pedestrian flows. A weakness with the existing model is that it is best suited to urban rather than suburban areas, but the inclusion of an urban density measure may help improve its flexibility. The current model is also not sensitive to estimating temporal changes in pedestrian volume over a 24-hour period. Research has commenced though that is adapting the pedestrian model to be sensitive to pedestrian volume differences by modeling these volumes for each one-hour interval.

The creation of the pedestrian model used in this research was for the particular purpose of monitoring pedestrian changes as a result of the introduction of London's congestion charging zone. London is therefore fairly unique in its requirement, hence why many other areas in the UK have yet to be treated to similar modeling. However, this is slowly changing as planners and transport providers begin to better recognise the important role that walking has as a mode of transport, and thus commission models similar to that carried out in London. Indeed, spaces in the cities of Melbourne (Australia), High Wycombe (UK), Kingston upon Thames (UK), Duisburg (Germany) and Boston (USA) have already been treated to pedestrian modeling. Interest from this street crime research has also stimulated the requirement to better consider on-street pedestrian populations in the analysis of crime and has already resulted in another similar study into violent crime Haringey (UK).

Most importantly though, this research has highlighted the need to consider pedestrian volumes for helping to qualify and define crime hotspots. While not all areas may be in the position to commission a pedestrian model, simply visiting the crime study area during different times of the day will help to put into context the drivers behind crime that has been committed.

REFERENCES

Benham, J., Patel, B. (1976). A method for estimating pedestrian volume in a central business district. *Transportation Research Record, 629*, 22-26.

Brantingham, & Brantingham (1995). Criminality and place: Crime generators and crime attractors. *European Journal on Criminal Policy and Research, 3*, 3.

Chainey, S. P., & Desyllas, J. (2004). Measuring, identifying and analysing . In *Proceedings at the 2004 UK National Crime Mapping Conference, University of London, London.* Retrieved August 28, 2007, from www.jdi.ucl.ac.uk/news_events/conferences/index.php

Chainey, S. P., & Ratcliffe, J. H (2005). *GIS and crime mapping.* London: Wiley.

Clarke, R. V., & Eck, J. (2003). *Become a problem-solving crime analyst in 55 small steps.* London: Jill Dando Institute, University College. Retrieved August 28, 2007, from http://www.jdi.ucl.ac.uk/publications/other_publications/55steps

Clarke, R. V., & Mayhew, P. (1998). Preventing crime in parking lots: What we know and what we need to know. In M. Felson, & R. B. Peiser (Eds.), *Reducing crime through real estate development and management.* Urban Land Institute.

Duxbury, E., & Desyllas, J. (2000). *St. Giles circus: Report on current patterns of pedestrian movement and accidents.* London: Intelligent Space Partnership.

Duxbury, E., & Desyllas, J. (2001). *Shoreditch triangle pedestrian movement and safety report.* London: Intelligent Space Partnership.

Eck, J. E., Chainey, S. P., Cameron, J. G., Leitner, M., & Wilson, R. E. (2005). *Mapping crime: Understanding hot spots.* USA: National Institute of Justice. Retrieved August 28, 2007, from www.ojp.usdoj.gov/nij

Fruin, J. (1971.) *Pedestrian and planning design.* Alabama: Elevator World Inc.

Harries, K. (1999). *Mapping crime: Principle and practice.* United States National Institute of Justice. Retrieved August 28, 2007, from http://www.ojp.usdoj.gov/nij/maps/pubs.html

Hillier, B. (1996). *Space is the machine.* Cambridge University Press.

Hillier, B., Penn, A., Hanson, J., Grajewski, T., & Xu, J. (1993). Natural movement: Or configuration and attraction in urban pedestrian movement. *Environment and Planning B: Planning and Design, 20*, 29-66.

Home Office (2005). *Crime mapping: Improving performance, a good practice guide for front line officers*. London: Home Office. Retrieved August 28, 2007, from http://www.jdi.ucl.ac.uk/downloads/publications/other_publications/crime_mapping_guide.pdf

Home Office (2003). *Crime in England and Wales: April 2002 to March 2003*. London: Home Office. Retrieved August 28, 2007, from www.homeoffice.gov.uk

Home Office (2001a). *Crime reduction toolkit: Focus area and hotspots. Crime reduction unit.* Retrieved August 28, 2007, from www.crimereduction.gov.uk/toolkits/fa00.htm

Home Office (2001b). *Crime reduction toolkit: Street crime and robbery. Crime reduction unit.* Retrieved August 28. 2007, from www.crimereduction.gov.uk/toolkits/sc00.htm

Lee, D. B. (1973). Requiem for large-scale models. *Journal of the American Institute of Planners, 39*, 163-178.

MVA (2000). *Profiling the populations available for stops and searches.* Police Research Series Paper 131. London: Home Office.

Newman, P., & Kenworthy, J. (1999). *Sustainability and cities, Overcoming automobile dependance.* Island Press.

Pushkarev, B., & Zupan, J. (1975). Urban space for pedestrians. Cambridge: MIT Press.

Ratcliffe, J. (2004). *HotSpot detective for mapInfo Helpfile* (Version 2.0).

Robinson, W. S. (1950). Ecological correlations and the behavior of individuals. *American Sociological Review, 15*(3), 351-357.

Transport for London (2003). *Analysis of the transport programme to support the draft London plan.* Technical Report. Transport for London.

UTMC (2000). *Review of current and future data requirements and detector technologies and the implications for UTMC.* Urban Traffic Management and Contol, Deliverable 2. Retrieved August 28, 2007, from http://www.utmc.gov.uk/research/index.htm

Willumsen, L. G., & Ortúzar, J. de D. (1994). *Modeling transport.* Sussex, England: John Wiley and Sons Ltd.

ADDITIONAL READING

The references that are listed provide many valuable starting points for further reading. In particular we identify to the reader the Brantingham and Brantingham article that discusses the different definitions of hotspots (and why rates are important for understanding crime) and the MVA UK Home Office publication that makes many important points about why it is important to understand on-street pedestrian levels in policing.

The papers listed below provide useful further reading in the background to pedestrian modeling, and further information on the pedestrian model that was calibrated for central London:

Desyllas, J., Duxbury, E., Ward, J., & Smith, A. (2003). *Pedestrian demand modeling of large cities.* Centre for Advanced Spatial Analysis, University College London. Working Paper Series 62. Retrieved August 28, 2007, from http://www.intelligentspace.com/

Desyllas, J., & Duxbury, E. (2004). *Pedestrian modeling of central London.* Intelligent Space paper written for Transport for London. Retrieved August 28, 2007, from http://www.intelligentspace.com/

This work was previously published in Artificial Crime Analysis Systems: Using Computer Simulations and Geographic Information Systems, edited by L. Liu and J. Eck, pp. 71-91, copyright 2008 by Information Science Reference (an imprint of IGI Global).

Chapter 16
A Framework for Data Warehousing and Mining in Sensor Stream Application Domains

Nan Jiang
Cedarville University, USA

ABSTRACT

The recent advances in sensor technologies have made these small, tiny devices much cheaper and convenient to use in many different applications, for example, the weather and environmental monitoring applications, the hospital and factory operation sites, sensor devices on the traffic road and moving vehicles and so on. The data collected from sensors forms a sensor stream and is transferred to the server to perform data warehousing and mining tasks for the end user to perform data analysis. Several data preprocessing steps are necessary to enrich the data with domain information for the data warehousing and mining tasks in the sensor stream applications. This chapter presents a general framework for domain-driven mining of sensor stream applications. The proposed framework is able to enrich sensor streams with additional domain information that meets the application requirements. Experimental studies of the proposed framework are performed on real data for two applications: a traffic management and an environmental monitoring site.

INTRODUCTION

Networks of thousands of sensors present a feasible and economic solution to some of our most challenging problems, such as real-time traffic modeling, weather and environmental monitoring, and military sensing and tracking. Recent advances in sensor technology have made possible the development of relatively low cost and low-energy-consumption micro sensors, which can be integrated in a wireless sensor network. These devices - Wireless Integrated Network Sensors (WINS) - will enable fundamental changes in applications spanning the home, office, clinic, factory, vehicle, metropolitan area, and the global environment.

Concerning the needs of the user for knowledge discovery from sensor streams in these application domains, new data warehousing, data mining techniques have to be developed to extract meaningful, useful and understandable patterns for the end users to perform data analysis. Many research projects have been conducted by different organizations regarding wireless sensor networks; however, few of them discuss the sensor stream processing infrastructure, and the data warehousing and data mining issues need to be addressed in the sensor network application domains. There is a need for new methodologies in order to extract interesting patterns in a sensor stream application domain. Since the semantics of sensor stream data is application dependent, the extraction of interesting, novel, and useful patterns from stream data applications becomes domain dependent.

Some data warehousing and data mining methods have been recently proposed to mine stream data, for example in (Manku 2002, Chang 2003, Li 2004, Yang 2004, Yu 2004, Dang 2007), the authors proposed algorithms to find frequent patterns over the entire history of data streams. In (Giannella 2003, Chang 2004, Lin 2005, Koh 2006, Mozafari 2008), the authors use different sliding window models to find recently frequent patterns in data streams. These algorithms focus on mining frequent patterns with one scan over the entire data stream.

In (Chi, 2004), Chi et al considers the problem of mining closed frequent itemsets over a data stream sliding window in the Moment algorithm, and in (Li, 2006), the authors proposed the New-Moment algorithm which uses a bit-sequence representation of items to reduce the time and memory needed. The CFI-Stream algorithm in (Jiang, 2006) directly computes the closed itemses online and incrementally without the help of any support information. In (Li, 2008), Li et al proposed to improve the CFI-stream algorithm with bitmap coding named CLIMB (Closed Itemset Mining with Bitmap) over data stream's sliding window to reduce the memory cost.

Besides pattern mining in data stream applications, as the number of data streaming applications grows, there is also an increasing need to perform association mining in data streams. One example application is to estimate missing data in sensor networks (Halatchev, 2005). Another example application is to predict frequency of Internet packet streams (Demaine, 2002). In the MAIDS project (Cai, 2004), an association mining technique is used to find alarming incidents from data streams. Association mining can also be applied to monitor manufacturing flows (Kargupta, 2004) to predict failures or generate reports based on accumulated web log streams. In (Yang, 2004), (Halatchev, 2005), and (Shin, 2007), the authors proposed using two, three, and multiple frequent pattern based methods to perform association rule mining.

In general, these approaches have focused on mining patterns and associations in data streams, without considering an application domain. As a consequence, these methods tend to discover general patterns, which for specific applications can be useless and uninteresting. Stream patterns are usually extracted based on the concept of pattern frequency. With no semantic or domain information, the discovered patterns cannot be applied directly to a specific domain.

In this book chapter, we present a data warehousing and mining framework where the users give to the data the semantics that is relevant for the application, and therefore the discovered patterns will refer to a specific domain. We will also discuss the issues needed to be considered in the data warehousing and mining components of this framework for sensor stream applications.

The remaining of the chapter is organized as follows: in the background section, we present some basic concepts about sensor data applications and the data warehousing and mining issues needed to be considered in stream data applications. Following with the background information, we present a framework for domain-driven data warehousing and mining from sensor streams. The case study section shows the experimental results with real data in two application domains. Finally, we conclude the chapter and discuss the future trends.

BACKGROUND

In this section, the background information is reviewed and discussed for two main areas: the sensor stream application domain and the data warehousing and mining issues needed to be considered in this specific domain. These are covered in the following two subsections, respectively.

Sensor Stream Application Domain

A data stream is a sequence of items that arrive in a timely order. Different from data in traditional static databases, data streams are continuous, unbounded, usually come with high speed, and have a data value distribution that often changes with time (Guha, 2001). A data stream is represented mathematically as an ordered pair (r, Δ) where: r is a sequence of tuples, Δ is the sequence of time intervals (i.e. rational or real numbers) and each $\Delta_i > 0$.

Applications that rely on data streams can be classified into offline and online streaming. Offline streaming applications are characterized by regular bulk arrivals (Manku, 2002). Generating reports based on accumulated web log streams is an example of mining offline data streams because most of reports are made based on log data that is collected over a relatively large period of time. Online streaming applications are characterized by real-time updated data that needs to be quickly processed as the data arrives. Predicting frequency of Internet packet streams is an application of mining online data streams because the prediction needs to be made in real time. Other potential online data streaming applications include stock tickers, network measurements, and evaluation of sensor data. In online data streaming applications, data is often discarded soon after it arrives and has been processed, because of the high update rate and huge resulting amount of data.

Figure 1 shows an example of wireless sensor stream applications. In this figure, the sensors generate a series of sensor streams with domain information, such as the sensor identifiers, sensor locations, time stamps, sensed values at a particular time, and power left for the particular sensor, etc. The information is reported to the server through single-hop or multiple-hop rout-

Figure 1. Wireless sensor stream application

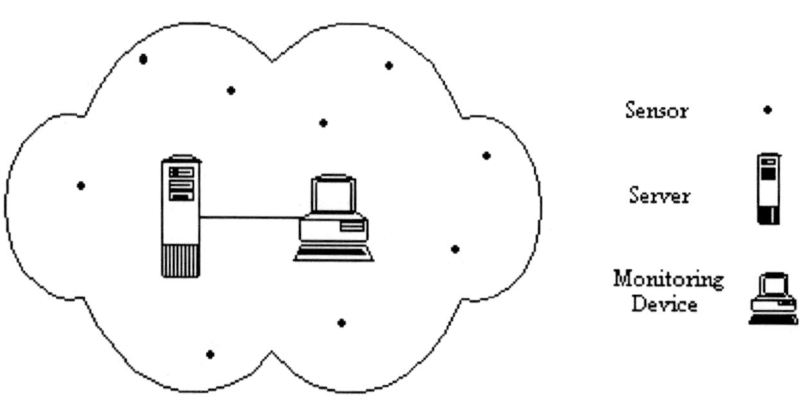

ings. General stream mining methodologies do not have a mechanism to connect the domain information of each application to the reported sensor values; correspondingly the discovered patterns and their relationships with each sensor could not be connected together.

Moreover, concerning the needs of the user for queries about particular sensors, mining the whole set of information from all the sensors may not always be necessary and effective. Connecting the sensors' domain information with reported values will make it easy when the mining task only associated with part of potential sensors. It will also save computing power and memory spaces, which consequently provide higher quality of service in the wireless sensor networks. More data warehousing and mining issues in the sensor stream application domain will be discussed in the following subsection.

Data Warehousing and Mining Issues

As shown in Figure 2, there are several stages in the sensor stream processing infrastructure, namely the data collecting stage, the data preprocessing stage, the data storage and processing stage, and the data reporting and analysis stage. In this subsection, we discuss the data warehousing and mining issues needed to be considered in each of these stages, respectively.

Data Collecting Issues

In the sensor stream application domain, data is gathered through wireless sensors and is delivered to the server or multiple servers for centralized or distributed processing and computing through single-hop or multiple-hop routing. Different data warehousing and mining techniques need to be considered regarding to the different computing models, for example, the centralized computing, distributed or parallel computing, etc. In this book chapter, we focus our discussion to the centralized computing, in which there is one server to collect and process all the delivered data from multiple sensors.

Data Preprocessing Issues

After the data is delivered to the server, before it can be processed, several data preprocess steps need to be performed to get the data ready for the processing unit and connect it with domain information. We discuss each of the steps in the following paragraphs.

Data Extraction

The first step of the data preprocessing stage is data extraction. In this step, the package sent from a particular sensor is extracted and information

Figure 2. Sensor stream processing infrastructure

associated with the data warehousing and mining tasks is collected.

Questions need to be considered in this step include: which information we need to extract? Why we need to collect the information? How the collected data can work with the data warehousing and mining tasks and process? After the needed information is extracted and collected from the data transportation package, some Meta data information regarding the application domain can be gathered from the end user or from the collected information. For example, the data description of the application domain, the data ranges, and data type, etc. All these associated information can be loaded to the framework after being transferred to the required format.

Data Transformation

In the data transformation step, data is transformed to the formats that are able to be processed by the processing unit. The Meta data is used to generate indicator flags which reflect the data's effectiveness, i.e. missing, corrupted or not. Each set of information from a particular sensor is connected together with the sensor's identifier, so that it is easier to identify the sensor relationships and perform data analysis. More of the detail information will be discussed in the framework section.

Questions need to be considered in this step include: What are the source data formats? What are the target data formats? How to perform the data transformation? What kind of data can be regarded as corrupted? How the transformed data can work with the data warehousing and mining task and process? What is the Meta data? What are the usage and constraints of the data? After the data transformation task is complete, all the information can be loaded to the processing system and is ready to be processed.

Data Loading

Data loading is the last step of the data preprocessing stage. In this step, the extracted and transformed data is loaded into the processing system.

Things need to be considered in this step are: If the data coming rate can be kept with the data processing rate? The speed of the mining system should be faster than the data coming rate, otherwise data approximation techniques, such as sampling and load shedding, need to be applied. If so, the questions needed to be considered include: Which kind of speed adjusting techniques are suitable to use? How it will affect the accuracy of the mining result? What is the maximum allowable error rate specified by the end user? How we can apply the selected technique to the application system?

Data Storage and Processing Issues

After the data is being preprocessed, the next fundamental issue we need to consider is how to optimize the storage and process the collected information to perform the data warehousing and mining tasks. We will discuss the related issues in the following subsections.

One important thing in the data stream application domain different from the traditional application domains is that due to the continuous, unbounded, high speed characteristics of data streams, there is a huge amount of data in both offline and online data stream applications. Thus, there is not enough space to store and accumulate all the stream data and wait for bulk offline processing as in traditional database applications. One scan processing of data and compact data storage structure are preferable in this environment.

Data Storage

In this stage, data is stored in different data structures associated with the relative data warehousing and mining tasks. Efficient and compact data structure is needed to store, update and retrieve the collected information. This is done to the bounded memory size and huge amounts of data stream characteristics. Furthermore, the data structure needs to be incrementally maintained since it is not possible to rescan the entire input due to the huge amount of stream data and requirement of rapid online querying speed.

Questions need to be considered in this stage include: What information we need to store to process the data warehousing and mining tasks? What data structures we use to store the data and/or Meta data? How and when we update the stored information? Is the data structure efficient to perform the data mining and retrieving tasks? The data storage stage is directly related with the data processing stage, where the data warehousing and mining tasks are performed.

Data Processing

In the data processing stage, different data warehousing and mining tasks are performed based on the user's query. This is the main stage of the data processing system where different data warehousing and data mining methodologies are used to discovery knowledge or potential important information.

The first issue need to be addressed in this stage is the data processing model. According to (Zhu, 2002), there are three stream data processing models, Landmark, Damped and Sliding Windows. The Landmark model mines all collected information over the entire history of stream data from a specific time point called landmark to the present. The Damped model, also called the Time-Fading model mines information in stream data in which each transaction has a weight and this weight decreases with age. Older transactions contribute less weight towards the mining results. The Sliding Windows model finds and maintains most recent information in sliding windows. Only part of the data streams within the sliding window are stored and processed at the time when the stream data flows in. The size of the sliding window may be decided according to the applications and system resources. All these three models can be converted to one another. Choosing which kind of data processing model to use largely depends on the application needs.

The next issue need to be considered in this stage is which data warehousing and data mining methods are suitable in the application environment? A number of questions need to be considered: Should we use an exact or approximate algorithm to perform the mining task? Can the error rate be guaranteed if it is an approximate algorithm? How to reduce and guarantee the error? What is the tradeoff between the accuracy and processing speed? Is the data processed within one pass? Can this method handle a large amount of data? What are the mechanism to maintain and update the data structure and mining results? Is the method resource aware? Can the processing methods handle timeline queries and/or multidimensional stream data?

Data Reporting and Analysis Issues

Data reporting and analysis is the last stage in the sensor stream processing infrastructure, where the mining results are monitored, analyzed and reported. What monitoring and visualization techniques and devices to use is largely application dependent.

Questions need to be considered in this stage include: What is the suitable layout and structure of the reporting interface? How the end users set parameters in the queries? For example, from the pick lists or drop down menus? What is the maximum query response time requested by the end users? What visualization techniques are appropriate to use in a particular application domain?

After discussing the issues needed to be considered in the data warehousing and mining in sensor stream application domain, in the next section we present a framework where the domain-driven data warehousing and mining tasks are used for knowledge discovery in sensor steam applications.

A FRAMEWORK FOR DOMAIN-DRIVEN DATA WAREHOUSING AND MINING FROM SENSOR STREAMS

As shown in Figure 3, raw data is collected from the sensor stream application, which includes the data from wireless sensor networks and data profile from a particular application domain. It is known that raw sensor stream data requires a lot of work to be transformed into data that is ready to be processed (processed data level in Figure 3). The processed data is generated for any application domain, and therefore is application independent. They can be used, for instance, to build applications of transportation management, environmental monitoring, urban planning, etc. On the other hand, the domain information is application dependent, and therefore, will contain only the entities of processed sensor stream data that are relevant to the sensor stream application, as shown in the application level in Figure 3.

Data warehousing and data mining are on the top level, and are also application dependent. In data warehousing and data mining tasks, the user is interested in patterns and/or their relationships about a specific problem or application. For instance, transportation managers are interested in patterns about traffic conditions, accidents, etc., but are not interested in, for instance, patterns of weather conditions.

Figure 3. Proposed framework for sensor stream application domain

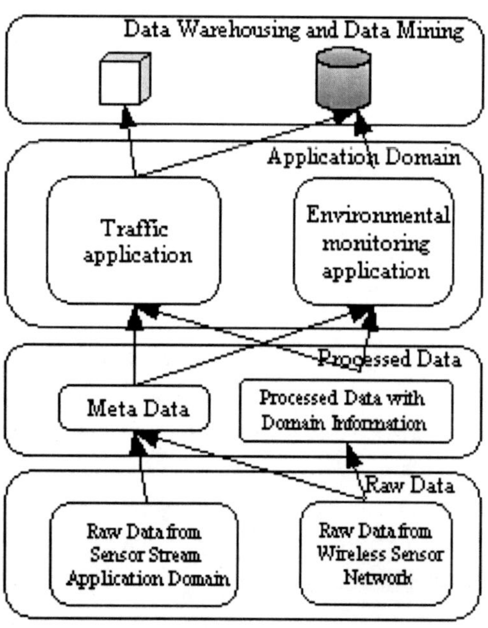

In data stream mining, in general, mining has been directly performed over raw stream data or sensor databases, and the background sensor network information has not been integrated to the mining process. For data warehousing and mining in sensor stream applications, sensor streams need to be integrated with domain information, the discovered pattern information will not be meaningful to the end user or data analyst until it connects with the sensor stream domain information.

In our framework, shown in Figure 3, in the Application Domain level, we proposed to generate domain-enriched sensor data streams, using the connected sensor identifier and sensor value information to find important patterns and relationships among sensors. Let $D = \{d_1, d_2, \ldots, d_n\}$ be a set of n sensor identifiers, and $V = \{v_1, v_2, \ldots, v_m\}$ be a set of m sensor values. A sensor value with domain information is a combination of sensor identifier D and sensor value V, denoted as J = D.V. For example, $d_n.v_m$ means that a sensor with identifier d_n has the value v_m.

In this framework, there are basically four abstraction levels as shown in Figure 4. On the bottom is the data: raw data from wireless sensor network and from sensor stream application domain. Raw data is obtained from the integration of sensor package information from server and sensor stream application domain. Data from the wireless sensor networks contains the information from wireless sensors, such as sensor locations, time stamps, sensor values, etc. Data from sensor stream application domain contains data description and constraints, such as data description, data range, data type, etc.

Figure 4. Architecture of a domain-driven framework for sensor stream application

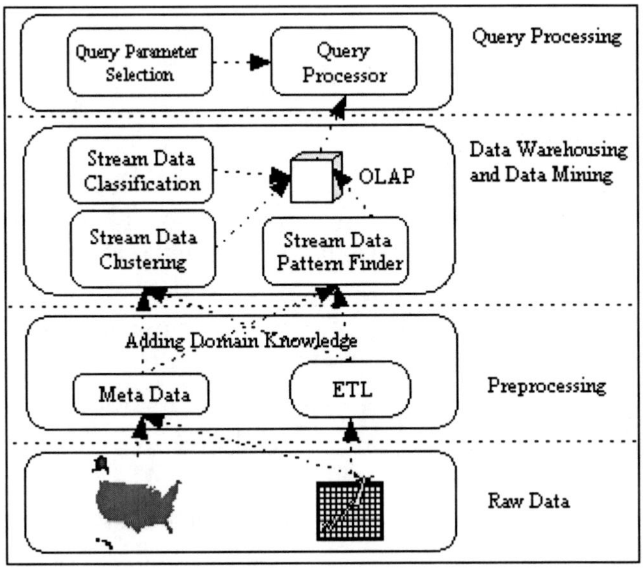

The second level is the preprocessing tasks, where the raw data are preprocessed and integrated with the domain information, and the sensor identifier are connected with sensor values. For data warehousing and mining, another fundamental task is the transformation of the data in different granularity levels. This is essentially important for sensor stream data, where both space and time need to be aggregated in order to allow the discovery of patterns. The user may aggregate both space and time in different granularities.

On the third level of the framework are the data warehousing and data mining tasks. Once the raw data are preprocessed and transformed, different data mining tasks can be applied. For example, data classification, data clustering, pattern and association discovering, and online data analytic processing, etc. These tasks can be combined together to get different mining results requested by the end users.

On the top of the framework is the query processing component, where the end users select query parameters and their queries can be answered based on the mining results. Different end users' requests can be answered at the same time according to their specified query parameters.

CASE STUDY

In this section we describe two different applications and present some experiments to show the usability of the framework for sensor stream application domain. Our experiments were performed with different data warehousing and data mining methods.

The Traffic Management Sensor Stream Application

The simulation data of the traffic management application was collected in year 2000 at various locations throughout the city of Austin, Texas. The data represents the current location, the time interval, and the number of vehicles detected during this interval. All sensor nodes report to a single server. The sensors are deployed on city streets, collect and store the number of the vehicles detected for a given time interval. The vehicle counts taken as sensor readings that are used as input for our simulation experiments are traffic data provided by (Austin, 2003).

Framework Setup and Implementation

As discussed in the framework for domain-driven data warehousing and mining from sensor streams, raw data was collected from the traffic sensor stream application and being transformed into data that was ready to be processed. The data was then enriched with the domain information from the traffic sensor network, and ready to be processed by different data warehousing and mining processes. The mining results were used to answer different user queries at requests. Below, we describe the process in each stage of the framework setup and implementation for the traffic management sensor stream application.

Raw Data Stage

At the raw data stage, which is on the bottom level of our proposed framework for sensor stream application domain, raw data is collected from the traffic management sensor stream application network. This information includes each sensor's current location, sensor identifier, the time interval for the reported value, and the number of vehicles detected during this time interval. The collected data was then transferred to the data preprocessor to be cleaned, integrated and added with domain information for the traffic management sensor stream application.

Data Preprocessing Stage

At the data preprocessing stage, the collected raw data was first integrated into a single sensor stream sequenced by their arrival time stamps. It was then cleaned to map the actual vehicle counts to ten different traffic states which represent the heavy or light levels of current traffic status at each sensor location during the reported time interval. The cleaned data was then enriched with the domain information; in this case, the sensor identifier was integrated with associated sensor values reported during different time intervals. The Meta data is used to generate indicator flags which reflect the data's effectiveness, i.e. missing, corrupted or not. In this case we define the data which no reporting value is missing, and the data reporting value is not within the specified vehicle count range is corrupted. After the preprocessing each sensor reading is connected by a particular sensor identifier, a sequence number represents the time interval the sensor reading was reported, and the Meta data contains the locations of each sensor represented by each sensor identifier.

Data Warehousing and Data Mining Stage

At the data warehousing and data mining stage, which is on the third level of the domain-driven framework, different data warehousing and data mining tasks can be performed on the preprocessed data enriched with domain information. In this case, we perform data clustering and classification tasks based on the sensor locations, and perform data association mining task in each sensor cluster to find out the interesting patterns and associations between the sensor readings. We then use the discovered relationships between these sensor readings to perform missing sensor data estimation based on the sensor readings related with the missing sensor reading. Please refer to (Jiang, 2007) for the detailed algorithm description of this approach.

Query Processing Stage

At the query processing stage, which is on the top level of the domain-driven framework, different users' query can be fulfilled at the users' specified query criteria at the same time. The end user will specify their query parameters on their online queries, in this case, the support, confidence threshold, and the sliding window size of their online association queries, or specify they would use the system recommend parameter values. The query processor then use these specified parameters as query criteria to pass the request to the data warehousing and mining components and retrieve the query results for different end users' requests.

Performance Study

The performance of our proposed framework is studied by means of simulation. Several different data warehousing and data mining experiments are conducted in order to evaluate the proposed framework while using the Average Window Size (AWS) approach, the linear interpolation approach, the linear trend approach, the Window based Association Rule Mining (WARM) approach (Halatchev, 2005), and the Closed itemsets based Association Rule Mining (CARM) approach (Jiang, 2007).

All these methods are applied to our proposed framework to answer the user's request for missing sensor value estimation. The AWS approach uses the average value of the missing sensor readings in the current sliding window as the estimated value. The linear interpolation approach uses the linear interpolation of the missing sensor's neighbor readings as the estimated value. The linear trend uses the linear regression trend for the missing sensor readings as the estimated value. The WARM approach (Halatchev, 2005) is a revised data warehousing technique for sensor network database based on the relationship between two sensors. And the CARM approach is a pattern and association mining approach for stream data application based on multiple sensor relationships. We compared the estimation accuracy, running time and memory space usage when applying each method to our proposed framework.

Performance Study of Estimation Accuracy

The evaluation of the estimation accuracy of the missing values is done by using the average Root Mean Square Error (RMSE):

$$RMSE = \frac{1}{numStates}\sqrt{\frac{\sum_{i=1}^{\#\,estimations}(Xa_i - Xe_i)^2}{\#\,estimations}}$$

where X_{ai} and X_{ei} are the actual value and the estimated value, respectively; #estimations is the number of estimations performed in a simulation run; and numStates is the number of states in which the actual readings are distributed.

The expression $\sqrt{\frac{\sum_{i=1}^{\#\,estimations}(Xa_i - Xe_i)^2}{\#\,estimations}}$ represents the standard error and is an estimate of the standard deviation under the assumption that the errors in the estimated values (i.e. X_{ai} - X_{ei}) are normally distributed. Thus, the RMSE allows the construction of confidence intervals describing the performance of different candidate missing value estimators. The smaller the RMSE (the standard deviation), the better the estimated results.

From Figure 5, we can see that CARM gives the best result of the above approaches in the proposed framework regarding the estimation accuracy, followed by the WARM approach. The linear interpolation, AWS, and linear trend approaches perform no better than WARM and CARM approaches. The main reason might be that it only considers the relationship between the neighbor nodes, while CARM and WARM find out more of the relationships between the existing sensors. Also from this figure, we can see that the CARM approach provides better estimation accuracy than the WARM approach does. This is because CARM performs the estimation based on the associations from multiple sensors derived by a compact and complete set of information, while WARM performs the estimation based on the associations derived by two sensors as frequent patterns in the data structure.

Performance Study of Running Time

Figure 6 illustrates the running time in seconds of AWS, linear interpolation, linear trend, WARM

Figure 5. Performance study of average and maximum estimation accuracy

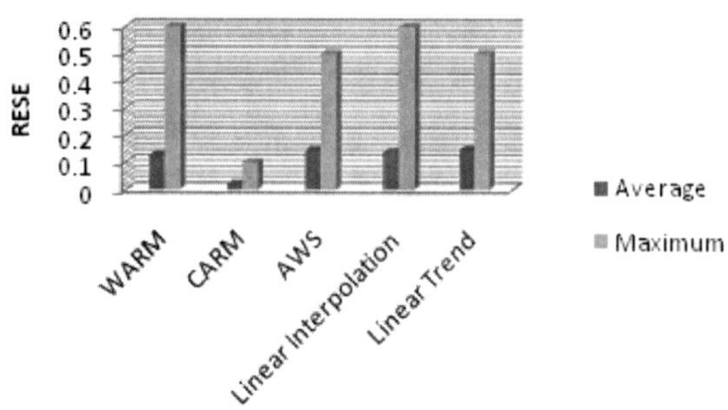

Figure 6. Performance study of running time

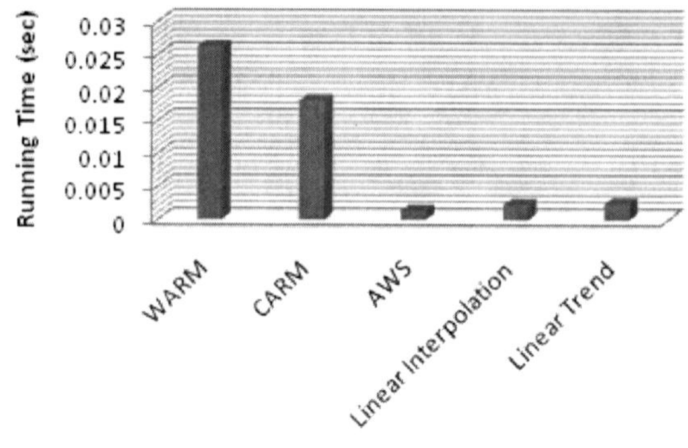

and CARM approaches in the proposed framework. The experimental results show that in terms of running time, the WARM and CARM approach are outperformed by AWS, linear interpolation and linear trend approaches. The CARM approach is faster than the WARM technique. This is because in the AWS, linear interpolation, linear trend approaches, the calculation is based on relationships in a certain sensors reading trend, while for WARM and CARM the calculation is based on relationships between not only the missing sensors previous readings but also the other sensors' readings that related with the missing sensor reading.

Performance Study of Memory Usage

Figure 7 illustrates the memory usage of AWS, linear interpolation, linear trend, WARM and CARM approaches in the proposed framework. The experimental results show that in terms of memory space, the WARM approach is outperformed by all other four approaches. The results of the simulation experiments show that for all processed sensors the needed memory space using WARM, is higher than that using CARM. This is because the lattice data structure in CARM uses less memory space than the cube data structures in WARM, and it stores only the condensed closed patterns. And for the AWS, linear interpolation, and linear trend approach, they only need to store and process information regarding the missing sensor's readings.

Analysis Summary

In this section, we discussed how to setup and implement a traffic management sensor stream application in the domain-driven framework for data warehousing and mining from sensor streams. We described the process of the framework setup in each implementing stage: the raw data stage, data preprocessing stage, data warehousing and data mining stage, and the query processing stage.

The performance of our proposed framework is studied by means of simulation. Several different data warehousing and data mining experiments are conducted in order to evaluate the proposed framework while using the Average Window

Figure 7. Performance study of memory usage

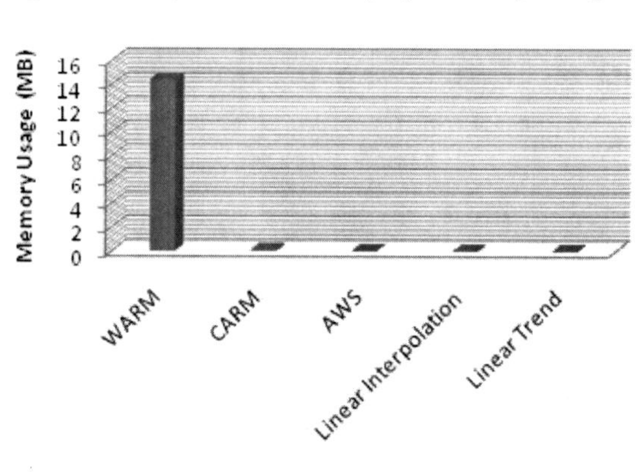

Size (AWS) approach, the linear interpolation approach, the linear trend approach, the WARM approach (Halatchev, 2005), and the CARM approach (Jiang, 2007). All these methods are applied to our proposed framework to answer the user's request for missing sensor value estimation. We compared the estimation accuracy, running time and memory space usage when applying each method to our proposed framework.

Our performance study shows that the proposed domain-driven framework can be applied to different data warehousing and mining tasks. For the traffic management sensor stream application, the closed pattern based association mining to estimate missing sensor data online is an area worth to explore.

The Environmental Monitoring Sensor Stream Application

A second experiment was performed over sensor data collected in the Huntington Botanical Garden in Sam Marino, California. The sensor reports the air temperature of several places in the gardens for different time intervals.

The simulation data of the environmental monitoring application was collected in year 2008 at various locations throughout the sensor network in Huntington Botanical Garden. The data represents the current location, the time interval, and the air temperature of detected environment during this interval. All sensor nodes report to a single server. The sensors are deployed on different places of the botanical garden, collect and store the air temperature detected for a given time interval. The actual air temperatures are taken as sensor readings that are used as input for our simulation experiment.

Framework Setup and Implementation

As discussed in the domain-driven framework for data warehousing and mining from sensor streams, raw data was collected from the environmental monitoring sensor stream application and being transformed into data that was ready to be processed. The data was then enriched with the domain information from the environmental monitoring sensor network, and ready to be processed by different data warehousing and mining processes. The mining results were used to answer different user queries at requests. Below, we describe the process in each stage of the framework setup and implementation for the environmental monitoring sensor stream application.

Raw Data Stage

At the raw data stage, which is on the bottom level of our proposed framework for sensor stream application domain, raw data is collected from the environmental monitoring sensor stream application network. This information includes each sensor's current location, sensor identifier, the time interval for the reported value, and the air temperature detected during this time interval. The collected data was then transferred to the data preprocessor to be cleaned, integrated and added with domain information for the environmental monitoring sensor stream application.

Data Preprocessing Stage

At the data preprocessing stage, the collected raw data was first integrated into a single sensor stream sequenced by their arrival time stamps. It was then cleaned to round up the detected air temperature to its nearest integer value which represents the temperature at each sensor location during the reported time interval. The cleaned data was then enriched with the domain information; in this case, the sensor identifier was integrated with associated sensor values reported during different time intervals. The Meta data is used to generate indicator flags which reflect the data's effectiveness, i.e. missing, corrupted or not. In this case we define the data which no reporting value is missing, and the data reporting value is beyond the specified temperature range is corrupted. After the preprocessing each sensor reading is connected by a particular sensor identifier, a sequence number represents the time interval the sensor reading was reported, and the Meta data contains the locations of each sensor represented by each sensor identifier.

Data Warehousing and Data Mining Stage

At the data warehousing and data mining stage, which is on the third level of the domain-driven framework, different data warehousing and data mining tasks can be performed on the preprocessed data enriched with domain information. In this case, we perform data association mining task in each sensor cluster from the Huntington Botanical Garden sensor network application to find out the interesting patterns and associations between the sensor readings. We then use the discovered relationships between these sensor readings to perform missing sensor data estimation based on the sensor readings related with the missing sensor reading.

Query Processing Stage

At the query processing stage, which is on the top level of the domain-driven framework, different users' query can be fulfilled at the users' specified query criteria at the same time. In this case, the temperatures at different locations during the specified time interval. If the query information is not missing, it can be directly retrieved from the sensor network database. Otherwise, the request was send through the data estimation component in the data warehousing and mining level and the estimated results are retrieved for different end users' requests.

Performance Study

Several different data mining techniques are conducted in order to evaluate the proposed framework using the Average Window Size (AWS) approach, the linear interpolation approach, the linear trend approach, and the CARM approach (Jiang, 2007). All these methods are applied to our proposed framework to answer the user's request for missing sensor air temperature value. We compared the estimation accuracy, running time and memory space usage when applying each method to our proposed framework.

Performance Study of Estimation Accuracy

The evaluation of the estimation accuracy of the missing values is done by using the average Root Mean Square Error (RMSE).

From Figure 8, we can see that CARM gives the best result of the above approaches regarding the estimation accuracy. The AWS, and linear series approaches perform no better than CARM approaches. The main reason might be that it only considers the relationship between the neighbor nodes, while CARM find out all of the relationships between the existing sensors. The linear interpolation method fails to report the missing temperature value, because it predicates the missing value based on the missing readings neighbor reported values. Therefore, when there are two consecutive missing values on the same sensor, this method couldn't calculate the estimated result.

Figure 8. Performance study of average estimation accuracy

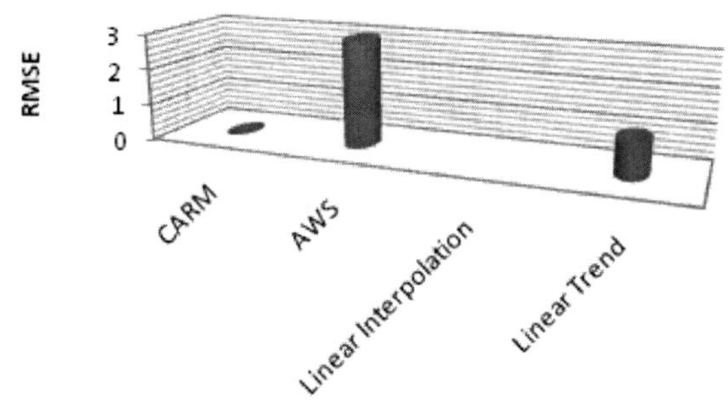

Performance Study of Running Time

Figure 9 illustrates the running time in seconds of AWS, linear interpolation, linear trend, and CARM approaches. The experimental results show that in terms of running time, the CARM approach are outperformed by AWS, linear interpolation and linear trend approaches. This is because in the AWS, linear interpolation, linear trend approaches, the calculation is based on relationships in a certain sensors reading trend, while for CARM the calculation is based on relationships between multiple sensor readings of the missing senor and the other related sensors.

Performance Study of Memory Usage

Figure 10 illustrates the memory usage of AWS, linear interpolation, linear trend, and CARM approaches in MB. The experimental results show that in terms of memory space, the CARM approach is outperformed by all other three approaches. This is because CARM stores the relationships of multiple sensor readings, and for the AWS, linear interpolation, and linear trend approach, they only need to store and process information regarding the missing sensor's readings.

Analysis Summary

In this section, we discussed how to setup and implement an environmental monitoring sensor stream application in the domain-driven framework for data warehousing and mining from sensor streams. We described the process of the framework setup in each implementing stage: the raw data stage, data preprocessing stage, data warehousing and data mining stage, and the query processing stage.

The performance of our proposed framework is studied by means of simulation. Several different data mining techniques are conducted in order to evaluate the proposed framework using the Average Window Size (AWS) approach, the linear interpolation approach, the linear trend approach, and the CARM approach (Jiang, 2007). All these methods are applied to our proposed framework to answer the user's request for missing sensor air temperature value. We compared the estimation accuracy, running time and memory space usage when applying each method to our proposed framework.

Our performance study shows that the proposed domain-driven framework can be applied to the environmental monitoring sensor stream application for different data warehousing and mining tasks. For the missing sensor value application component in the sensor network applica-

Figure 9. Performance study of running time

Figure 10. Performance study of memory usage

tion, the closed pattern based association mining approach is an area worth to explore.

CONCLUSION AND FUTURE TRENDS

Sensor stream applications are becoming very common with the advances in technologies for sensor devices. There data are normally available as stream data with very little or no semantics. This makes their analysis and knowledge extraction very complex from an application point of view. Most sensor stream mining works have focused on the streaming properties without considering the background of sensor domain information.

In this chapter we have addressed the problem of mining sensor stream system from an application point of view. We presented a framework to preprocess sensor streams for domain-driven data mining. The objective is to integrate domain information that is relevant for stream data warehousing and data mining.

We have evaluated the framework with real data from two different stream data application domains, which shows that the framework is general enough to be used in different application scenarios. This is possible because the user can choose the domain information that is important for data warehousing and mining tasks.

Domain-driven data mining is an open research field, especially for spatial, temporal sensor stream data. We believe that in the future new data warehousing and mining algorithms that consider data semantics and domain information have to be developed in order to extract more interesting patterns and associations in different application domains.

REFERENCES

Austin, F. I. (2000). *Austin Freeway ITS Data Archive*. Retrieved January, 2003 from http://austindata.tamu.edu/default.asp.

Cai, Y. D., Pape, G., Han, J., Welge, M., & Auvil, L. (2004). MAIDS: Mining alarming incidents from data streams. *International Conference on Management of Data.*

Chang, J. H., Lee, & W. S. (2004). A sliding window method for finding recently frequent itemsets over online data streams. *Journal of Information Science and Engineering.*

Chang, J. H., Lee, W. S., & Zhou, A. (2003). Finding recent frequent itemsets adaptively over online data streams. *ACM SIGKDD International Conference on Knowledge Discovery and Data Mining.*

Chi, Y., Wang, H. X., Yu, P. S., & Muntz, R. R. (2004). Moment: Maintaining closed frequent itemsets over a stream sliding window. *IEEE International Conference on Data Mining.*

Dang, X. H., Ng, W. K., & Ong, K. L. (2007). Online mining of frequent sets in data streams with error guarantee. *Knowledge and Information Systems.*

Demaine, E. D., Ortiz, A. L., & Munro, J. I. (2002). Frequency estimation of internet packet streams with limited space. *European Symposium on Algorithms.*

Giannella, C., Han, J. W., Pei, J., Yan, X. F., & Yu, P. S. (2003). Mining frequent patterns in data streams at multiple time granularities. *Data Mining: Next Generation Challenges and Future Directions*, AAAI/MIT.

Guha, S., & Koudas, N. (2002). Approximating a data stream for querying and estimation: Algorithms and performance evaluation. *International Conference on Data Engineering.*

Halatchev, M., & Gruenwald, L. (2005). Estimating missing values in related sensor data streams. *International Conference on Management of Data.*

Jiang, N., & Gruenwald, L. (2006). CFI-Stream: Mining Closed Frequent Itemsets in Data Streams.

ACM SIGKDD international conference on knowledge discovery and data mining.

Jiang, N., & Gruenwald, L. (2007). Estimating missing data in data streams, *the International Conference on Database Systems for Advanced Applications.*

Kargupta, H., Bhargava, R., Liu, K., Powers, M., Blair, P., Bushra, S., et al. (2004). VEDAS: A mobile and distributed data stream mining system for real-time vehicle monitoring. *SIAM International Conference on Data Mining.*

Koh, J. L., & Shin, S. N. (2006). An approximate approach for mining recently frequent itemsets from data streams. *The 8th International Conference on Data Warehousing and Knowledge Discovery.*

Li, H. F., & Cheng, H. (2008). Improve frequent closed itemsets mining over data stream with bitmap. *Ninth ACIS International conference on software engineering, artificial intelligence, networking, and parallel/distributed computing.*

Li, H. F., Ho, C. C., Kuo, F. F., & Lee, S. Y. (2006) A new algorithm for maintaining closed frequent itemsets in data streams by incremental updates. *Six IEEE International Conference on Data Mining Workshop.*

Li, H. F., Lee, S. Y., & Shan, M. K. (2004). An efficient algorithm for mining frequent itemsets over the entire history of data streams. *The International Workshop on Knowledge Discovery in Data Streams.*

Lin, C. H., Chiu, D. Y., Wu, Y. H., & Chen, A. L. P. (2005). Mining frequent itemsets from data streams with a time-sensitive sliding window. *SIAM International Conference on Data Mining.*

Manku, G. S., & Motwani, R. (2002). Approximate frequency counts over data streams. *International Conference on Very Large Databases.*

Mozafari, B., Thakkar, H., & Zaniolo, C. (2008). Verifying and mining frequent patterns from large windows over data streams. *IEEE International Conference on Data Engineering.*

Shin, S. J., & Lee, W. S. (2007). An online interactive method for finding assoication rules data streams. *ACM 16th Conference on Information and Knowledge Management.*

Yang, L., & Sanver, M. (2004). Mining short association rules with one database scan; *International Conference on Information and Knowledge Engineering.*

Yu, J. X., Chong, Z. H., Lu, H. J., & Zhou, A. Y. (2004). False positive or false negative: Mining frequent itemsets from high speed transactional data streams. *International Conference on Very Large Databases.*

Zhu, Y. Y., & Shasha, D. (2002). StatStream: Statistical monitoring of thousands of data streams in real time. *International Conference on Very Large Databases.*

This work was previously published in Evolving Application Domains of Data Warehousing and Mining: Trends and Solutions, edited by P. N. San-Bento Furtado, pp. 113-128, copyright 2010 by Information Science Reference (an imprint of IGI Global).

Compilation of References

Aadland, D. M., Caplan, A. J., & Phillips, O. R. (2007). A Bayesian examination of information and uncertainty in contingent valuation. *Journal of Risk and Uncertainty, 35*, 149–178.

Aarts, H., & Dijksterhuis, A. (2000). The automatic activation of goal-directed behavior: the case of travel habit. *Journal of Environmental Psychology, 20*, 75–82.

Aarts, H., Verplanken, B., & van Knippenberg, A. (1997). Habit and information use in travel mode choices. *Acta Psychologica, 96*, 1–14.

Aarts, H., Verplanken, B., & van Knippenberg, A. (1998). Predicting behavior from actions in the past: repeated decision making or a matter of habit? *Journal of Applied Social Psychology, 28*, 1355–1374.

Abowd, G., Atkeson, C.G., Hong, J., Long, A., Kooper, R. & Pinkertin, M. (1997). Cyberguide: A mobile context-aware tour guide. *Wireless Networks*, 421-433.

Adams, P., Ashwell, G., & Baxter, R. (2003). Location Based Services – an overview of the standards. *BT Technology Journal, 21*(1), 31–42. doi:10.1023/A:1022572210026

Aggarwal, J. K., & Cai, Q. (1994). Human Motion Analysis: A Review. *Proceedings of the IEEE Workshop on Motion of Non-Rigid and Articulated Objects,* (pp. 2-14), Austin, Texas: IEEE.

Ahas, R., Aasa, A., Mark, Ü., Pae, T., & Kull, T. (2007). Seasonal tourism spaces in Estonia: case study with mobile positioning data. *Tourism Management, 28*, 898–910.

Ahas, R., Aasa, A., Silm, S., & Tiru, M. (2007). Mobile positioning data in tourism studies and monitoring: case study in Tartu, Estonia . In Sigala, M., Mich, L., & Murphy, J. (Eds.), *Information and Communication Technologies in Tourism 2007* (pp. 119–128). Vienna: Springer. doi:10.1007/978-3-211-69566-1_12

Ahas, R., Aasa, A., Silm, S., Aunap, R., Kalle, H., & Mark, Ü. (2007). Mobile positioning in space-time behavior studies: Social Positioning Method experiments in Estonia. *CaGIS, 34*, 259–273.

Ahrend, C. (2002). *Mobilitätsstrategien zehnjähriger Jungen und Mädchen als Grundlage städtischer Verkehrsplanung. Internationale Hochschulschriften (Vol. 381)*. Berlin: Waxmann. (in German)

Alvares, L. O., Bogorny, V., Kuijpers, B., Macedo, J., Moelans, B., & Vaisman, A. (2007). A model for enriching trajectories with semantic geographical information. In *Proceedings of the 15th Annual ACM International Symposium on Advances in Geographic Information Systems,* (pp. 162-169).

American Bar Association. (1996). *Digital Signature Guidelines*. Chicago: Author.

American Bar Association. (2001). *PKI Assessment Guidelines*. Chicago: Author.

Anderson, N. H. (1981). Foundations of information integration theory. New York: Academic Press.

Andersson, M., Gudmundsson, J., Laube, P., & Wolle, T. (2008). Reporting leaders and followers among trajectories of moving point objects. *GeoInformatica, 12*(4), 497–528. doi:10.1007/s10707-007-0037-9

Andirenko, N., Andirenko, G., Wachowicz, M., & Orellana, D. 2008. Uncovering Interactions between Moving Objects. In T.J. Cova, H.J. Miller, K. Beard, A. Frank & M.F. Goodchild, (Eds.), *Geographic Information Science, Extended Abstracts of 5th International Conference, GIScience 2008,* Park City, UT, September 23-26, 2008 (pp. 16-27).

Andrienko, G., & Andrienko, N. (2008). A visual analytics approach to exploration of large amounts of movement data. In M. Sebillo, G. Vitiello & G. Schaefer (Eds.), Visual information systems, 10th international conference, VISUAL 2008 (pp. 1-4). Berlin: Springer-Berlin. doi:10.1007/978-3-540-85891-1_1

Andrienko, G., Andrienko, N., Kopanakis, I., Ligtenberg, A., & Wrobel, S. (2008). Visual analytics methods for movement data. In F. Giannotti, & D. Pedreschi (Eds.), Mobility, data mining and privacy (pp. 375-410). Berlin: Spinger-Verlag.

Andrienko, N., Andrienko, G., Pelekis, N., & Spaccapietra, S. (2008). Basic concepts of movement data. In F. Giannotti, & D. Pedreschi (Eds.), Mobility, data mining and privacy (pp. 15-38). Berlin: Spinger-Verlag.

Antonini, G., Bierlaire, M., & Weber, M. (2006). Discrete choice models of pedestrian walking behavior. *Transportation Research Part B: Methodological, 40*(8), 667–687. doi:. doi:10.1016/j.trb.2005.09.006

Arentze, T. A., & Timmermans, H. J. P. (2005). Representing mental maps and cognitive learning in micro-simulation models of activity-travel choice dynamics. *Transportation, 32,* 321–341.

Arentze, T. A., & Timmermans, H. J. P. (2005). Information gain, novelty seeking and travel: a model of dynamic activity-travel behavior under conditions of uncertainty. *Transportation Research, 39A,* 125–145.

Argo. (2006). *Five years of progress, five decades of potential.* Retrieved September 26, 2006, from http://www.argo.ucsd.edu/FrAbout_Articles.html

Arikawa, M., Konomi, S., Ohnishi, K., & Navitime, S. (2007). Supporting pedestrian navigation in the real world. *IEEE Pervasive Computing / IEEE Computer Society [and] IEEE Communications Society, 6,* 21–29.

Arikawa, M., Tsuruoka, K. I., Hideyuki, F., & Ome, A. (2007). Place-tagged podcasts with synchronised maps on mobile media players. *Cartography and Geographic Information Science, 34,* 293–304.

Arvind, D. K., (2005). Speckled computing. In *Proc. Nanotech 2005, 3,* 351-354.

Asakura, Y., & Iryo, T. (2007). Analysis of tourist behavior based on the tracking data collected using a mobile communication. *Transportation Research Part A, Policy and Practice, 41,* 684–690.

Ashbrook, D., & Starner, T. (2003). Using GPS to Learn Significant Locations and Predict Movement Across Multiple Users. *Personal and Ubiquitous Computing,* 7(5), 275–286. doi:10.1007/s00779-003-0240-0

Auld, J., Williams, C., Mohammadian, A. & Nelson, P. (2009). An automated GPS-based prompted recall survey with learning algorithms. *Transportation letters, 1,* 58-79.

Austin, F. I. (2000). *Austin Freeway ITS Data Archive.* Retrieved January, 2003 from http://austindata.tamu.edu/default.asp.

Avineri, E., & Prashker, J. N. (2003). *Sensitivity to uncertainty: the need for a paradigm shift.* Paper presented at the 82nd meeting of the Transportation Research Board, Washington, D.C.

Avineri, E., & Prashker, J. N. (2003). *The impact of travel time information on travelers' learning under uncertainty.* Paper presented at the 10th International Conference of Travel Behavior Research, Lucerne.

Axhausen, K. W., Schönfelder, S., Wolf, J., Oliveira, M., & Samaga, U. (2004). Eighty weeks of GPS traces: approaches to enriching trip information. In *Proceedings of the 83rd Annual Meeting of the Transportation Research Board,* Washington, D.C., USA.

Axhausen, K. W., Zimmermann, A., Schonfelder, S., Rindsfuser, G., & Haupt, Th. (2002). Observing the rhythms of daily life: A six-week travel diary. *Transportation, 29*, 95–124.

Azuma, R., Baillot, Y., Behringer, R., Feiner, S., Julier, S., & MacIntyre, B. (2001). Recent advances in augmented reality. *IEEE Computer Graphics & Applications, 21*(6), 34-47.

Baader, F., Calvanese, D., McGuinness, D., Nardi, D., & Patel-Schneider, P. (2003). The description logic handbook: Theory, implementation and applications. Cambridge, UK: Cambridge University Press.

Bachu, P. K., Dudala, T., & Kothuri, S. M. (2001). Prompted recall in Global Positioning System survey – proof-of-concept study. *Transportation Research Record, 1768*, 106–113.

Backhouse, J., Hsu, C., Baptista, J., & Tseng, J. (2003). The key to trust? Signalling quality in the PKI market. In *Proceedings of the 11 European Conference on Information Systems* (ECIS 2003), Naples.

Badenoch, D. (2005). *Interdisciplinary Curriculum Design and Teaching for Integrative Learning and Development of Graduate Qualities: A case study of reconceptualising Physical Education: University of South Australia.* Paper presented to the Division of Education, Arts & Social Sciences' Teaching & Learning Colloquium, University of South Australia.

Baglioni, M., Macedo, J., Renso, C., & Wachowicz, M. (2008). *An ontology-based approach for the semantic modelling and reasoning on trajectories.* Paper presented at the ER '08: Proceedings of the ER 2008 Workshops (CMLSA, ECDM, FP-UML, M2AS, RIGiM, SeCoGIS, WISM) on Advances in Conceptual Modeling, Barcelona, Spain (pp. 344-353). doi:0.1007/978-3-540-87991-6_41

Baglioni, M., Macedo, J., Renso, C., Trasarti, R., & Wachowicz, M. (2009). Towards a semantic interpretation of movement behaviour. In *Proceedings of 12th AGILE Conference, Lecture Notes in Geoinformation and Cartography.*

Batarfi, O., & Marshall, L. (2007). Conformance testing a set of criteria for assessing trust. *Journal of Computers, 2*(1). doi:10.4304/jcp.2.1.44-49

Battelle (1997). *Lexington area travel data collection test.* Final Report. Prepared for Federal Highway Administration, September, 1997.

Batty, M. (2003). Agent-based pedestrian modelling. In M. Batty, & P. A. Longley (Eds.), *Advanced spatial analysis: The CASA book of GIS* (pp. 81-106). Redlands, CA: Esri Press. Retrieved from http://www.casa.ucl.ac.uk/working_papers/paper61.pdf

Batty, M., DeSyllas, J., & Duxbury, E. (2003). The discrete dynamics of small-scale spatial events: Agent-based models of mobility in carnivals and street parades. *International Journal of Geographical Information Science, 17*(7), 673-697. Retrieved from http://www.casa.ucl.ac.uk/working_papers/paper56.pdf

Baus, J. R., Chevherst, K., & Kray, C. (2005). A survey of map-based mobile guides. In *Map-Based Mobile Services*, (pp. 197-216).

Beardsley, P., Van Baar, J., Raskar, R., & Forlines, C. (2005). Interaction using a handheld projector. *IEEE Computer Graphics and Applications, 25*(1), 39-43.

Belcher, D., Bilinghurst, M., Hayes, S. E., & Stiles, R. (2003). Using augmented reality for visualising complex graphs in three dimensions. In *Proceedings of the International Symposium on Mixed and Augmented Reality* (ISMAR 2003) (pp. 84-92).

Bellemans, T., Kochan, B., Janssens, D., Wets, G., & Timmermans, H. J. P. (2008). *In the field evaluation of the impact of a GPS-enabled personal digital assistant on activity-travel diary data quality.* Presented at 87th Annual Meeting of the Transportation Research Board, Washington, DC.

Belzer, D., & Autler, G. (2002). *Transit-Oriented Development, Moving From Rhetoric to Reality.* Berkeley, CA: Strategic Economics. Retrieved July 1, 2009, from http://www.reconnectingamerica.org/public/reports/90

Ben-Akiva, M., & Boccara, B. (1995). Discrete choice models with latent choice-sets. *International Journal of Marketing, 12*, 9–24.

Benham, J., Patel, B. (1976). A method for estimating pedestrian volume in a central business district. *Transportation Research Record, 629*, 22-26.

Benkert, M., Gudmundsson, J., Hübner, F., & Wolle, T. (2008). Reporting flock patterns. *Computational Geometry, 41*(3), 111–125. doi:. doi:10.1016/j.comgeo.2007.10.003

Benyon, D. R. (2001). The new HCI? Navigation of information space. *Knowledge-based Systems, 14*(8) 425-430.

Benyon, D. R. (2005). Information space. In C. Ghaoul (Ed.), *The encyclopedia of human-computer interaction*. Hershey, PA: Idea Group Reference.

Benyon, D. R. (2006). Navigating information space: Web site design and lessons from the built environment. *Psychnology, 4*(1), 7-24.

Benyon, D., & Höök, K. (1997). Navigation in information spaces: Supporting the individual. In S. Howard, J. Hammond, & G. Lindgaard (Eds.), *Human-computer interaction: INTERACT'97*, (pp. 39-46). UK: Chapman & Hall.

Beresford, A., & Stajano, F. (2003). Location Privacy in Pervasive Computing. *Pervasive Computing, IEEE, 2*(1), 46–55. doi:10.1109/MPRV.2003.1186725

Bierlaire, M., Antonini, G., & Weber, M. (2003). *Behavioral dynamics for pedestrians.* Paper presented at the Moving through Nets: The Physical and Social Dimensions of Travel, Lucerne.

Birnbaum, M. H. (1983). Base rates in Bayesian inference: signal detection analysis for the cab problem. *The American Journal of Psychology, 96*, 85–94.

Blivice, S. (1974). *Pedestrian Route Choice: A Study of Walking to Work in Munich.* Unpublished Ph.D. Dissertation, University of Michigan.

Blue, V. J., & Adler, J. L. (2001). Cellular automata microsimulation for modeling bi-directional pedestrian walkways. *Transportation Research Part B: Methodological, 35*(3), 293–312. doi:. doi:10.1016/S0191-2615(99)00052-1

Blythe, P. W., Miller, G. F., & Todd, P. M. (1996). Human simulation of adaptive behavior: Interactive studies of pursuit, evasion, courtship, fighting, and play. Paper presented at the *From Animals to Animats 4: Proceedings of the Fourth International Conference on Simulation of Adaptive Behavior.*

Bohte, W., & Maat, K. (2008)., Deriving and validating trip destinations and modes for multi-day GPS-based travel surveys: An application in the Netherlands. In *Proceedings 87th Annual Meeting of the Transportation Research Board*, Washington DC.

Boivin, S., & Gagalowicz, A. (2001). Image-based Rendering of Diffuse, Specular and Glossy Surface from a Single Image. *SIGGRAPH*, (pp. 107-116).

Bonsall, P., Firmin, P., & Beale, J. (2004). *Perception of modal attributes: how accurate and how susceptible to change?* Paper presented at the 83rd meeting of the Transportation Research Board, Washington, DC.

Boone, G. (2004). Reality mining: Browsing reality with sensor networks. *Sensors Mag, 21*(9). Retrieved September 25, 2006, from http://www.sensorsmag.com/articles/0904/14/main.shtml

Bottino, A., & Laurentini, A. (2001). A Silhouette Based Technique for the Reconstruction of Human Movement. *Computer Vision and Image Understanding, 83*, 75-95.

Botts, M., & Robin, A. (Eds.). (2007). OpenGIS Sensor Model Language (SensorML) Implementation Specification 1.0.0 (OGC Document Number: 07-000). Wayland, MA: OGC.

Botts, M., Percivall, G., Reed, C., & Davidson, J. (Eds.). (2007). *OGC White Paper - OGC Sensor Web Enablement: Overview and High Level Architecture (OGC Document Number: 07-165).* Wayland, MA: OGC.

Bouguet, J-Y. (nd). *Camera calibration toolkit for Matlab and in OpenCV.* Retrieved May 12, 2008, from http://www.vision.caltech.edu/bouguetj/calib_doc/

Bouquet, P., Ghidini, C., Giunchiglia, F., & Blanzieri, E. (2003). Theories and uses of context in knowledge representation and reasoning. *Journal of Pragmatics, 35*(3), 455–484. doi:10.1016/S0378-2166(02)00145-5

Bourdieu, P. (1976). *Entwurf einer Theorie der Praxis auf der ethnologischen Grundlage der kabylischen Gesellschaft*. Frankfurt am Main, Germany: Suhrkamp. (in German)

Brantingham, & Brantingham (1995). Criminality and place: Crime generators and crime attractors. *European Journal on Criminal Policy and Research, 3*, 3.

Bregler, C., Malik, J., & Pullen, K. (2004). Twist Based Acquisition and Tracking of Animal and Human kinematics. *International Journal of Computer Vision, 56*, 179-194.

Briggs, R. (1973). *Urban Cognitive Distance. Image and Environment* (pp. 361–388). Chicago: Aldine Publishing Co.

Brin, D. (1998). *The Transparent Society*. Reading, MA: Addison-Wesley.

Brown, D.L., & Wheatley, G.H. (1989).Relationship between spatial Knowledge. In C. Maher, G. Goldin, & R. Davis (Eds.), *Proceedings of the 11th Annual Meeting, North America Chapter of the International Group for the Psychology of Mathematic Education* (pp. 143-148). New Brunswick, NJ: International Group for the Psychology of Mathematics Education.

Burrell, J., Broke, T., & Beckwith, R. (2004). Vineyard computing: Sensor networks in agricultural production. *IEE Pervasive Computing, 3*(1), 38-45.

Burstedde, C., Klauck, K., Schadschneider, A., & Zittartz, J. (2001). Simulation of pedestrian dynamics using a two-dimensional cellular automaton. *Physica A: Statistical Mechanics and its Applications, 295*(3-4), 507-525.

Byon, Y., Abdulhai, B., & Shalaby, A. (2007). Impact of Sampling Rate of GPS-enabled Cell Phones on Mode Detection and GIS Map Matching Performance. In *Proceedings of the 86th Annual Transportation Research Board Meeting*, Washington D.C.

Cai, Y. D., Pape, G., Han, J., Welge, M., & Auvil, L. (2004). MAIDS: Mining alarming incidents from data streams. *International Conference on Management of Data*.

Calthorpe, P. (1993). The Next American Metropolis, Ecology, Community and the American Dream. New York: Princeton Architectural Press.

Calvanese, D., Lembo, D., Lenzerini, M., Poggi, A., & Rosati, R. (2007). *Ontology-based database access*. Paper presented at the Proc. of the 15th Italian Conf. on Database Systems SEBD, 324-331.

Camenisch, J., & Groß, T. (2008). Efficient attributes for anonymous credentials. In *Proceedings of the 15th ACM Conference on Computer and Communications Security, Alexandria, VA, October 27 - 31, CCS '08*, 345-356. New York: ACM.

Camenisch, J., Groß, T., & Heydt-Benjamin, T. S. (2008). Rethinking accountable privacy supporting services: extended abstract. In *Proceedings of the 4th ACM Workshop on Digital Identity Management*, Alexandria, Virginia, USA, October 31 - 31.

Cameron, K. (2005). *The Laws of Identity*. Redmond, WA: Microsoft Corporation.

Card, S. K., MacKinlay, J. D., & Shneiderman, B. (1999). *Readings in information visualisation: Using vision to think*. San Diego, CA: Academic Press.

Carranza, J., Theobalt, C., Magnor, M., & Seidel, H-P. (2003). Free-Viewpoint Video of Human Actors. *SIGGRAPH*, (pp. 569-577).

Cervero, R. (1997). Paratransit in America: Redefining Mass Transportation. Santa Barbara, CA: Praeger Publishers.

Chainey, S. P., & Desyllas, J. (2004). Measuring, identifying and analysing . In *Proceedings at the 2004 UK National Crime Mapping Conference, University of London, London*. Retrieved August 28, 2007, from www.jdi.ucl.ac.uk/news_events/conferences/index.php

Chainey, S. P., & Ratcliffe, J. H (2005). *GIS and crime mapping*. London: Wiley.

Chang, J. H., Lee, & W. S. (2004). A sliding window method for finding recently frequent itemsets over online data streams. *Journal of Information Science and Engineering*.

Chang, J. H., Lee, W. S., & Zhou, A. (2003). Finding recent frequent itemsets adaptively over online data streams. *ACM SIGKDD International Conference on Knowledge Discovery and Data Mining.*

Chaum, D. L. (1981). Untraceable electronic mail, return addresses, and digital pseudonyms. *Communications of the ACM, 24*(2), 84–90. doi:10.1145/358549.358563

Chen, R. B., & Mahmassani, H. S. (2004). *Travel time perception and learning mechanisms in traffic networks.* Paper presented at the 83rd meeting of the Transportation Research Board, Washington, DC.

Chernoff, H. (1973). The use of faces to represent points in k-dimensional space graphically. *Journal of the American Statistical Association, 68*(342), 361-368.

Cheung, G., Baker, S., & Kanade, T. (2003). Shape-from-silhouette for articulated objects and its use for human body kinematics estimation and motion capture. *Proceedings of the IEEE Conference on Computer Vision and Pattern Recognition,* (pp. 77-84), Madison, Wisconsin: IEEE.

Cheung, G., Baker, S., & Kanade, T. (2003). Visual Hull Alignment and Refinement Across Time: A 3D Reconstruction Algorithm Combining Silhouette with Stereo. *Proceedings of the IEEE Conference on Computer Vision and Pattern Recognition,* (pp. 375-382), Madison, Wisconsin: IEEE.

Chi, Y., Wang, H. X., Yu, P. S., & Muntz, R. R. (2004). Moment: Maintaining closed frequent itemsets over a stream sliding window. *IEEE International Conference on Data Mining.*

Chorus, C. G., & Timmermans, H. J. P. (In Press). Personal Intelligent Travel Assistants. In A. de Palma, R. Lindsey, E. Quinet, & R. Vickerman (Eds.), Handbook in Transport Economics. London: Edward Elgar.

Chorus, C. G., Arentze, T. A., & Timmermans, H. J. P. (2008). *A general model of Bayesian travel time learning.* Paper presented at the 87th annual meeting of the *Transportation Research Board*, Washington, D.C.

Chorus, C. G., Arentze, T. A., & Timmermans, H. J. P. (2009). Traveler compliance with advice: a Bayesian utilitarian perspective. *Transportation Research Part E, Logistics and Transportation Review, 36*, 538–551.

Chorus, C. G., Walker, J. L., & Ben-Akiva, M. E. (in press). The value of travel information: A search-theoretic approach. *Journal of Intelligent Transportation Systems.*

Chorus, C.G., Arentze, T.A. & Timmermans, H.J.P. (2008). A random regret minimization model of travel choice. *Transportation Research, 42*(Part B), 1-18

Chorus, C.G., Arentze, T.A., Molin, E.J.E. & Timmermans, H.J.P. (2006). The value of travel information: Decision-strategy specific conceptualizations and numerical examples. *Transportation Research, 40*(Part B), 504-519.

Chung, E., & Shalaby, A. (2005). A Trip Reconstruction Tool for GPS-Based Personal Travel Surveys. *Journal of Transportation Planning and Technology, 28*(5), 381–401. doi:10.1080/03081060500322599

Chung, E., & Shalaby, A. (2005). Development of a trip reconstruction tool for GPS-based personal travel surveys. *Journal of Transportation Planning and Technology, 28*, 381–401.

Civilis, A., Jensen, C. S., & Pakalnis, S. (2005). Techniques for Efficient Road-Network-Based Tracking of Moving Objects. *IEEE Transactions on Knowledge and Data Engineering, 17*(5), 698–712. doi:10.1109/TKDE.2005.80

Clarke, R. V., & Eck, J. (2003). *Become a problem-solving crime analyst in 55 small steps.* London: Jill Dando Institute, University College. Retrieved August 28, 2007, from http://www.jdi.ucl.ac.uk/publications/other_publications/55steps

Clarke, R. V., & Mayhew, P. (1998). Preventing crime in parking lots: What we know and what we need to know. In M. Felson, & R. B. Peiser (Eds.), *Reducing crime through real estate development and management.* Urban Land Institute.

Clarkson, B. (2002). *Life Patterns: structure from wearable sensors*. Cambridge, MA: Massachusetts Institute of Technology.

Cohn, A. G. (1995). The Challenge of Qualitative Spatial Reasoning. *ACM Computing Surveys, 27*(3), 323–325. doi:10.1145/212094.212112

Commission of the European Communities. (2001). *European transport policy for 2010: time to decide*. Policy White Paper. Retrieved July 1, 2009, from http://ec.europa.eu/transport/white_paper/documents/doc/lb_com_2001_0370_en.pdf

Committee on Sporting Singapore (CoSS) (2000). Retrieved May 12, 2008, from http://app.mcys.gov.sg/web/sprt_towards_committeesporting.asp

Conroy Dalton, R. (2003). The secret is to follow your nose: Route path selection and angularity. *Environment and Behavior, 35*(1), 107–131. doi:10.1177/0013916502238867

Corazza, S., Mundermann, L., & Andriacchi, T. (2006). The Evolution of Methods for the Capture of Human Movement Leading to Markerless Motion Capture for Biomechanical Applications. *Journal of Neuroengineering and Rehabilitation, 3*, 6.

Cox, S. (Ed.). (2007). Observations and Measurements - Part 1 - Observation schema 1.0 (OGC Document Number: 07-022r1). Wayland, MA: OGC.

Craik, F. I. M., & Lockhart, R. S. (1972). Levels of Processing: A Framework for Memory Research. *Journal of Verbal Learning and Verbal Behavior, 11*, 671–684. doi:10.1016/S0022-5371(72)80001-X

Cyberware (n.d.). Retrieved May 12, 2008, from http://www.cyberware.com

Cyrus, F. (2006).Vodafone, TomTom partner to create real-time traffic data network. *Engadget* October 2006. Retrieved July 23, 2008: http://www.engadget.com/2006/10/27/vodafone-tomtom-partner-to-create-real-time-traffic-data-networ/

Daamen, W., & Hoogendoorn, S. P. (2003). Controlled experiments to derive walking behaviour. *European Journal of Transport and Infrastructure Research, 3*(1), 39–59.

Daamen, W., & Hoogendoorn, S. P. (2003). Experimental research of pedestrian walking behavior. *Transportation Research Record, 1828*, 20–30. doi:10.3141/1828-03

Daamen, W., & Hoogendoorn, S. P. (2003b). Research on pedestrian traffic flows in the Netherlands. Proceedings Walk 21 IV, (pp. 101-117). Portland, OR: Walk 21 conference.

Dablanc, L. (2007). Goods Transport in large European Cities: Difficult to Organize, Difficult to Modernize. *Transportation Research Part A, Policy and Practice, 41*, 280–285. doi:10.1016/j.tra.2006.05.005

Dang, X. H., Ng, W. K., & Ong, K. L. (2007). Online mining of frequent sets in data streams with error guarantee. *Knowledge and Information Systems.*

Dantu, K., Rahimi, M., Shah, H., Babel, S., Dhariwal, A., & Sukhatme, G. S. (2005). Robomote: Enabling Mobility in Sensor Networks. In *IEEE/ACM Fourth International Conference on Information Processing in Sensor Networks (IPSNSPOTS)* (pp. 404 – 409), New York.

Davis, R., Shrobe, H., & Szolovits, P. (1993). What is a knowledge representation? *AI Magazine, 14*(1), 17–33.

de Vries, B., Lin, Y., & Jessurun, J. (2008). Sense of the City. In *Proceedings DDSS Conference*, Leende, The Netherlands.

Dee, H. M. & Hogg, D. C. (2006). *Navigational strategies and surveillance.* Paper presented at the Sixth IEEE International Workshop on Visual Surveillance, Graz, Austria.

Delft University of Technology. (2008). *Delft - Urbanism on Track.* Retrieved July 1, 2009 from: http://www.bk.tudelft.nl/urbanismontrack

Demaine, E. D., Ortiz, A. L., & Munro, J. I. (2002). Frequency estimation of internet packet streams with limited space. *European Symposium on Algorithms.*

Dementhon, D.F., & Davis, L. (1995). Model-based Object Pose in 25 Lines of Code. *Intl. Journal of Computer Vision, 15*, 123-141.

Dey, A. K. (2001). Understanding and using context. *Personal and Ubiquitous Computing*, *5*(1), 4–7. doi:10.1007/s007790170019

Dijst, M. (2004). ICT and accessibility: An action space perspective on the impact of new information and communication technologies. In Beuthe, M., Reggiani, A., & Zamarini, L. (Eds.), *Transport Developments and Innovation in an Evolving World* (pp. 27–46). Berlin: Springer Verlag.

Dillemuth, J. (2007). Choosing the scale and extent of maps for navigation with mobile computing systems. *Journal of LBS*, *1*(1), 46–62.

Dittmar, H., & Ohland, G. (2003). The New Transit Town, Best Practices In Transit-Oriented Development. Washington, DC: Island Press.

Dobson, J., Fisher, P. (2003, Spring). GeoSlavery. *IEEE Technology and Society Magazine*, (pp. 47-52).

Dodge, S., Weibel, R., & Lautenschütz, A. K. (2008). Towards a taxonomy of movement patterns. *Information Visualization*, *7*, 240–252. doi:. doi:10.1057/palgrave.ivs.9500182

Doherty, S. T., Noel, N., Lee-Gosselin, M., Sirois, C., & Ueno, M. (2001). Moving Beyond Observed Outcomes: Integrating Global Positioning Systems and Interactive Computer-Based Travel Behaviour Surveys. In Transportation Research Circular Number E-C026: Personal Travel: The Long and Short of It, (pp. 449-446). Washington, DC: Transportation Research Board, National Research Council, Washington, D.C.

Doherty, S. T., Papinski, D., & Lee-Gosselin, M. (2006). An Internet-based Prompted Recall Diary with Automated GPS Activity-trip Detection: System Design. In *85th Annual Meeting of the Transportation Research Board*, January 22-26.

Dorai, C., & Venkatesh, S. (2002). Media Computing: Computational Media Aesthetics. (The International Series in Video Computing). *Springer*, Edition 1.

Downs, R., & Stea, D. (1973). *Cognitive Maps and Spatial Behavior: Process and Products. Image and Environment* (pp. 8–26). Chicago: Aldine Publishing Co.

Draijer, G., Kalfs, N., & Perdok, J. (2000). Global positioning system as data collection method for travel research. *Transportation Research Record*, *1719*, 147–153.

Du, J., & Aultman-Hall, L. (2007)... *Transportation Research Part A, Policy and Practice*, *41*, 220–232. Increasing the accuracy of trip rate information from passive multi-day GPS travel datasets: Automatic trip end identification issues

Durlach, N., Allen, G., Darken, R., Garnett, R.L., Loomis, J., Templeman, J., & von Wiegand, T. E. (2000). Virtual environments and the enhancement of spatial behavior: Towards a comprehensive research agenda. *PRESENCE-Teleoperators and Virtual Environments, 9*, 593-615.

Duxbury, E., & Desyllas, J. (2000). *St. Giles circus: Report on current patterns of pedestrian movement and accidents*. London: Intelligent Space Partnership.

Duxbury, E., & Desyllas, J. (2001). *Shoreditch triangle pedestrian movement and safety report*. London: Intelligent Space Partnership.

Eck, J. E., Chainey, S. P., Cameron, J. G., Leitner, M., & Wilson, R. E. (2005). *Mapping crime: Understanding hot spots*. USA: National Institute of Justice. Retrieved August 28, 2007, from www.ojp.usdoj.gov/nij

Edwards, W., Lindman, H., & Savage, L. J. (1963). Bayesian statistical research for psychological research. *Psychological Review*, *70*, 193–242.

Egenhofer, M. J., & Al-Taha, K. K. (1992). Reasoning about Gradual Changes of Topological Relations. In A. Frank, I. Campari & U. Formentini (Eds.), Theory and Methods of Spatio-Temporal Reasoning in Geographic Space (LNCS 639, pp. 196-219). Berlin: Springer-Verlag.

Egenhofer, M. J., & Mark, D. M. (1995). Modeling Conceptual Neighborhoods of Topological Line-Region Relations. *International Journal of Geographical Information Systems*, *9*(5), 555–565. doi:10.1080/02693799508902056

Eijck, J v and Orzan, S. (2007). Epistemic Verification of Anonymity. *Electronic Notes in Theoretical Computer Science*, (168), 159-174.

Einhorn, H. J., & Hogarth, R. M. (1981). Behavioral decision theory: processes of judgment and choice. *Annual Review of Psychology, 32*, 53–88.

El-Gamal, M. A., & Grether, D. M. (1995). Are people Bayesian? Uncovering behavioral strategies. *Journal of the American Statistical Association, 90*, 1137–1145.

Emmerink, R. H. M., Axhausen, K. W., Nijkamp, P., & Rietveld, P. (1995). Effects of information in road transport networks with recurrent congestion. *Transportation, 22*, 21–53.

Emmerink, R. H. M., Nijkamp, P., Rietveld, P., & Van Ommeren, J. N. (1996). Variable message signs and radio traffic information: an integrated empirical analysis of drivers' route choice behavior. *Transportation Research, 30A*, 135–153.

ESRI. (2008). ArcGIS 9.2 [Computer software]

Esser, H. (1985). *Befragtenverhalten als rationales Handeln Zur Erklärung von Antwortverzerrungen in Interviews*. ZUMA-Arbeitsbericht Nr. 85/01 (in German).

Estevez, E., Janowski, T., & Ojo, A. (2007). *Planning for e-Government – A Service-Oriented Agency Survey*. United Nations University, UNU-IIST Report No. 361.

ETSI TS 101 042 (2002). *Policy requirements for certification authorities issuing public key certificates*, Sophia Antipolis.

ETSI TS 101 456 V1.2.1 (2004). *Policy requirements for certification authorities issuing qualified certificates*, Sophia Antipolis.

ETSI TS 102 231 V1.2.1 (2005). *Provision of harmonized Trust Service Provider status information*, Sophia Antipolis.

Eurostat (2008). *Modal split of passenger transport - % in total inland passenger-km: tran_typ > Passenger cars.* Retrieved July 1, 2009, from http://epp.eurostat.ec.europa.eu/tgm/table.do?tab=table&init=1&language=en&pcode=tsdtr210&plugin=1

Eymeri, J. (2001). The electronic identification of citizens and organisations in the European Union: State of Affairs. In *37th Meeting of the Directors-General of the Public Service of the Member States of the European Union*, 26-27 November 2001, Bruges.

Fennema, E., & Sherman, J. A. (1977). Sex-related differences in mathematics achievement, spatial visualization, and affective factors. *American Educational Research Journal, 23*(1), 51-71.

Finocchiaro, G. (Ed.). (2008). *Diritto all'anonimato.* Padova: Cedam.

Forrest, T. L., & Pearson, D. F. (2005). *A comparison of trip determination methods in GPS enhanced household travel surveys.* Presented at 84th Annual Meeting of the Transportation Research Board, Washington, D.C., 2005.

Fruin, J. (1971.) *Pedestrian and planning design.* Alabama: Elevator World Inc.

Fudenberg, D., & Tirole, J. (1991). Game theory. Cambridge, MA: MIT Press.

Fujii, S., & Gärling, T. (2003). Development of script-based travel mode choice after forced change. *Transportation Research, 6F*, 117–124.

Fujii, S., & Kitamura, R. (2003). What does a one-moth free bus ticket do to habitual drivers? *Transportation, 30*, 81–95.

Fujii, S., Gärling, T., & Kitamura, R. (2001). Changes in drivers' perceptions and use of public transport during a freeway closure. *Environment and Behavior, 33*, 796–808.

Fusco, G., Tatarelli, L., & Valentini, M. (2003). Last-Mile: A procedure to set-up an Optimized Delivery Scheme. In *Proceedings of the 3rd International Conference on City Logistics,* (pp. 147-161). Madeira, Portugal: Elsevier.

Gagalowicz, A. (1990). Collaboration between Computer Graphics and Computer Vision. *Proceedings of the International Conference on Computer Vision,* (pp. 733-737), Osaka, Japan: IEEE.

Gagalowicz, A., & Gerard, P. (2000). Three Dimensional Object Tracking using Analysis/Synthesis Techniques. In A. Leonardis, F. Solina, R. Bajcsy, & F. Solina (Eds.), *Confluence of Computer Vision and Computer Graphics* (pp. 307-330), Dordrecht, NL: Kluwer.

Gagalowicz, A., & Mathieu, M. (2006). Modeling 3D Humans from Uncalibrated Wide Baseline Views. Paper presented at *SAE Digital Human Modeling Conference*, Lyon, France.

Gajek, S., Sadeghi, A. R., Stüble, C., & Winandy, M. (2007) Compartmented security for browsers or how to thwart a phisher with trusted computing. In *2nd Intl. Conference on Availability, Reliability and Security (ARES 2007)*, 120-127.

Galton, A. (1995, September). *Towards a Qualitative Theory of Movement*. Paper presented at the Spatial Information Theory: A Theoretical Basis for GIS - International Conference COSIT'95, Semmering, Austria.

Gandon, F. L., & Sadeh, N. M. (2004). Context-awareness, privacy and mobile access: a web semantic and multiagent approach. In *Proceedings of the 1st French-Speaking Conference on Mobility and Ubiquity Computing,* Nice, France, June 01 - 03, UbiMob '04, vol. 64. 123-130. New York: ACM. DOI= http://doi.acm.org/10.1145/1050873.1050901

Garcia, F. D., Hasuo, I., Pieters, W., & van Rossum, P. (2005). Provable anonymity. In *Proceedings of the 2005 ACM Workshop on Formal Methods in Security Engineering* (Fairfax, VA, USA, November 11 - 11, 2005).

Gärling, T., Eek, D., Loukopoulos, P., Fujii, S., Johansson-Stenman, O., Kitamura, R., et al. (2002). A conceptual analysis of the impact of travel demand management on private car-use. *Transport Policy, 9*, 59–70.

Gartner, G., & Hiller, W. (2008). Impact of restricted display size on spatial knowledge acquisition in the context of pedestrian navigation . In Gartner, G., & Rehrl, K. (Eds.), *LBS and TeleCartography II*. Heidelberg: Springer.

Gavrila, D.M. (1999). The Visual Analysis of Human Movement: A Survey. *Computer Vision and Image Understanding, 73*, 82-98.

Gedik, B., & Liu, L. (2005). Locations Privacy in Mobile Systems, A personalized Anonymization Model. In *Proceedings of the 25th IEEE International Conference on Distributed Computing Systems.* (pp. 620-629). Los Alamitos, CA: IEEE Press.

Gerard, P., & Gagalowicz, A. (2003). Human Body Tracking using a 3D Generic Model Applied to Golf Swing Analysis. Paper presented at *MIRAGE 2003Conference,* Rocquanecort, France.

Giannella, C., Han, J. W., Pei, J., Yan, X. F., & Yu, P. S. (2003). Mining frequent patterns in data streams at multiple time granularities. *Data Mining: Next Generation Challenges and Future Directions*, AAAI/MIT.

Giannotti, F., & Pedreschi, D. (Eds.). (2008). *Mobility, Data Mining and Privacy.* Heidelberg: Springer. doi:10.1007/978-3-540-75177-9

Giannotti, F., Nanni, M., Pinelli, F., & Pedreschi, D. (2007). *Trajectory pattern mining.* Paper presented at the KDD '07: Proceedings of the 13th ACM SIGKDD International Conference on Knowledge Discovery and Data Mining, San Jose, CA. (pp. 330-339). doi:10.1145/1281192.1281230

Gigerenzer, G., & Hoffrage, U. (1995). How to improve Bayesian reasoning without instruction: Frequency formats. *Psychological Review, 102*, 684–704.

Girardin, F., Calabrese, F., Dal Fiore, F., Ratti, C., & Blat, J. (2008). Digital footprinting: Uncovering tourists with user-generated content. *IEEE Pervasive Computing / IEEE Computer Society [and] IEEE Communications Society, 7*(4), 36–43. doi:10.1109/MPRV.2008.71

Giunchiglia, F., & Bouquet, P. (1997). Introduction to contextual reasoning: an artificial intelligence perspective. *Perspectives on Cognitive Science, 3*, 138–159.

Goeree, J. K., & Holt, C. A. (2004). A theory of noisy introspection. *Games and Economic Behavior, 46*, 365–382.

Goh, W.B., & Chan, K.Y. (2007). The Multiresolution Gradient Vector Field Skeleton. *Pattern Recognition, 40*, 1255-1269.

Golledge, R. G. (1995). *Defining the criteria used in path selection (Technical Report UCTC No. 78)*. University of California Transportation Center.

Goncalves, L., Di Bernardom, E., Ursella, E., & Perona, P. (1995). Monocular Tracking of the Human Arm in 3D. *Proceedings of the International Conference on Computer Vision,* (pp. 764-770), Cambridge, Massachusetts: IEEE.

González, M. C., Hidalgo, C. A., & Barabási, A. L. (2008). Understanding individual human mobility patterns. *Nature, 453*, 779–782. doi:10.1038/nature06958

Gopnik, A., Glymour, C., Sobel, D. M., Schulz, L. E., Kushnir, T., & Danks, D. (2004). A theory of causal learning in children: Causal maps and Bayes nets. *Psychological Review, 111*, 3–32.

Gortmaker, J., Janssen, M., & Wagenaar, R. (2007). Requirements on cross-agency processes in eGovernment: the need for a reference model . In Mitrakas, A., Hengenveld, A., Polemi, D., & Gamper, J. (Eds.), *Secure eGovernment web services*. Hershey, PA: IGI Publishing.

Götz, K., Loose, W., Schmied, M., & Schubert, S. (2003). *Mobility Styles in Leisure Time*. Paper presented at the 10th International Conference on Travel Behaviour Research, Lucerne, Switzerland.

Grest, D., Woetzel, J., & Koch, R. (2005). Nonlinear Body Pose Estimation from Depth Images. In W. Kropatsch (Ed.), *Proceedings of the DAGM - German Association for Pattern Recognition Conference*, (pp. 285-292). Vienna, Austria: DAGM.

Gruber, T. (2008). Ontology. In L. Liu, & M. T. Özsu (Eds.), Encyclopedia of database systems. Berlin: Springer-Verlag.

Guay, R.B., & McDaniel, E. (1977). The relationship between mathematics achievement and spatial abilities among elementary school children. *Journal for Research in Mathematics Education, 8*(2), 211-215.

Gudmundsson, J., van Kreveld, M., & Speckmann, B. (2004). *Efficient detection of motion patterns in spatio-temporal data sets*. Paper presented at the Proceedings of the 12th Annual ACM International Workshop on Geographic Information Systems, (pp. 250-257).

Guensler, R., & Wolf, J. (1999). Development of a handheld electronic travel diary for monitoring individual trip making behavior. Washington, DC: TRB, National Research Council.

Güting, R. H., & Schneider, M. (2005). Moving Objects Databases, (Morgan Kaufmann Series in Data Management Systems). San Francisco, CA: Morgan Kaufmann.

Guha, S., & Koudas, N. (2002). Approximating a data stream for querying and estimation: Algorithms and performance evaluation. *International Conference on Data Engineering.*

Gundotra, V. (2009). See where your friends are with Google Latitude. *The official Google blog.* Retrieved March 1, 2009 from: http://googleblog.blogspot.com/2009/02/see-where-your-friends-are-with-google.html

Hägerstrand, T. (1970). What about people in regional science? *Papers in Regional Science, 24*(1), 6–21. doi:10.1007/BF01936872

Hakes, J., & Viscusi, W. K. (1997). Mortality risk assessments: A Bayesian reassessment. *Journal of Risk and Uncertainty, 15*, 135–150.

Halatchev, M., & Gruenwald, L. (2005). Estimating missing values in related sensor data streams. *International Conference on Management of Data.*

Hamamatsu (n.d.). Retrieved May 12, 2008, from http://usa.hamamatsu.com

Han, Q., & Timmermans, H. J. P. (2006). Interactive learning in transportation networks with uncertainty, bounded rationality, and strategic choice behavior: quantal response model. *Transportation Research Record, 1964*, 27–34.

Han, Q., & Timmermans, H. J. P. (2007). Towards models of strategic spatial choice behavior: theory and application issues. *GeoJournal, 67*, 195–206.

Han, Q., Arentze, T. A., Timmermans, H. J. P., Janssens, D., & Wets, G. (2008). *Modeling context-sensitive dynamic activity-travel behavior under conditions of uncertainty incorporating reinforcement learning, habit formation, and behavioral and cognitive adaptation strategies*. Paper presented at the 87th Annual Meeting of the Transportation Research Board, Washington D.C.

Harries, K. (1999). *Mapping crime: Principle and practice*. United States National Institute of Justice. Retrieved August 28, 2007, from http://www.ojp.usdoj.gov/nij/maps/pubs.html

Hartmann, R. (1984). Ein „Touristentag“ in der Stadt – Methodische Betrachtungen zum Städtetourismus, dargestellt am Beispiel der Europareisen junger nordamerikanischer Touristen. [in German]. *Zeitschrift fur Wirtschaftsgeographie*, *28*(3/4), 145–156.

Hartmann, R. (1988). Combining Field Methods in Tourism Research. *Annals of Tourism Research*, *15*, 88–105. doi:10.1016/0160-7383(88)90073-4

Hato, E., & Asakura, Y. (2001). New approach for collection of activity diary using mobile communication systems. In *Proceedings 80th Annual Meeting of the Transportation Research Board*, Washington DC.

Hato, E., Itsubo, S., & Mitani, T. (2006). Development of MoALs (Mobile Activity Loggers supported by GPS-phones) for travel behavior analysis. In *Proceedings 85th Annual Meeting of the Transportation Research Board*, Washington DC.

Hato. (2006). Development of behavioral context addressable loggers in the shell for travel-activity analysis. In *Proceedings IATRB Conference*, Kyoto, Japan.

Havens, S. (Ed.). (2007). OpenGIS Transducer Markup Language (TML) Implementation Specification 1.0.0 (OGC Document Number: 06-010r6). Wayland, MA: OGC.

Heinzelman, W.B., Murphy, A.L., Carvalho, H.S., & Perillo, M.A.. (2004, January). Middleware to support sensor network applications. *IEEE Network Magazine Special Issue*.

Helbing, D. (1991). A mathematical model for the behavior of pedestrians. *Behavioral Science*, *36*, 298–310. doi:10.1002/bs.3830360405

Helbing, D., Buzna, L., Johansson, A., & Werner, T. (2005). Self-organized pedestrian crowd dynamics: Experiments, simulations, and design solutions. *Transportation Science*, *39*(1), 1–24. doi:. doi:10.1287/trsc.1040.0108

Helbing, D., Molnár, P., Farkas, I. J., & Bolay, K. (2001). Self-organizing pedestrian movement. *Environment and Planning B*, *28*(3), 361–384. doi:10.1068/b2697

Helbing, D., Twining, C. J., Bulsari, A., Oliker, S., Furst, M., & Maimon, O. (1993). A fluid-dynamic model for the movement of pedestrians. *Complex Systems*, *6*, 391–415.

Henricksen, K., Indulska, J., & Rakotonirainy, A. (2002). Modeling context information in pervasive computing systems. In F. Mattern & M. Naghshineh (Eds.), *Proceedings of 1st International Conference on Pervasive Computing* (LNCS 2414, pp. 167-180). Berlin: Springer.

Herrmann, T., Schweizer, K., Janzen, G., & Katz, S. (1998). Routen- und Überblickswissen – konzeptuelle Überlegungen. [in German]. *Kognitionswissenschaft*, *7*, 145–159. doi:10.1007/s001970050067

Hill, M. (1984). Stalking the Urban Pedestrian: A Comparison of Questionnaire and Tracking Methodologies for Behavioral Mapping in Large-Scale Environments. *Environment and Behavior*, *16*, 539–550. doi:10.1177/0013916584165001

Hillier, B. (1996). *Space is the machine*. Cambridge University Press.

Hillier, B., Penn, A., Hanson, J., Grajewski, T., & Xu, J. (1993). Natural movement: Or, configuration and attraction in urban pedestrian movement. *Environment and Planning. B, Planning & Design*, *20*, 29–66. doi:10.1068/b200029

Hillier, B., Penn, A., Hanson, J., Grajewski, T., & Xu, J. (1993). Natural movement: Or configuration and attraction in urban pedestrian movement. *Environment and Planning B: Planning and Design, 20*, 29-66.

HitLab-Washington. (2006). *ARToolkit Homepage.* Retrieved September 24, 2006, from http://www.hitl.washington.edu/artoolkit/

Hodgson, G. M. (2004). Reclaiming habit for institutional economics. *Journal of Economic Psychology, 25*, 651–660.

Hoeven, F.D. van der (2005). Connected Cities. *Nova Terra*, December, 3.

Hoeven, F.D. van der (2008). Connected Cities: the Final Cut. *Nova Terra*, April, 3-6.

Hoffman, T. (2003). Smart dust: Mighty motes for medicine, manufacturing, the military and more. *Computer World: Mobile/Wireless*. Retrieved September 24, 2007, from http://www.computerworld.com/mobiletopics/mobile/story/0,10801,79572,00.html

Holden, A. (2000). *Environment and Tourism.* London: Routledge.

Holmquist, L. E., Gellersen, H-W, Kortuem, G., Schmidt, A., Strohbach, M., Antifakos, S., Michahelles, F., Schiele, B., Beigl, M., & Mazé, R. (2004). Building intelligent environments with Smart-Its. *IEEE Computer Graphics & Applications, 24*(1), 56-64.

Home Office (2001). *Crime reduction toolkit: Focus area and hotspots. Crime reduction unit.* Retrieved August 28, 2007, from www.crimereduction.gov.uk/toolkits/fa00.htm

Home Office (2001). *Crime reduction toolkit: Street crime and robbery. Crime reduction unit.* Retrieved August 28. 2007, from www.crimereduction.gov.uk/toolkits/sc00.htm

Home Office (2003). *Crime in England and Wales: April 2002 to March 2003.* London: Home Office. Retrieved August 28, 2007, from www.homeoffice.gov.uk

Home Office (2005). *Crime mapping: Improving performance, a good practice guide for front line officers.* London: Home Office. Retrieved August 28, 2007, from http://www.jdi.ucl.ac.uk/downloads/publications/other_publications/crime_mapping_guide.pdf

Hoogendoorn, S. P., Bovy, P. H. L., & Daamen, W. (2002). Microscopic pedestrian wayfinding and dynamics modelling. In M. Schreckenberg, & S. Sharma (Eds.), Pedestrian and evacuation dynamics, (pp. 123-155). Berlin: Springer.

Hornsby, K., & Egenhofer, M. J. (2002). Modeling moving objects over multiple granularities. *Annals of Mathematics and Artificial Intelligence, 36*, 177–194. doi:10.1023/A:1015812206586

Horowitz, J. L. (1984). The stability of stochastic equilibrium in a two-link transportation network. *Transportation Research, 18B*, 13–28.

Hu, P., Robinson, R., & Indulska, J. (2007). Sensor Standards: Overview and Experiences. In *Proceedings of the 3rd International Conference on Intelligent Sensors, Sensor Networks and Information Processing ISSNIP'07*, Melbourne, Australia, December 3-6.

Hu, Y., & Wang, H. (2005). A framework for location privacy in wireless networks . In *ACM SIGCOMM.* Asiba Workshop.

Hurtig, T. (2006). A mobile multimodal dialogue system for public transportation navigation evaluated. In *Proceedings of the 8th conference on Human-computer interaction with mobile devices and services*, September 12-15, Helsinki, Finland.

IAIS. (2008). *CommonGIS* [Computer software].

Ishikawa, T., Fujiwara, H., Imai, O., & Okabe, A. (2008). Wayfinding with a GPS-Based Mobile Navigation System: A Comparison with Maps and Direct Experience. *Journal of Environmental Psychology, 28*, 74–82. doi:10.1016/j.jenvp.2007.09.002

Ito, T., Hayano, M., Naito, T., Asakura, Y., Hato, E., & Wada, T. (2004). Empirical Analysis on hazardous material transportation using road traffic sensors and accident data. In *Proceedings of the 3rd International Conference on City Logistics,* (pp. 235-250). Madeira, Portugal: Elsevier.

Itsubo, S., & Hato, E. (2006). A study of the effectiveness of a household travel survey using GPS-equipped cell phones and a WEB diary through a comparative study with a paper based travel survey. In *Proceedings 85th Annual Meeting of the Transportation Research Board*, Washington DC.

Jackson, C., Boneh, D., & Mitchell, J. (2007). Transaction generators: Root kits for web. In *2nd USENIX Workshop on Hot Topics in Security (HotSec '07)*.

Jagadeesh, G. R., Srikanthan, T., & Zhang, X. D. (2004). A Map Matching Method for GPS Based Real-Time Vehicle Location. *Journal of Navigation, 57*, 429–440. doi:10.1017/S0373463304002905

Jakob, A. (2001). Möglichkeiten und Grenzen der Triangulation quantitativer und qualitativer Daten am Beispiel der (Re-) Konstruktion einer Typologie erwerbsbiographischer Sicherheitskonzepte. *Forum: Qualitative Social Research, 2*(1). Retrieved April 5, 2009, from http://www.qualitative-research.net (in German)

Janssen, M., & Wagenaar, R. W. (2004). Developing Generic Shared Services for e-Government. *Electronic . Journal of E-Government, 2*(1), 31–38.

Janssens, D., Hannes, E., & Wets, G. (2007). *Planning interventions in the interactions between individual activity patterns, patterns of functions and infrastructure*. Paper presented at Urbanism On Track – Expert meeting on the application in urban design and planning of GPS-based and other tracking-based research, Delft, The Netherlands.

Jha, M., Madanat, S., & Peeta, S. (1998). Perception updating and day-to-day travel choice dynamics in traffic networks with information provision. *Transportation Research Part C, Emerging Technologies, 6*, 189–212.

Jiang, N., & Gruenwald, L. (2006). CFI-Stream: Mining Closed Frequent Itemsets in Data Streams. *ACM SIGKDD international conference on knowledge discovery and data mining.*

Jiang, N., & Gruenwald, L. (2007). Estimating missing data in data streams, *the International Conference on Database Systems for Advanced Applications.*

Joh, C.-H., Arentze, T. A., Harry, J. P., & Timmermans, H. J. P. (2001). Understanding activity scheduling and rescheduling behaviour: Theory and numerical illustration. *GeoJournal, 53*, 359–371. doi:10.1023/A:1020145214526

Kakadiaris, I. A., & Metaxas, D. (1998). 3D Human Body Acquisition from Multiple Views. *International Journal of Computer Vision, 30*, 191-218.

Kalfs, N., & Saris, W. E. (1997). New data collection methods in travel surveys . In Ettema, D. F., & Timmermans, H. J. P. (Eds.), *Activity-Based Approaches to Travel Analysis*. New York: Pergamon Press.

Kalnis, P., Mamoulis, N., & Bakiras, S. (2005). On discovering moving clusters in spatio-temporal data. In C. Bauzer-Medeiros, M. J. Egenhofer & E. Bertino (Eds.), Advances in spatial and temporal databases (pp. 364-381). Berlin: Springer. doi:10.1007/11535331

Kargupta, H., Bhargava, R., Liu, K., Powers, M., Blair, P., Bushra, S., et al. (2004). VEDAS: A mobile and distributed data stream mining system for real-time vehicle monitoring. *SIAM International Conference on Data Mining.*

Karimi, H. A., & Liu, X. (2003, November). *A Predictive Location Model for Location-Based Services*. Paper presented at the 11th ACM International Symposium on Advances in Geographic Information Systems, New Orleans, USA.

Kaysi, I. (1991). *Framework and models for provision of driver information system*. PhD-thesis. Cambridge, MA: Massachusetts Institute of Technology.

Kehl, R., & van Gool, L. (2006). Markerless Tracking of Complex Human Motions from Multiple Views. *Computer Vision and Image Understanding, 104*, 190-209.

Kelly, D. J., Raines, R. A., Grimaila, M. R., Baldwin, R. O., & Mullins, B. E. (2008). A survey of state-of-the-art in anonymity metrics. In *Proceedings of the 1st ACM Workshop on Network Data Anonymization,* Alexandria, VA, October 31 - 31, (NDA '08) (pp. 31-40). New York: ACM.

Kerr, Ian (ed.). (2009) Lessons from the Identity Trail: Anonymity, Privacy, and Identity in a Networked Society, *Oxford University Press.*

Keul, A &, Kühberger, A. (1996). *Die Strasse der Ameisen. Beobachtungen und Interviews zum Salzburger Städtetourismus.* München/Wien, Germany: Profil-Verlag (in German).

Keul, A., & Kühberger, A. (1997). Tracking the Salzburg Tourist. *Annals of Tourism Research, 24,* 1008–1012. doi:10.1016/S0160-7383(97)00038-8

Kirk, B., Fagan, K., & Renner, R. (2005). Development and Demonstration of a System for Using Cell Phones as Traffic Probes. In *ITS Society of Canada Annual Conference and General Meeting.*

Kitazawa, K., Zhao, H., & Shibasaki, R. (2003). A Study for Agent-based Modeling of Migration Behavior of Shoppers. In *Proceedings of the 8th International Conference on Computers in Urban Planning and Urban Management*, Sendai, Japan.

Koehler, U. (2004). New ideas for the City Logistics Project in Kassel. In *Proceedings of the 3rd International Conference on City Logistics,* (pp. 321-332). Madeira, Portugal: Elsevier.

Koh, J. L., & Shin, S. N. (2006). An approximate approach for mining recently frequent itemsets from data streams. *The 8th International Conference on Data Warehousing and Knowledge Discovery.*

Koh, M., & Anwari, K. (2004). Integrating video and computer technology in teaching – an example in gymnastics initial PE teacher-training in Singapore. *British Journal of Teaching Physical Education*, 35(3), 43 -46.

Koike, H., Morimoto, A., Inoue, T., & Kawano, T. (2003). Studies on the Characteristic Differences of Pedestrian Behaviors among Cities and Between City Districts such as Downtown and Suburban Shopping Centers [Portland, OR.]. *Walk, 21,* IV.

Krijgsman, S., Nel, J., & De Jong, T. G. (2008). *Deriving transport data with cellphones: Methodological lessons from South Africa.* Paper presented at the *8th International Conference on Survey Methods in Transport*, Annecy, France.

Krüger, A., Aslan, I., & Zimmer, H. (2004). The Effects of Mobile Pedestrian Navigation Systems on the Concurrent Acquisition of Route and Survey Knowledge . In Brewster, S., & Dunlop, M. (Eds.), *Mobile Computer Interaction -- MobileHCI 2004* (pp. 446–450). Berlin: Springer.

Kuner, C. (2007). *European Data Protection Law: Corporate Regulation and Compliance.* Oxford, UK: Oxford University Press.

Kunze, O. (2004). A new interactive Approach on Route Planning with Tight Delivery Time Windows. In *Proceedings of the 3rd International Conference on City Logistics,* (pp. 83-96). Madeira, Portugal: Elsevier.

Laasonen, K. (2005, October). *Clustering and Prediction of Mobile User Routes from Cellular Data.* Paper presented at the PKDD'05 - 9th European Conference on Principles and Practice of Knowledge Discovery in Databases, Porto, Portugal.

Lafortune, M.A., Cavanagh, P.R., Sommer, H.J., & Kalenak, A. (1992). Three-Dimensional Kinematics of the Human Knee during Walking. *Journal of Biomechanics, 25,* 347-357.

Larson, P., & Van Rooyen, A. (1998). The Virtual Reality Mental Rotation Spatial Skills Project. *CyberPsychology and Behavior, 1,* 113-120.

Laube, P., Imfeld, S., & Weibel, R. (2005). Discovering relative motion patterns in groups of moving point objects. *International Journal of Geographical Information Science, 19*(6), 639. doi:. doi:10.1080/13658810500105572

Lave, J. (1988). *Cognition in practice, mind, mathematics and culture in everyday life.* Cambridge, UK: CUP

Lee, D. B. (1973). Requiem for large-scale models. *Journal of the American Institute of Planners, 39,* 163-178.

Lee, M. D., Reilly, R. E., & Butavicius, M. A. (2003). An empirical evaluation of Chernoff faces, start glyphs, and spatial visualizations for binary data. In *Proceedings of the Asia-Pacific symposium on Information Visualisation, 24,* 1-10.

Lee-Gosselin, M. E. H., & Harvey, A. S. (2006). Non-web technologies . In Stopher, P. R. (Ed.), *Travel Survey Methods: Quality and Future Directions* (pp. 561–568). Hoboken, NJ: Elsevier.

Lerasle, F., Rives, G., & Dhome, M. (1999). Tracking of Human Limbs by Multicular Vision. *Computer Vision and Image Understanding, 75*, 229-246.

Lerman, S. R., & Mahmassani, H. S. (1985). The econometrics of search. *Environment and Planning, 17A*, 1009–1024.

Li, H. F., & Cheng, H. (2008). Improve frequent closed itemsets mining over data stream with bitmap. *Ninth ACIS International conference on software engineering, artificial intelligence, networking, and parallel/distributed computing.*

Li, H. F., Ho, C. C., Kuo, F. F., & Lee, S. Y. (2006) A new algorithm for maintaining closed frequent itemsets in data streams by incremental updates. *Six IEEE International Conference on Data Mining Workshop.*

Li, H. F., Lee, S. Y., & Shan, M. K. (2004). An efficient algorithm for mining frequent itemsets over the entire history of data streams. *The International Workshop on Knowledge Discovery in Data Streams.*

Li, H., Guensler, R., Ongle, J., & Wang, J. (2005). Using GPS data to understand the day-to-day dynamics of the morning commute behavior. In *Proceedings 84th Annual Meeting of the Transportation Research Board*, Washington, DC.

Li, Y. (2003). Evaluating the urban commute experience: a time perception approach. *Journal of Public Transportation, 6*, 41–66.

Li, Z., & Shalaby, A. (2008). Web-based GIS system for prompted recall of GPS-assisted personal travel surveys: system development and experimental study. In *Proceedings 87th Annual Meeting of the Transportation Research Board*, Washington, DC.

Liang, S. H. L. (2008). A New Fully Decentralized Scalable Peer-to-Peer GIS Architecture. *Int. Arch. Photogram Rem. Sens. Spatial Inform. Sci., 37*, 687–692.

Lim, W.Y., & Koh, M. (2006). Effectiveness of Learning Technologies in the Teaching and Learning of Gymnastics. *Pacific Asian Education, 18*(2), 69–77.

Lin, C. H., Chiu, D. Y., Wu, Y. H., & Chen, A. L. P. (2005). Mining frequent itemsets from data streams with a time-sensitive sliding window. *SIAM International Conference on Data Mining.*

Lozano, A., Antun, J. P., Santos, C., Alarcon, R., Granados, F., & Hernandez, R. (2005). Basis for a policy for the development of Logistics Platforms in the Metropolitan Zone of Mexico City. In *Proceedings of the 4rd International Conference on City Logistics,* (pp. 147-161). Langkawi, Malaysia: Elsevier.

Lutz, M., & Klien, E. (2006). Ontology-based retrieval of geographic information. *International Journal of Information Science, 20*(3), 233–260.

Lyons, G. D. (2001). Towards integrated traveler information. *Transport Reviews, 21*, 217–235.

Ma, C. M., & Sonka, M. (1996). A Fully Parallel 3D Thinning Algorithm and Its Applications. *Computer Vision and Image Understanding*, 64, 420-433.

Mahmassani, H. S., & Chang, G. (1987). On boundedly rational user equilibrium in transportation systems. *Transportation Science, 22*, 89–99.

Mahmassani, H. S., & Jayakrishnan, R. (1991). System performance and user response under real-time information in a congested traffic corridor. *Transportation Research, 25A*, 293–307.

Manco, G., Baglioni, M., Giannotti, F., Kuijpers, B., Raffaetà, A., & Renso, C. (2008). Querying and reasoning for spatiotemporal data mining. In F. Giannotti, & D. Pedreschi (Eds.), Mobility, data mining and privacy (pp. 335-374). Berlin: Springer-Verlag. doi.:10.1007/978-3-540-75177-9_13

Mandelbrot, B. (1967). How long is the coast of Britain? Statistical self-similarity and fractional dimension. *Science, 156*, 636–638. doi:10.1126/science.156.3775.636

Manku, G. S., & Motwani, R. (2002). Approximate frequency counts over data streams. *International Conference on Very Large Databases.*

Marca, J., Rindt, C. R., & McNally, M. (2002). Collecting activity data from GPS readings. In *Proceedings 81st Annual Meeting of the Transportation Research Board*, Washington DC.

Marca, J., Rindt, C. R., & McNally, M. (2002). The Tracer data collection system: Implementation and operational experience. In *Proceedings 81st Annual Meeting of the Transportation Research Board*, Washington, DC.

Marmasse, N., & Schmandt, C. (2002). A User-Centered Location Model. *Personal and Ubiquitous Computing*, *6*, 318–321. doi:10.1007/s007790200035

Mastrogiannidou, C., Kozanidis, G., & Ziliaskopoulos A.K. (2006). Paratransit systems, a case study for the Mount Pelion region in Greece. *Nova Terra*, June, 18-22.

Matsuyama, T., Wu, X., Takai, T., & Nobuhara, S. (2004). Real-time 3D Shape Reconstruction, Dynamic 3D Mesh Deformation, and High Fidelity Visualization for 3D Video. *Computer Vision and Image Understanding*, *96*, 393-434.

Matthew, C. E. (2005). Calibration of Accelerometer Output for Adults. *Medicine and Science in Sports and Exercise*, *37*(11), S512–S522. doi:10.1249/01.mss.0000185659.11982.3d

McCreadie, C., Raper, J., Gunesh, A., Wood, J., Carey, K., Petrie, H., et al. (2006). Persuasive technology for leisure and health: development of a personal navigation tool. In *Persuasive Technology*, (pp. 187-190).

McFadden, D. (1999). Rationality for economists? *Journal of Risk and Uncertainty*, *19*, 73–105.

McFadden, D. (2001). Disaggregate behavioral travel demand's RUM side: a 30 year retrospective. In D. Hensher (Ed.), Travel behavior research: the leading edge. Amsterdam: Pergamon.

McFerren, G., Roos, S., & Terhorst, A. (2006). *Fire alerts on the geospatial semantic web.* Paper presented at the Workshop Terra Cognita 2006, Athens, GA.

McKenna, S. J., Jabri, S., Duric, Z., Rosenfeld, A., & Wechsler, H. (2000). Tracking Groups of People. *Computer Vision and Image Understanding*, *80*, 42-56.

McNamara, L., Mascolo, C., & Capra, L. (2008). Media sharing based on collocation prediction in urban transport. In Proceed. ACM MobiCom '08, San Francisco, CA, Sept. 14 - 19, 58-69.

Meriall, W. M., Newberg, F., Sohrabi, K., Kaiser, W., & Pottie, G. (2003). Collaborative networking requirements for unattended ground sensor systems. In *Proc. IEEE Aerospace Conference*, 5.

Mikic, I., Trivedi, M., Hunter, E., & Cosman, P. (2003). Human Body Model Acquisition and Tracking using Voxel Data. *International Journal of Computer Vision*, *53*, 199-223.

Miller, H. (2005). A measurement theory for time geography. *Geographical Analysis*, *37*, 17–45. doi:10.1111/j.1538-4632.2005.00575.x

Miller, S. (1995). *Civilizing Cyberspace: Policy, Power and the Information Superhighway.* Reading, MA: Addison-Wesley.

Millonig, A. & Schechtner, K. (2006). Developing Landmark-based Pedestrian Navigation Systems. *IEEE Transactions on Intelligent Transport Systems (ITS), Special Selection on ITSC'05, 8*(1), 43–49.

Millonig, A., & Gartner, G. (2007). *Monitoring Pedestrian Spatio-Temporal Behaviour* (Technical Report 42). Bremen, Germany: Universität Bremen, TZI Technologie-Zentrum Informatik.

Millonig, A., & Schechtner, K. (2007). Decision Loads and Route Qualities for Pedestrians – Key Requirements for the Design of Pedestrian Navigation Services. In Waldau, N., Gattermann, P., Knoflacher, H., & Schreckenberg, M. (Eds.), *Pedestrian and Evacuation Dynamics 2005* (pp. 109–118). Berlin: Springer. doi:10.1007/978-3-540-47064-9_10

Mitrakas, A. (2005). Policy Frameworks for Secure Electronic Business. In Khosrow-Pour, M. (Ed.), *Encyclopedia of Information Science and Technology* (*Vol. I-V*). Hershey, PA: Idea Group Publishing.

Mitrakas, A. (2005). Policy-driven signing frameworks in open electronic transactions. In Soft Law constraints in eGovernment, BILETA 2005, (British Irish Law Education & Technology Association), Belfast, 7 April 2005.

Mitrakas, A. (2008). Heidi: an electronic identity intermediary model in eGovernment . In Bouwman, H., Bons, R., Hoogeweegen, M., Janssen, M., & Pronk, H. (Eds.), *Let a Thousand Flowers Bloom: Essays in Commemoration of Prof. Dr. René Wagenaar*. Amsterdam: IOS Press.

Mitrakas, A., & Bos, J. (1998). The ICC ETERMS Repository to support Public Key Infrastructure. *Jurimetrics, 38*(3).

Mitrakas, A., Hengenveld, A., Polemi, D., & Gamper, J. (2007). Towards secure eGovernment . In Mitrakas, A., Hengenveld, A., Polemi, D., & Gamper, J. (Eds.), *Secure eGovernment web services*. Hershey, PA: IGI Publishing.

Moehring, M., Lessig, C., & Bimber, O. (2004). Optical tracking and video see-rhrough AR on consumer cell phones. In *Proceedings of Workshop on Virtual and Augmented Reality of the GI-Fachgruppe AR/VR* (pp. 193-204).

Moeslund, T.B., & Granum, E. (2001).A Survey of Computer Vision-based Human Motion Capture. *Computer Vision and Image Understanding, 81*, 231-268.

Moiseeva, A., Arentze, T. A., & Timmermans, H. J. P. (2009). *Imputing relevant information from multi-day GPS tracers for retail planning and management using fusion and context-sensitive learning.* To be presented at the 16th RARSS conference, Niagara Falls, Canada, July 6 – 9, 2009.

Molz, J. (2007). Cosmopolitans on the Couch: Mobile Hospitality and the Internet . In Molz, J., & Gibson, S. (Eds.), *Mobilizing Hospitality, the ethics of social relations in a mobile world* (pp. 65–80). Burlington, VT: Ashgate Publishing Ltd.

Moselund, T.B., Hilton, A., & Kruger, V. (2006). A Survey of Advances in Vision-based Human Motion Capture and Analysis. *Computer Vision and Image Understanding, 104*, 90-126.

Motion Analysis (nd). Retrieved May 12, 2008, from http://www.motionanalysis.com

Mozafari, B., Thakkar, H., & Zaniolo, C. (2008). Verifying and mining frequent patterns from large windows over data streams. *IEEE International Conference on Data Engineering.*

Münzer, S., & Stahl, C. (2007). Providing Individual Route Instructions for Indoor Wayfinding in Complex, Multi-Level Buildings. In F. Probst & C. Keßler (Eds.), GI-Days 2007 Young Researchers Forum (pp. 241-246). Münster, Germany: ifgi Print series.

Münzer, S., Zimmer, H. D., Schwalm, M., Baus, J., & Aslan, I. (2006). Computer-Assisted Navigation and the Acquisition of Route and Survey Knowledge. *Journal of Environmental Psychology, 26*, 300–308. doi:10.1016/j.jenvp.2006.08.001

Munuzuri, J., Larraneta, J., Onieva, L., & Cortes, P. (2004). Estimation of an origin destination matrix for urban freight transportation: Application to the city of Seville. In *Proceedings of the 3rd International Conference on City Logistics,* (pp. 67-82). Madeira, Portugal: Elsevier.

Murakami, E., & Wagner, D. P. (1999). Can using Global Positioning System (GPS) improve trip reporting? *Transportation Research Part C, Emerging Technologies, 7*, 149–165.

Murakami, E., Wagner, D. P., & Neumeister, D. M. (1999) Using Global Positioning Systems and Personal Digital Assistants for Personal Travel Surveys in the United States. In TRB Transportation Research Circular E-C008: Transport Surveys: Raising the Standard.

MVA (2000). *Profiling the populations available for stops and searches.* Police Research Series Paper 131. London: Home Office.

Na, A., & Priest, M. (Eds.). (2006). Sensor Observation Service 1.0 (OGC Document Number: 06-009r6). Wayland, MA: Open Geospatial Consortium Inc.

Nams, V. O. (1996). The VFractal: a new estimator for fractal dimension of animal movement paths. *Landscape Ecology, 11*(5), 289–297. doi:10.1007/BF02059856

Nams, V. O. (2005). Using animal movement paths to measure response to spatial scale. *Oecologia, 143,* 179–188. doi:10.1007/s00442-004-1804-z

Nams, V. O., & Bourgeois, M. (2004). Fractal dimension measures habitat use at different spatial scales: an example with marten. *Canadian Journal of Zoology, 82,* 1738–1747. doi:10.1139/z04-167

Nanni, M., Kuijpers, B., Korner, C., May, M., & Pedreschi, D. (2008). Spatiotemporal data mining. In F. Giannotti, & D. Pedreschi (Eds.), Mobility, data mining and privacy, (pp. 267-296). Berlin: Springer.

NASA. (2008). *Semantic Web for Earth and Environmental Terminology (SWEET).* Retrieved February 20, 2009, from http://sweet.jpl.nasa.gov/ontology/

Nath, S., Lie, J., & Zhao, F. (2006). *Challenges in Building a Portal for Sensors World-Wide.* Microsoft Technical Report MSR-TR-2006-133. Redmond, WA: Microsoft Research.

Nebert, D., Whiteside, A., & Vretanos, P. (Eds.). (2007). OpenGIS Catalogue Services Specification 2.0.2 (OGC Document Number: 07-006r1). Wayland, MA: OGC.

Neisser, U. (1976). *Cognition and Reality.* San Francisco: WH Freeman.

Nemoto, T. Browne, M. Visser, J. & Castro, J. (2005). Intermodal transport and City Logistics policies. In *Proceedings of the 4rd International Conference on City Logistics,* (pp. 15-29). Langkawi, Malaysia: Elsevier.

Neuman, B. C. (1993). Proxy-Based Authorization and Accounting for Distributed Systems. In *Proceedings of 13th International Conference on Distributed Computing Systems.*

Newman, P., & Kenworthy, J. (1999). *Sustainability and cities, Overcoming automobile dependance.* Island Press.

Nicoll, C.-P. Corien -Van Dellen, M. J. M. (eds.). (2003). Digital Anonymity and the Law, Cambridge University Press, TMC Asser Press,The Hague.

Nisbett, R. E., & Wilson, T. D. (1977). Telling more than We can Know: Verbal Reports on Mental Processes. *Psychological Review, 84,* 231–259. doi:10.1037/0033-295X.84.3.231

Nobuhara, S., & Matsuyama, T. (2006). Deformable Mesh Model for Complex Multi-Object 3D Motion Estimation from Multi-Viewpoint Video. In M. Pollefeys (Ed.), *Proceedings of the 3rd Intl. Symposium on 3D Processing, Visualization and Transmission (3DPVT 2006),* (pp. 264-271). Chapel Hill: University of North Carolina.

O'Connor, A., Zerger, A., & Itami, R. (2005). Geo-Temporal Tracking and Analysis of Tourist Movement. *Mathematics and Computers in Simulation, 69,* 135–150. doi:10.1016/j.matcom.2005.02.036

O'Neill, E., Kostakos, V., & Kindberg, T. gen. Schieck, T., Penn, A. et al. (2006). Instrumenting the city: Developing methods for observing and understanding the digital cityscape. In Ubicomp, 2006.

Oduntan, G. (2004). The evidentiary issues arising from the proposed use of satellite based vehicle monitoring system and electronic logbooks in the FishCAM Project of the European Union. *International Journal of law and Information Technology, 12*(1).

OECD. (2005). *Fishing vessel monitoring systems: past present and future, OECD High seas task force.* Paris: Author.

Ohmori, N., Nakazato, M., & Harata, N. (2005). GPS mobile phone-based activity diary survey. In. *Proceedings of the Eastern Asia Society of Transportation Studies, 5,* 1104–1115.

Ohmori, N., Nakazato, M., Harata, N., Sasaki, K., & Nishii, K. (2006). Activity diary surveys using GPS mobile phones and PDA. In *Proceedings 85th Annual Meeting of the Transportation Research Board,* Washington DC.

Olander, F. (1973). Search behavior in non-simultaneous choice-situations: satisficing or maximizing? In *Selected proceedings of an interdisciplinary research conference,* Rome.

Orellana, D., Wachowicz, M., Andrienko, N., & Andrienko, G. (2009). Uncovering interaction patterns in mobile outdoor gaming. Paper presented at the *Advanced Geographic Information Systems & Web Services, 2009. GEOWS '09.* (pp. 177-182). doi:10.1109/GEOWS.2009.13

Ortale, R., Manco, G., Ritacco, E., Pelekis, N., Trasarti, R., Costa, G., et al. (2008). DAEDALUS: Towards progressively querying and mining movement dataICAR-CNR.

Peak Performance. (nd). Retrieved May 12, 2008, from http://www.peakperform.com

Pedreschi, D., Bonchi, F., Turini, F., Verykios, V. S., Atzori, M., & Malin, B. (2008). Privacy Protection: Regulations and Technologies, Opportunities and Threats . In Giannotti, F., & Pedreschi, D. (Eds.), *Mobility, Data Mining and Privacy* (pp. 101–119). Berlin: Springer-Verlag. doi:10.1007/978-3-540-75177-9_5

Peissl, W., Raguse, M., Meints, M., & Langfeldt, O. (2008). Criteria for privacy enhancing security technologies. In Deliverable 6.2 of the PRISE project, Vienna, May 2008.

Penn, A., & Turner, A. (2001). Space syntax based agent simulation. In M. Schreckenberg, & S. Sharma (Eds.), Pedestrian and evacuation dynamics (pp. 99-114). Berlin: Springer.

Perry, M. (2003). Distributed cognition. In J. Carroll (Ed.), *HCI models, theories and frameworks*. Boston, MA: Morgan Kaufman.

Pfoser, D., & Jensen, C. S. (2001). *Querying the trajectories of on-line mobile objects*. Paper presented at the MobiDe '01: Proceedings of the 2nd ACM International Workshop on Data Engineering for Wireless and Mobile Access, Santa Barbara, CA, (pp. 66-73). doi:10.1145/376868.376902

Pichler, R. (2000). *Trust and Reliance – Enforcement and Compliance: Enhancing Consumer Confidence in the Electronic Marketplace*. Unpublished doctoral dissertation, Stanford Law School, Palo Alto, CA.

Pierce, B., Casas, J., & Giaimo, G. (2003). Estimating trip rate under-reporting: preliminary results from the Ohio household travel survey. In *Proceedings 83th Annual Meeting of the Transportation Research Board*, Washington DC.

Pirolli, P. (2003). Exploring and finding information. In J. Carroll (Ed.), *HCI models, theories and frameworks*. Boston, MA: Morgan Kaufman,

Planker, R., & Fua, P. (2001). Tracking and Modeling People in Video Sequences. *Computer Vision and Image Understanding, 81*, 285-302.

Platzer, E. (2005). *Spatial Cognition research: The human navigation process and its comparability in complex real and virtual environments*. Germany: PhD, Universität der Bundeswehr München.

Polak, J., & Jones, P. (1993). The acquisition of pre-trip information: a stated preference approach. *Transportation, 20*, 179–198.

Pushkarev, B., & Zupan, J. (1975). Urban space for pedestrians. Cambridge: MIT Press.

Quack, H., Van Duin, R., & Visser, J. (2006). City Logistics over the Years. Lessons learnt, Research Directions and Interests. In *5th International Conference on City Logistics*.

Quah, C. K., Gagalowicz, A., Roussel, R., & Seah, H S. (2005) 3D modeling of humans with skeletons from uncalibrated wide baseline views. In A.Gagalowicz & W. Philips (Eds.), *Proceedings of the International Conference on Computer Analysis of Images and Patterns*, (pp. 379-389), Versailles, France: Springer.

Quah, C.K., Gagalowicz, A., & Seah, H.S. Marker-less 3D Video Motion Capture in Cluttered Environments. In F.K. Fuss (Ed.), *Proceedings of the Asia-Pacific Congress on Sports Technology*, (pp. 121-126). Singapore: Nanyang Technological University.

Qualysis. (nd). Retrieved May 12, 2008, from http://www.qualysis.com

Radoczky, V. (2003). *Kartographische Unterstützungsmöglichkeiten zur Routenbeschreibung von*

Fußgängernavigationssystemen im In- und Outdoorbereich. MTh. TU Vienna. (in German)

Raento, M., Oulasvirta, A., & Eagle, N. (2009). Smartphones: An Emerging Tool for Social Scientists. *Sociological Methods & Research, 37*(3), 426–454. doi:10.1177/0049124108330005

Raiffa, H., & Schlaifer, R. (1961). Applied statistical decision theory. Boston: Harvard University Press.

Ratcliffe, J. (2004). *HotSpot detective for mapInfo Helpfile* (Version 2.0).

Rattenbury, T., Good, N., & Naaman, M. (2007). Towards automatic extraction of event and place semantics from flickr tags. In Proceed. ACM SIGIR '07, Amsterdam, July 23-27, 103-110.

Ratti, C., Pulselli, R. M., Williams, S., & Frenchman, D. (2006). Mobile landscapes: Using location data from cell-phones for urban analysis. *Environment & Planning., 33*(5), 727–748. doi:10.1068/b32047

Ray, M., & Schrom-Feiertag, H. (2007). *Cell-based Finding and Classification of Prominent Places of Mobile Phone Users.* Paper presented at The 4th International Symposium on Location Based Services & TeleCartography, Vienna, Austria.

Reades, J., Calabrese, F., Sevtsuk, A., & Ratti, C. (2007). Cellular census: Explorations in urban data collection. *IEEE Pervasive Computing / IEEE Computer Society [and] IEEE Communications Society, 6*(3), 30–38. doi:10.1109/MPRV.2007.53

Redman, R.W. (2006). The challenge of interdisciplinary teams. *Research and Theory for Nursing Practice, 20*(2), 105-107.

Rehl, K., Bruntsch, S., & Mentz, H.-J. (2007). Assisting multimodal travelers: design and prototypical implementation of a personal travel companion. *IEEE Transactions on Intelligent Transportation Systems, 8*, 31–42.

Renso, C., Puntoni, S., Frentzos, E., Mazzoni, A., Moelans, B., Pelekis, N., et al. (2008). Wireless network data sources: Tracking and synthesizing trajectories. In F. Giannotti, & D. Pedreschi (Eds.), Mobility, data mining and privacy, (pp. 73-100). doi:10.1007/978-3-540-75177-9_4

Resnick, P., Zeckhauser, R., Swanson, J., & Lockwood, K. (2006). The value of reputation on eBay: A controlled experiment. *Experimental Economics, 9*, 79–101. doi:10.1007/s10683-006-4309-2

RFC 2560. (1999). Internet X.509 Public Key Infrastructure: Online Certificate Status Protocol – OCSP.

RFC 3039. (2001). Internet X.509 Public Key Infrastructure - Qualified Certificates Profile.

RFC 3280. (2002). *Internet X.509 Public Key Infrastructure: Certificate and Certificate Revocation List (CRL) Profile,* (obsoletes RFC 2459).

Rinzivillo, S., Pedreschi, D., Nanni, M., Giannotti, F., Andrienko, N., & Andrienko, G. (2008). Visually–driven analysis of movement data by progressive clustering. *Information Visualization, 7*(3/4), 225–239. doi:. doi:10.1057/palgrave.ivs.9500183

Robinson, W. S. (1950). Ecological correlations and the behavior of individuals. *American Sociological Review, 15*(3), 351-357.

Rodriguez, D. A., Brown, A. L., & Troped, P. J. (2005). Portable Global Positioning Units to Complement Accelerometry-Based Physical Activity Monitors. *Medicine and Science in Sports and Exercise, 37*(11), S572–S581. doi:10.1249/01.mss.0000185297.72328.ce

Romer, K., & Mattern, F. (2004). The design space of wireless sensor networks. *IEEE Wireless Communications, 11*(6), 54-61.

Rosales, R., & Sclaroff, S. (2006). Combining Generative and Discriminative Models in a Framework for Articulated Pose Estimation. *International Journal of Computer Vision, 63*, 251-276.

Rosenhahn, B., He, L., & Klette, R. (2005). Automatic Human Model Generation. In A.Gagalowicz & W. Philips (Eds.), *Proceedings of the International Conference on Computer Analysis of Images and Patterns,* (pp. 41-48), Versailles, France: Springer.

Royal Institution of Chartered Surveyors. Policy Unit (2002). *Transport Development Areas, Guide to good practice*. Retrieved July 1, 2009, from http://connectedcities.eu/downloads/3rdparty/rics_tda_mainreport.pdf

Sander, H., & Hoeven, F.D. van der (2008). Transport as a field for local anarchy. *Nova Terra*, April, 7-10.

Santos, M. Y., & Moreira, A. (2007, May). *Topological Spatial Relations between a Spatially Extended Point and a Line for Predicting Movement in Space*. Paper presented at the 10th AGILE Conference on Geographic Information Science, Aalborg, Denmark.

Santos, M. Y., & Moreira, A. (2009, July). *How a Circular Spatially Extended Point is topological related with a Line?* Paper presented at the 2009 International Conference of Computational Intelligence and Intelligent Systems, London, England.

Santos, M. Y., & Moreira, A. (2009, July). *Conceptual neighborhood graphs for topological spatial relations*. Paper presented at the 2009 International Conference of Computational Intelligence and Intelligent Systems, London, England.

Saponas, S., Lester, J., Hartung, C., Sameer, A., & Tadayoshi, K. (2007). Devices that tell on you: privacy in consumer ubiquitous computing. In *Proceedings of the 16th USENIX Security Symposium*. (Boston, MA, August 06 - 10, 2007). N. Provos, Ed. USENIX Association, Berkeley, CA, 1-16.

Schlich, R., & Axhausen, K. W. (2003). Habitual travel behavior: evidence from a six-week travel diary. *Transportation, 30*, 13–36.

Schlieder, C. & Werner, A. (2003). *Interpretation of intentional behavior in spatial partonomies*.

Schmid, F. (2007). *Personalized Maps for Mobile Wayfinding Assistance*. Paper presented at The 4th International Symposium on Location Based Services and Telecartography, Vienna, Austria.

Schmidt, A. (2005). Interactive context-aware systems interacting with ambient intelligence. In G. Riva, F. Vatalaro, F. Davide, & M. Alcañiz (Eds.), *Ambient intelligence: The evolution of technology, communication and cognition towards the future of human-computer interaction*. IOS Press.

Schofer, J. L., Khattak, A., & Koppelman, F. S. (1993). Behavioral issues in the design and evaluation of advanced traveler information systems. *Transportation Research Part C, Emerging Technologies*, 107–117.

Schönfelder, S., Axhausen, K. W., Antille, N., & Bierlaire, M. (2002). Exploring the potentials of automatically collected GPS data for travel behavior analysis - a Swedish data source. In J. Möltgen & A. Wytzisk (Eds.), *GI-Technologien für Verkehr und Logis-tik, IfGIprints*, (132, 20, pp.155-179). Münster, Germany: Institut für Geoinformatik, Universität Münster.

Schuessler, N., & Axhausen, K. (2008). *Identifying trips and activities and their characteristics from GPS raw data without further information*. Paper presented at the *8th International Conference on Survey Methods in Transport*, Annecy, France.

Schweiger, C. L. (1996). Deployment of technology for paratransit: Examples of innovation. In *Vehicle Navigation and Information Systems Conference, VNIS '96* (Vol. 7, pp. 197- 204). Retrieved July 1, 2009 from http://ieeexplore.ieee.org/xpl/freeabs_all.jsp?tp=&arnumber=1623750&isnumber=34087

Schweizer, K. (1997). *Räumliche oder zeitliche Wissensorganisation? Zur Repräsentation der Blickpunktsequenzen bei räumlichen Anordnungen*. Lengerich, Germany: Pabst Science Publishers Verlag.

Schweizer, K., & Horn, M. (2006). Raumwahrnehmung und Raumvorstellung. [in German]. *Geographie und Schule, 164*, 21–26.

Shankland, S. (2008). Cameras with built-in geotagging on horizon. *CNET*. Retrieved March 1, 2009 from http://news.cnet.com/8301-13580_3-9868159-39.html

Sharma, J., Flewelling, D. M., & Egenhofer, M. J. (1994, September). *A Qualitative Spatial Reasoner*. Paper presented at the 6th International Symposium on Spatial Data Handling, Edinburgh, Scotland.

Shaw, S., & Yu, H. (2009). A GIS-based time-geographic approach of studying individual activities and interactions in a hybrid physical–virtual space. *Journal of Transport Geography*. doi:.doi:10.1016/j.jtrangeo.2008.11.012

Shaw, S., Yu, H., & Bombom, L. (2008). A space-time GIS approach to exploring large individual-based spatiotemporal datasets. *Transactions in GIS*, *12*(4), 425–441. doi:10.1111/j.1467-9671.2008.01114.x

Shenzhen Urban Transport Planning Center. (SUTPC). (2009). *2008 Shenzhen Traffic Condition Annual Report* 45-48.

Shin, S. J., & Lee, W. S. (2007). An online interactive method for finding assoication rules data streams. *ACM 16th Conference on Information and Knowledge Management*.

Shneiderman, (1988). *Designing the user interface* (3rd ed.). Reading, MA: Addison-Wesley.

Shoval, N., & Isaacson, M. (2006). The application of tracking technologies to the study of pedestrian spatial behavior. *The Professional Geographer*, *58*, 172–183.

Shoval, N., & Isaacson, M. (2007). Tracking Tourists in the Digital Age. *Annals of Tourism Research*, *34*(1), 141–159. doi:10.1016/j.annals.2006.07.007

Shoval, N., & Isaacson, M. (2007).Tracking tourists in the digital age. *Annals of Tourism Research*, *34*, 141–159.

Simon, H. A. (1955). A behavioral model of rational choice. *The Quarterly Journal of Economics*, *69*, 99–118.

Simonis, I. (Ed.). (2006). OGC Sensor Alert Service Candidate Implementation Specification 0.9 (OGC Document Number: 06-028r5). Wayland, MA: OGC.

Simonis, I., & Dibner, P. C. (Eds.). (2007). OpenGIS Sensor Planning Service Implementation Specification 1.0 (OGC Document Number: 07-014r3). Wayland, MA: OGC.

Simonis, I., & Echterhoff, J. (2006). *Draft OpenGIS Web Notification Service Implementation Specification 0.0.9 (OGC Document Number: 06-095)*. Wayland, MA: OGC.

Sirin, E., Parsia, B., Grau, B. C., Kalyanpur, A., & Katz, Y. (2007). Pellet: A practical owl-dl reasoner. *Web Semantics: Science, Services and Agents on the World Wide Web*, *5*(2), 51-53. Retrieved from http://pellet.owldl.com/papers/sirin05pellet.pdf

Smith, B. L., Zhang, H., Fontaine, M., & Green, M. (2006). Cellphone Probes as an ATMS Tool. In *Research Project Report for the National ITS Implementation Research Center and U.S. DOT University Transportation Center*. Retrieved March 3rd 2006 from http://ntl.bts.gov/lib/23000/23400/23431/CellPhoneProbes-final.pdf

Smith, S. E., & Vela, E. (2001). Environmental Context-Dependent Memory: A Review and Meta-Analysis. *Psychonomic Bulletin & Review*, *8*, 203–220.

Smyth, C. S. (2001). Mining mobile trajectories. In H. J. Miller & J. Han (eds.), Geographic Data Mining and Knowledge Discovery, (pp. 337-361). London: Taylor and Francis.

Souvignier, E. (2001). Training räumlicher Fähigkeiten. [Training spatial abilities.]. In K.J. Klauer (Ed.), *Handbuch Kognitives Training* (pp. 293-319). Göttingen: Hogrefe.

Spaccapietra, S., Parent, C., Damiani, M. L., Macêdo, J., Porto, F., & Vangenot, C. (2008). A conceptual view on trajectories. *Data & Knowledge Engineering*, *65*(1), 126–146. doi:10.1016/j.datak.2007.10.008

SpeckNet (2006). *Speckled Computing Consortium*. Retrieved April 28, 2006, from http://www.specknet.org/

Spence, R. (2001). *Information visualization*. Essex: Pearson Education Limited.

Stanford Center for Biomedical Informatics Research. (2008). *Protégé* [Computer software].

Starck, J., & Hilton, A. (2003). Model-based Multiple View Reconstruction of People. *Proceedings of the International Conference on Computer Vision,* (pp. 915-922), Nice, France: IEEE.

Statistikbanken. (n.d.). Retrieved from http://www.statistikbanken.dk/statbank5a/default.asp?w=1440

Stefanidis, A. (2006). The Emergence of GeoSensor Networks. *Directions Magazine*. Retrieved February 20, 2009, from http://www.directionsmag.com/article.php?article_id=2100&trv=1

Stopher, P., & Collins, A. (2005). Conducting a GPS prompted recall survey over the internet. *TRB Annual Meeting, CD-ROM,* Transportation Research Board, National Research Council, Washington, DC.

Stopher, P., Bullock, P. J., & Horst, F. N. H. (2003). Conducting a GPS survey with a time-use diary. In *Proceedings 83th Annual Meeting of the Transportation Research Board*, Washington, DC.

Stopher, P., Greaves, S., & Fitzgerald, C. (2005). Developing and deploying a new wearable gps device for transport applications. In *Proceedings 18th Australasian Transport Research Forum*, Sydney, Australia.

Suleyman, K. (2008). A nested decomposition approach for solving the paratransit vehicle scheduling problem. *Transportation Research Part B: Methodological, 43*(4), 448–465.

Sun, Z., Arentze, T. A., & Timmermans, H. J. P. (2005a). Modeling the impact of travel information on activity-travel rescheduling decisions under conditions of travel time uncertainty. *Transportation Research Record, 1926*, 79–87.

Sun, Z., Arentze, T. A., & Timmermans, H. J. P. (2005b). Modeling the impact of travel information on activity-travel rescheduling decisions under multiple uncertain events: distributed myopic decision heuristics. In *the Ninth International Conference on Computers in Urban Planning and Urban Management*, London, UK.

Svetsuk, A. (2007). *Experiments in urban mobility analysis in Rome using mobile phone data.* Paper presented at Urbanism On Track – Expert meeting on the application in urban design and planning of GPS-based and other tracking-based research, Delft, The Netherlands.

Swait, J., & Ben-Akiva, M. (1987). Incorporating random constraints in discrete models of choice set generation. *Transportation Research, 21B*, 91–102.

Sweeney, L. (2001). Information Explosion . In Zayatz, L., Doyle, P., Theeuwes, J., & Lane, J. (Eds.), *Confidentiality, Disclosure, and Data Access: Theory and Practical Applications for Statistical Agencies*. Washington, DC: Urban Institute.

Sweeney, L. (2007).*Technology Dialectics: Constructing Provably Appropriate Technology*. Retrieved October 7, 2008 from http://privacy.cs.cmu.edu/dataprivacy/projects/dialectics/index.html

Taniguchi, E., Thompson, R., & Yamada, T. (2001). Recent advances in modeling city logistics . In Taniguchi, E., & Thompson, R. G. (Eds.), *City Logistics II* (pp. 3–34). Kyoto, Japan: Institute of Systems Science Research.

Terzopolus, D. (1999). Visual Modeling for Computer Animation: Graphics with a Vision. *Computer Graphics, 33*, 42-45.

Theobalt, C., Aguiar, E., Magnor, M., Theisel, H., & Seidel H-P. (2004). Marker-free Kinematic Skeleton Esimation from Sequence of Voume Data. In R. Lad & G. Baciu (Eds.), *Proceedings of the ACM Virtual Reality Software and Technology Conference,* (pp. 57-64), Hong Kong: ACM.

Thomas, J. J., & Cook, K. A. (2005). Illuminating the path: The research and development agenda for visual analytics. Washington, DC: IEEE.

Thompson, R. G. (2004). Intelligent Vehicle Routing and Scheduling. In *Proceedings of the 3rd International Conference on City Logistics,* (pp. 97-110). Madeira, Portugal: Elsevier.

Thornton, P., Williams, A., & Shaw, W. G. (1997). Revisiting Time-Space Diaries: An Exploratory Case Study of Tourist Behavior in Cornwall, England. *Environment & Planning A, 29*, 1847–1867. doi:10.1068/a291847

Tillwick, H., & Olivier, M. (2005). Towards a framework for connection anonymity. In *Proceedings of the 2005 Annual Research Conference of the South African institute of Computer Scientists and information Technologists on IT Research in Developing Countries* (White River, South Africa, September 20 - 22, 2005). ACM International Conference Proceeding Series, vol. 150. South

African Institute for Computer Scientists and Information Technologists, 113-122.

Timmermans, H. J. P., Arentze, T. A., & Ettema, D. F. (2003). Learning and adaptation behavior: Empirical evidence and modelling issues. In *Proceedings ITS Conference*, April 1-4, Eindhoven, The Netherlands.

Transport for London (2003). *Analysis of the transport programme to support the draft London plan*. Technical Report. Transport for London.

Triandis, H. C. (1977). Interpersonal behavior. Monterey, CA: Brooks / Cole Publishing Company.

Tsui, S. Y. A., & Shalaby, A. S. (2006). (forthcoming). An Enhanced System for Link and Mode Identifications for GPS-Based Personal Travel Surveys. *Transportation Research Record*. doi:10.3141/1972-07

Tsui, S. Y. A., & Shalaby, A. S. (2006). An enhanced system for link and mode identification for GPS-based personal travel Surveys. In *Proceedings 85th Annual Meeting of the Transportation Research Board*, Washington, DC.

Turner, A., Doxa, M., O'Sullivan, D., & Penn, A. (2001). From isovists to visibility graphs: A methodology for the analysis of architectural space. *Environment and Planning. B, Planning & Design*, *28*(1), 103–121. doi:10.1068/b2684

Tversky, A., & Kahneman, D. (1974). Judgment under uncertainty: heuristics and biases. *Science*, *85*, 1124–1131.

Ueno, M., Noël, N., Doherty, S. T., Lee-Gosselin, M. E. H., Théberge, F., & Sirois, C. (1999). Extending the scope of travel surveys using differential GPS. In ION-GPS 1999, Nashville, Tennessee (CDROM).

University of California. (2006). *James Reserve: Data management systems*. Retrieved September 21, 2006, from http://dms.jamesreserve.edu/

UTMC (2000). *Review of current and future data requirements and detector technologies and the implications for UTMC*. Urban Traffic Management and Contol, Deliverable 2. Retrieved August 28, 2007, from http://www.utmc.gov.uk/research/index.htm

Vaidya, J., Clifton, C., & Zhu, M. (2006). *Privacy Preserving Data Mining*. Berlin: Springer.

Van Berkum, E. C., & Van der Mede, P. H. J. (1993). *The impact of traffic information*. PhD-thesis. Delft: Delft University of Technology.

van der Hoeven, F. D., Smit, M. G. J., & van der Spek, S. C. (Eds.). (2008). Street Level Desires: Discovering the city on foot, Pedestrian mobility and the regeneration of the European city centre. Charleston, SC: Booksurge Publishing.

van der Spek, S. (2008). Measuring and observing pedestrian activity: Tracking pedestrians in Norwich, Rouen and Koblenz. In *Proceedings Walk 21*.

van der Spek, S. C. (2007). *Legible City – Walkable City – Liveable City: Observation of Walking Patterns in City Centres*. Paper presented at Urbanism On Track – Expert meeting on the application in urban design and planning of GPS-based and other tracking-based research, Delft, The Netherlands.

van der Spek, S. C. (2008). Sensing Human Activity: GPS tracking . In Kooistra, L., & Ligtenberg, A. (Eds.), *Proceedings of International Workshop Sensing a Changing World 2008. CGI report CGI-08-005*. Wageningen, The Netherlands: Wageningen University and Research Centre.

van Marwijk, R., & Pitt, D. G. (2008). Where Dutch recreationists walk: Path design, physical features and walker usage. In *The Fourth International Conference on Monitoring and Management of Visitor Flows in Recreational and Protected Areas, Management for Protection and Sustainable Development*, Montecatini Terme, Italy.

van Schaick, J., & van der Spek, S. C. (Eds.). (2008). Urbanism on Track. Application of tracking technologies in Urbanism (Research in Urbanism Series, Vol. 1). Amsterdam: IOS Press.

Verplanken, B., Aarts, H., & van Knippenberg, A. (1997). Habit, information acquisition and the process of making modal choices. *European Journal of Social Psychology, 27*, 539–560.

Vicon. (nd). Retrieved May 12, 2008, from http://www.vicon.com

Viscusi, W. K., & Magat, W. A. (1992). Bayesian decisions with ambiguous belief aversion. *Journal of Risk and Uncertainty, 5*, 371–387.

Vishik, C., Johnson, S., & Hoffman, D. (2007). Infrastructure for Trusted Environment: In Search of a Solution. In ISSE/SECURE 2007 Securing Electronic Business Processes, 219-227. Wiesbaden, Germany: Vieweg.

Von Neumann, J., & Morgenstern, O. (1947). Theory of games and economic behavior. Princeton, NJ: Princeton University Press.

W3C Consortium. (2004). *OWL web ontology language.*

Wagner, D., & Schmalstieg, D. (2003). First steps towards handheld augmented reality. In *Proceedings of the 7th International Conference on Wearable Computers* (p 127).

Waller, S. T., & Ziliaskopoulos, A. K. (2003). On the Online Shortest Path Problem. *Networks, 40*(4), 182–202.

Weaver, C. (2008). *Cross-dimensional visual queries for interactive animated analysis of movement.* Paper presented at the Proceedings of Geospatial Visual Analytics Workshop at GIScience 2008, Utah. Retrieved from http://geoanalytics.net/GeoVisualAnalytics08/

Weibull, J. W. (1978). A search model for microeconomic analysis –with spatial applications. In A. Karlqvist (Ed.), Spatial interaction theory and planning models. Amsterdam: North-Holland Publishing Company.

Weis, S., Sarma, S., Rivest, R., & Engels, D. (2004). *Security and Privacy Aspects of Low-Cost Radio Frequency Identification Systems.* Security in Pervasive Computing.

Weiser, M. (2002). Mark Weiser (1952-1999), The founder of Ubiquitous Computing. *IEEE Pervasive Computing, 1*(1).

Whittaker, J. O., & Whittaker, S. J. (1972). A cross-cultured study of geocentrism. *Journal of Cross-Cultural Psychology, 3*, 417–421. doi:10.1177/002202217200300412

Wilhelm, U. G., Staamann, S. M., & Buttyan, L. (2005). A pessimistic approach to trust in mobile agent platforms. *IEEE Internet Computing, 4*(05), 40–48. doi:10.1109/4236.877485

Williams, H. C. W. L., & Ortuzar, J. D. (1982). Behavioral theories of dispersion and the mis-specification of travel demand models. *Transportation Research, 16B*, 167–219.

Willumsen, L. G., & Ortúzar, J. de D. (1994). *Modeling transport.* Sussex, England: John Wiley and Sons Ltd.

Wingbermuhle, J., Liedtke, C-E., & Solodenko, J. (2001). Automated Acquistion of Lifelike 3D Human Models from Multiple Posture Data. In W. Skarbek (Ed.), *Proceedings of the International Conference on Computer Analysis of Images and Patterns,* (pp. 400-409), Warsaw, Poland: Springer.

Winter, S., & Nittel, S. (2006). Ad-hoc shared-ride trip planning by mobile geosensor networks. *International Journal of Geographical Information Science, 20*, 899–916.

Wolf, J., Guensler, R., & Bachman, W. (2001). Elimination of the travel survey diary – experiment to derive trip purpose from Global Positioning System travel data. *Transportation Research Record, 1768*, 125–134.

Wolf, J., Guensler, R., Frank, L., & Ogly, J. (2001). The use of electronic travel diaries and vehicle instrumentation packages in the year 2000 Atlanta regional household travel survey: test results, package configurations, and deployment plans. In *Proceedings 9th International Association of Travel Behavior Research Conference*, Gold Coast, Queensland, Australia.

Wolf, J., Loechl, M., Myers, J., & Arce, C. (2001). Trip rate analysis in GPS-enhanced personal travel surveys.

In *International Conference on Transport Survey Quality and Innovation Kruger Park, South Africa*, August 2001.

Wolf, J., Oliviera, M., & Thompson, M. (2003). Impact of underreporting on mileage and travel time estimates: results from Global Positioning System-enhanced household travel survey. In *Transportation Research Record: Journal of the Transportation Research Board*, No. 1854 (pp. 189–198). Washington, DC: TRB, National Research Council.

Wolf, J., Schönfelder, S., Samaga, U., Oliveira, M., & Axhausen, K. W. (2004). Eighty weeks of GPS traces: approaches to enriching trip information. In *Proceedings 83th annual meeting of the transportation research board*, Washington DC.

Wong, K., Wong, S. C., & Yang, H. (2001). Modeling urban taxi services in congested road networks with elastic demand. *Transportation Research Part B: Methodological*, *35*(9), 819–842. doi:10.1016/S0191-2615(00)00021-7

Wong, K., Wong, S., Bell, M., & Yang, H. (2005). Modeling the bilateral micro-searching behavior for urban taxi services using the absorbing Markov chain approach. *Journal of Advanced Transportation*, *39*(1), 81–104. doi:10.1002/atr.5670390107

Wong, K., Wong, S., Yang, H., & Tong, C. (2002). A sensitivity-based solution algorithm for the network model of urban taxi services. In M.A.P. Taylor, (Ed.), *Proceedings of the 15th International Symposium on Transportation and Traffic Theory*, Amsterdam: Elsevier Science. 23–42.

Wong, S., & Yang, H. (1998). Network model of urban taxi services: improved algorithm. *Transportation Research Record*, *1623*, 27–30. doi:10.3141/1623-04

Wood, Z., & Galton, A. (2008). *Collectives and how they move: A tale of two classifications*. Paper presented at the Proceedings of the 2nd Workshop on Behaviour Monitoring and Interpretation, BMI'08, Kaiserslautern, (pp. 57-71).

Wright, P. Fields, R., & Harrison, M. (2000). Aanalyzing human-computer interaction as distributed cognition: the resources model. *Human-Computer Interaction*, *15*(1), 1 – 4.

OSIRIS Consortium. (2008). *Deliverable 2400 - Scenario Definition. Paris: OSIRIS Consortium.* Retrieved from http://osiris-fp6.eu/doc/OSIRIS-WP2400-DEL-0017-scenario%20definition.pdf

OSIRIS Consortium. (2008). *Deliverable 6001 Revision A - OSIRIS Architecture Specification and Justification. Paris: OSIRIS Consortium.* Retrieved from http://osiris-fp6.eu/doc/OSIRIS-WP6000-DEL-0037-Archi%20Spec%20and%20Justification.pdf

Xian, H., & Feng, D. (2007). Protecting mobile agents' data using trusted computing technology. *Journal of Communication and Computer*, *4*(3), 44–51.

Xu, F., & Tenenbaum, J. B. (2007). Word learning as Bayesian inference. *Psychological Review*, *114*, 245–272.

Yager, R. Entropy and Specificity in a Mathematical Theory of Evidence. (1983). *International Journal of General Systems*, (9), 249-260.

Yamada, T., Yoshimura, Y., & Mori, K. (2004). Road Network Reliability Analysis using Vehicle Routing and Scheduling Procedures. In *Proceedings of the 3rd International Conference on City Logistics*, (pp. 111-122). Madeira, Portugal: Elsevier.

Yan, W., & Forsyth, D. A. (2005). *Learning the Behavior of Users in a Public Space through Video Tracking*. Paper presented at the 7th IEEE Workshop on Applications of Computer Vision / IEEE Workshop on Motion and Video Computing (WACV/MOTION 2005), Breckenridge, CO, USA.

Yang, H., & Wong, S. (1998). A network equilibrium model of urban taxi services. *Transportation Research Part B: Methodological*, *32*(4), 235–246. doi:10.1016/S0191-2615(97)00042-8

Yang, H., Wong, K., & Wong, S. (2001). Modeling urban taxi services in road networks: progress, problem and prospect. *Journal of Advanced Transportation*, *35*(3), 237–258. doi:10.1002/atr.5670350305

Yang, H., Wong, S. C., & Wong, K. I. (2002). Demand-supply equilibrium of taxi services in a network under competition and regulation. *Transportation Research Part B: Methodological*, *36*(9), 799–819. doi:10.1016/S0191-2615(01)00031-5

Yang, H., Ye, M., Tang, W. H. C., & Wong, S. C. (2005). A multi-period dynamic model of taxi services with endogenous service intensity. *Operations Research*, *53*(3), 501–515. doi:10.1287/opre.1040.0181

Yang, H., Ye, M., Tang, W. H., & Wong, S. C. (2005). Regulating taxi services in the presence of congestion externality. *Transportation Research Part A, Policy and Practice*, *39*(1), 17–40. doi:10.1016/j.tra.2004.05.004

Yang, L., & Sanver, M. (2004). Mining short association rules with one database scan; *International Conference on Information and Knowledge Engineering.*

Yang, L., & Wang, F.-Y. (2007). Driving into Intelligent Spaces with Pervasive Communications. *IEEE Intelligent Systems*, *22*(1), 12–15. doi:10.1109/MIS.2007.8

Yu, H., & Shaw, S. (2007). Revisiting Hägerstrand's time-geographic framework for individual activities in the age of instant access. In H. J. Miller (Ed.), Societies and cities in the age of instant access (pp. 103-118). Dordrecht, Germany: Springer. doi:10.1007/1-4020-5427-0.

Yu, J. X., Chong, Z. H., Lu, H. J., & Zhou, A. Y. (2004). False positive or false negative: Mining frequent itemsets from high speed transactional data streams. *International Conference on Very Large Databases.*

Zhang, J. (1997). The nature of external representation in problem solving. *Cognitive Science*, *18*, 271 – 295.

Zhang, Z. (2000). A Flexible New Technique for Camera Calibration. *IEEE Trans. Pattern Analysis and Machine Intelligence, 22*, 1330-1334.

Zhao, S., & Harata, N. (2001). *Travel information, perceived travel time, and route diversion behavior.* Paper presented at the 9th World Conference on Transport Research, Seoul.

Zhou, J., & Golledge, R. (2007). Real-time tracking of activity scheduling/schedule execution within unified data collection framework. *Transportation Research Part A, Policy and Practice*, *41*, 444–463.

Zhu, Y. Y., & Shasha, D. (2002). StatStream: Statistical monitoring of thousands of data streams in real time. *International Conference on Very Large Databases.*

Zmud, J., & Wolf, J. (2003). Identifying the correlates of trip misreporting - results from the California statewide household travel survey GPS study. In *10th International conference on Travel Behavior Research.*

About the Contributors

Monica Wachowicz, after receiving a Ph.D. in Geography from the University of Edinburgh in 1995, she worked on spatio-temporal modelling for ubiquitous computing at the University of Cambridge, Computer Sciences Department in England, for two years. She then took a research fellow position at the Pennsylvania State University before joining the Wageningen University and Research (WUR) in 1999. In 2007 and 2008, she was a visiting Professor in the LatinGeo Group at the Technical University of Madrid, Spain. Her main research interests include the application of mobile sensor networks, the analysis of movement behaviour, and context modelling and reasoning over space and time. For the last decade, she has been working on policy, research and educational issues at European and national levels, aiming to address the needs in sustainable mobility and environmental management. Working in international funded research projects has provided her with the experience of operating in multidisciplinary teams from government, industry, and international organisations. In addition, one of her most rewarding experiences has been in teaching students from different cultural backgrounds and nationalities. She is currently the President of AGILE (Association of Geographic Information Laboratories for Europe) and the European Editor of the International Journal of Geographic Information Science edited by Taylor & Francis.

* * *

Clio Andris is a Ph.D. candidate in the Urban Information Systems Program at MIT. She wants to make GIS better with ideas from unsupervised classification, pattern recognition and machine learning so we can make the best decisions for the future. Her maps are usually pink.

Assaf Biderman is teaching at the Massachusetts Institute of Technology, where he is the Associate Director of the SENSEable City Laboratory. In 2004, Biderman joined Carlo Ratti to establish the SENSEable City Laboratory, an MIT research group that explores the "real-time city" by studying the increasing deployment of sensors and hand-held electronics, and their relationship to the built environment.

Francesco Calabrese received his M.S. and Ph.D. degrees in Computer and System Engineering from the University of Naples Federico II, Italy, in 2004 and 2007. He is currently Post-doctoral Associate at MIT in the SENSEable City Lab, and his research interests include ubiquitous computing, analysis of urban dynamics through sensor networks, analysis and design of distributed, hybrid, embedded control systems and CNC machines. He is a member of the IEEE and the IEEE Control Systems Society.

Caspar Chorus is a tenured Assistant Professor at Delft University of Technology (Section of Transport and Logistics). His main research interest concerns the development and testing of transport demand models that combine behavioural realism with formal tractability. He is co-editor-in-chief of the European Journal of Transport and Infrastructure Research, fellow of TRAIL research school, senior member of NETHUR research school, member of the Travel Survey Methods and Traveller Behaviour and Values committees of the Transportation Research Board, and member of the Netherlands Fulbright Alumni Association. His dissertation 'Traveller Response to Information' has been awarded the Eric Pas dissertation award of the International Association of Travel Behaviour Research, the best dissertation award of the Belgium-Netherlands-Luxembourg Interuniversity Association of Transport Economists, and the best dissertation award (second place) of the IEEE Intelligent Transportation Systems Society. In 2005, he was awarded a Fulbright scholarship.

Sean Doherty is an Associate Professor in the Department of Geography and Environmental Studies at Wilfrid Laurier University, Canada. He received a Ph.D. in Civil Engineering from the University of Toronto in 1998, and has a masters degree in Geography. His interests focus on human activity/mobility patterns and decision-making, including tracking methods, modelling, and the impacts on health, safety, and the environment. His current projects involve application of a Global Positioning System tracking system to the study of diabetic patients, elderly, children, tourists, and immigrants. His partners in these projects include Bloorview Kids Rehab, Toronto Rehab Institute, Research in Motion, Telus, and Innovations at the University of Toronto. His recent honours include receipt of the Fred Burggraf Award recognising excellence in transportation research. Alongside his professional career, Dr. Doherty enjoys an active family life and is an avid hockey player.

Giusella Finocchiaro is a Full Professor of Internet and Private Law at the University of Bologna, and owner of the Finocchiaro legal firm specialising in Internet law. She has been legal advisor for the EU Commission, UNCITRAL, the Russian and the Italian governments. She has served as legal consultant for many Italian and international companies. She is a member of ENISA (European Network and Information Security Agency) Permanent Stakeholders Group. She is also a member of the Scientific Advisory Committee of the European Privacy Institute. Since 2003 she has been the Scientific Director of the Master's programme on Information Security at the Alma Graduate School of the University of Bologna. She is also Professor of Information Technology Law, Management School, at the "Luigi Bocconi" University in Milan. She is the author of several books in the field of internet law and has been published in many books and international journals.

Georg Gartner is a Full Professor for Cartography at the Vienna University of Technology. He holds graduate qualifications in Geography and Cartography from the University of Vienna and received his Ph.D. and his Habilitation from the Vienna University of Technology. He was awarded a Fulbright grant to the University of Nebraska at Omaha in 1997 and a research visiting fellowship to the Royal Melbourne Institute of Technology in 2000, to South China Normal University in 2006 and to the University of Nottingham in 2009. He serves as Vice-President of the International Cartographic Association. He is Dean for Academic Affairs for Geodesy and Geoinformation at Vienna University of Technology. He is responsible organiser of the International Symposia on Location Based Services & TeleCartography and Editor of the Book Series "Lecture Notes on Geoinformation and Cartography" by Springer and Editor of the Journal on LBS by Taylor & Francis.

Frank van der Hoeven is Associate Professor Urban Design at the Faculty of Architecture, part of the Delft University of Technology (TU Delft). Since 2008 he has combined this task with the position of Research Director of the institute. As such he is responsible for the redevelopment of the research strategy and the portfolio of the Faculty of Architecture at TU Delft. He is also expert for the Netherlands Commission for Environmental Assessment and editor of the Dutch NIROV magazine Nova Terra. He conducted his Ph.D. research in the field of underground space technology and mixed-use development. Between 2005 and 2008 he was involved in European projects within the framework of both INTERREG IIIB and IIIC. He developed and managed as lead partner a successful INTERREG IIIC network, called Connected Cities. Connected Cities addressed the interaction between sustainable mobility and land use in cities and regions (connectedcities.eu).

Simon Jirka works as Community Leader for the Sensor Web group of the Open Source initiative 52° North. Furthermore he is employed as a research associate at the Institute for Geoinformatics (IfGI) and holds a master degree in Geoinformatics. He is currently working on his Ph.D. His research activities focus on discovery mechanisms for sensor networks and on the simulation of autonomous sensor behaviour. At IfGI, he is the internal project manager of the EU-funded projects OSIRIS and GENESIS, in which IfGI deals with the development of an interoperable architecture for sensor networks. Within the GENESIS project he is the work package leader for the technology research activities. In addition, he is also involved in the Sensor Web Enablement initiative of the Open Geospatial Consortium (OGC) which aims at developing standards for the interoperable integration of sensor networks.

Martin Lee-Gosselin was a Professor of Planning at Université Laval, Quebec, Canada, where he continues on an honorary basis after retiring in 2007. He is also Visiting Professor in the Centre for Transport Studies, Imperial College London, and advisor or consultant to a number of national and international organisations involved in transport, environmental and community energy efficiency strategies. Martin specialises in survey, analysis and evaluation methods to investigate activity/travel behaviour, both as it exists now, and in response to future changes in policy or the availability of alternative energy, vehicle and telecommunications technologies. With his wife, he also enjoys trying to understand the evolving telecommunications and activity behaviour of their two teenaged daughters. From 2000-2006, he led PROCESSUS, a major six-year international network research programme on the behavioural foundations of integrated land-use, transport and environment models, and directed related research to develop person-based GPS-aided survey methods. He co-chairs the International Steering Committee for Travel Survey Conferences and has published extensively on transport survey methods.

Liang Liu is a Post-doctoral Associate at the Massachusetts Institute of Technology's SENSEable City Lab. His research interests include using pervasive computing to understand urban mobility, intelligent transportation system, and new techniques for urban transportation planning and management.

Richard Mason is a Researcher with the Microsoft QUT eResearch Centre concentrating in environmental sensor networks. His research interests include distributed systems, peer-to-peer computing and the application of web based technologies to scientific research.

Alexandra Millonig graduated in Regional Planning from the Vienna University of Technology, Austria. At present she is a Project Assistant at the Cartography Research Group at the Vienna University of Technology. In her Ph.D., she is currently investigating pedestrian motion patterns and related determinants. Additionally, she is a researcher in the business field Dynamic Transportation Systems at the Mobility Department of the Austrian Institute of Technology, Vienna, Austria. Her research interests include monitoring and interpretation of human spatio-temporal behaviour, landmark orientation, mobile navigation tools for pedestrians, and transport information requirements of specific user groups.

Andreas Mitrakas is the Head of the Administration Department at the European Network and Information Security Agency (ENISA). He has been consulted in an array of private and public sector implementations of eBusiness and eGovernment with emphasis on identity management and transaction frameworks (e.g. national identities, tachographs, PKI etc.). He currently chairs the CEN/ISSS WG on Cyber-identity for Business. He holds a doctorate in law from Erasmus University of Rotterdam (Netherlands) and a Master's in computers and law from Queen's University of Belfast (UK). He is a member inter alia of ISACA, ACM and the Athens Bar Association. He has authored inter alia "Open EDI and Law in Europe: a regulatory framework", (Kluwer Law 1997) and he is co-editor of "Secure eGovernment Web Services" (IGI, 2007).

Adriano Moreira is an Associate Professor in the Information Systems Department, University of Minho, since 1996. He received the "Licenciatura" degree in Electronics and Telecommunications Engineering and the Ph.D. degree in Electrical Engineering, respectively in 1989 and 1997, from the University of Aveiro - Portugal. He has been a voting member of the IEEE 802.11 working group where he participated in the specification of the infrared physical layer. His research interests are in wireless local area networks, mobile and context-aware computing, urban computing, mobility and space modelling. He is a member of the IEEE Communications Society.

Daniel Orellana received his B.S. in Environmental Biology from Universidad del Azuay in 2002, his M.Sc. in Geographic Information Systems from Universitat Politècnica de Catalunya in 2006, and the Diploma of Advanced Studies in Geographical Engineering from Technical University of Madrid. He is currently conducting his Ph.D. thesis research at the University of Wageningen, The Netherlands. His research interests are in geographic information science, mainly focused on spatio-temporal knowledge representation and reasoning, as well as the analysis methods and models for pedestrian movement. He was a member of the scientific committee of 4th and 5th gvSIG conferences (2008 and 2009). He is currently a member of the Pool of Reviewers of the European Science Foundation.

Christine Outram is an Architect and Urban Designer, and her current research at MIT focuses on whether emerging technologies (cell-phone tracking, GPS etc) and the use of real-time data can improve our understanding of cities and be built into a new methodology for the practice of urban design.

Binh Pham has been an Emeritus Professor at the Queensland University of Technology since 2008. She was the Director of Research in The Faculty of IT from 2000-2007. She held the IBM Foundation Chair in Information Technology at the University of Ballarat from 1995-1999 and was an Associate Professor in the School of Computing and IT at Griffith University from 1993-1995. Her research interests cover computer graphics, image processing, multimedia, visualisation, data mining, intelligent systems, sensor networks, and their application to real world problems.

Carlo Ratti, an architect and engineer by education, he is an Associate Professor of the Practice of Urban Technologies at the Massachusetts Institute of Technology, where he directs the SENSEable City Laboratory. He is also founding partner and director of carlorattiassociati, a rapidly growing architectural practice that was established in Turin, Italy, in 2002; the practice is currently involved in a number of architectural schemes, both nationally and internationally. Carlo graduated in structural engineering from the Politecnico di Torino and the Ecole Nationale des Ponts et Chaussies, later specialising in architecture with M.Phil. and Ph.D. degrees from the University of Cambridge. He is a member of several professional organisations, including the Ordine degli Ingegneri di Torino, the Association des Anciens Eleves de l'Ecole Nationale des Ponts et Chaussees and the UK Architects Registration Board.

Chiara Renso received her M.Sc. and the Ph.D. in Computer Science from University of Pisa in 1992 and 1998 respectively. She holds a permanent research position at ISTI, Institute of National Research Council, Pisa, Italy. Her research interests range from spatio-temporal reasoning, semantic queries on geographical information systems, data mining query languages, semantic and conceptual aspects in data mining, data mining for e-Health and mobility data. She served as reviewer for several data mining conferences. She has been co-chair of the ICLP Workshop on Complex Reasoning on Geographical Data (2001), of the Spatial, Temporal, and Spatio-temporal Data Mining (SSTDM'08) and the First International Workshop on Semantic Aspects in Data Mining (SADM 08), both in conjunction with IEEE ICDM 2008. She also served as Local Organiser Chair for IEEE ICDM 2008. She is currently co-chairing the Second Workshop on Semantic Aspects in Data Mining (SADM09) to be held in conjunction with ICDM 2009.

Paul Roe received his MEng from the University of York in 1987 and his Ph.D. from the University of Glasgow in 1991. He is currently a Full Professor in the Faculty of Science and Technology at QUT, in Brisbane Australia. At QUT he leads the Microsoft-QUT eResearch Centre (www.mquter.qut.edu.au), a collaboration between the Queensland State Government, Microsoft Research and QUT which is investigating smart tools for eResearch. His research interests lie in mobile and distributed computing.

Maribel Yasmina Santos is an Assistant Professor in the Information Systems Department at the University of Minho in Portugal. She has a degree in Informatics and Systems Engineering from the University of Minho (1991), a M.Sc. in Informatics and a Ph.D. in Information Systems and Technologies, both from University of Minho (1996 and 2001, respectively). Her research interests include business intelligence, data mining, spatial data mining, spatio-temporal data models, and spatial reasoning.

Amer Shalaby is an Associate Professor of Transportation Engineering at the University of Toronto. He is specialised in the areas of urban transportation and transit planning and intelligent transportation systems. He is a member of the US Transportation Research Board committees on "Emerging and Innovative Public Transport and Technologies", "Bus Transit Systems" and "Light Rail Transit". He also serves on the editorial boards of the international Journal of Intelligent Transportation Systems and Journal of Advanced Transportation. He currently holds a visiting appointment at Carnegie Mellon University while on leave from the University of Toronto.

Jiro Sumitomo is a Research Assistant at Queensland University of Technology. His research interests include distributed parallel computing and sensor networks.

Harry Timmermans holds a Ph.D. degree in Geography/Urban and Regional Planning. He studied at the Catholic University of Nijmegen, The Netherlands. His dissertation concerned a theory of the functional and spatial structure and the dynamics of central place systems. Since 1976 he is affiliated with the Faculty of Architecture, Building and Planning of the Eindhoven University of Technology, The Netherlands. First as an Assistant Professor of Quantitative and Urban Geography, later as an Associate Professor of Urban Planning Research. In 1986 he was appointed chaired professor of Urban Planning at the same institute. In 1992 he founded the European Institute of Retailing and Services Studies (EIRASS) in Eindhoven, the Netherlands (a sister-institute of the Canadian Institute of Retailing and Services Studies). His main research interests concern the study of human judgment and choice processes, mathematical modelling of urban systems and spatial interaction and choice patterns and the development of decision support and expert systems for application in urban planning. He has published several books and many articles in journals in the fields of Marketing, Urban Planning, Architecture and Urban Design, Geography, Environmental Psychology, Transportation Research, Urban and Regional Economics, Urban Sociology, Leisure Sciences and Computer Science.

Michael Towsey is a Postdoctoral Fellow at the Queensland University of Technology. His research interests include applications of machine learning. He has predominantly worked in biological applications including bioinformatics. He is currently working in the QUT Microsoft eResearch Centre on the analysis of environmental acoustic recordings.

Claire Vishik's work at Intel Corporation UK focuses on hardware security, trusted computing, privacy enhancing technologies, some aspects of encryption and related policy issues. She is a member of the Permanent Stakeholders Group of ENISA, the European Network and Information Security Agency. She is active in standards development and is on the Board of Directors of TCG, the Trusted Computing Group. She received her Ph.D. from the University of Texas at Austin. Prior to joining Intel, she worked at Schlumberger Laboratory for Computer Science and AT&T Laboratories, studying security and other technologies, from electronic commerce and communication protocols to hardware/software systems and applications. Claire is the author of many papers and reports and 25 pending and granted US patents.

Jinglan Zhang is a Senior Lecturer in the Faculty of Science and Technology, Queensland University of Technology, Australia. She received the degree of Bachelor of Engineering in 1987 and Master of Engineering in 1990, both from Beijing University of Aeronautics and Astronautics, China. She received her Ph.D. from Queensland University of Technology in 2003. Her research interests include computer graphics, image processing, intelligent systems, mobile and distributed computing, and applications of sensor networks.

Index